Flowgraph Models for Multistate Time-to-Event Data

Flowgraph Models for Multistate Time-to-Event Data

APARNA V. HUZURBAZAR

University of New Mexico
Department of Mathematics and Statistics
Albuquerque, New Mexico

WILEY-
INTERSCIENCE

A JOHN WILEY & SONS, INC., PUBLICATION

Published by John Wiley & Sons, Inc., Hoboken, New Jersey.
Published simultaneously in Canada.

Library of Congress Cataloging-in-Publication Data:

Huzurbazar, Aparna V., 1966–
Flowgraph models for multistate time-to-event data / Aparna V. Huzurbazar.
p. cm.
Includes bibliographical references and index.
ISBN 0-471-26514-4 (cloth : acid-free paper)
1. Flowgraphs. 2. Stochastic processes. 3. Reliability (Engineering)—Statistical methods.
I. Title.

TA338.F5H89 2005
519.5′42—dc22

2004049225

10 9 8 7 6 5 4 3 2 1

Krishna

Contents

Preface

The main purpose of this book is to present an introduction to flowgraph models for time-to-event data. The focus is on stochastic models for censored time-to-event data with competing risks and recurrent events. The applications are geared to survival analysis and reliability. I view flowgraph models as providing a methodology for data analysis of semi-Markov processes that can be applied without becoming intimately familiar with the mathematical theory of stochastic processes. My early experience with stochastic processes left me with the impression that there were lots of nice models, but I could not think of how to analyze data with them except in the simplest cases. I found myself unhappy with the exponential assumption and limit theorems, which although they provided some approximation, did not reflect the real system. My early work in flowgraphs was on queues, but I soon found myself drawn to interesting applications in survival analysis. Consequently, I continued to work in both survival analysis and engineering systems. Flowgraphs bring together applied probability techniques such as transforms and saddlepoint methods and meld them with data analysis and statistical methods. Flowgraph models are analyzed using Bayesian methods, or if one prefers, maximum likelihood techniques.

This book is intended for students and practitioners of statistics who have some background at the level of a one-year graduate course in probability and statistics. Although background in survival analysis or systems reliability is not assumed, a one-semester course or some experience with standard methods in these areas would be helpful. The first seven chapters are structured for a one-semester topics course. The remaining two chapters are more advanced. A large number of worked examples are presented along with computer code as needed. For readers who are not as interested in computation, Section 3.5 and Chapter 6 can be skipped without loss of continuity.

I would like to thank the following people and institutions: Howard Snyder for my undergraduate course on aerospace systems analysis; my Ph.D. advisor, Ron Butler, who taught me about saddlepoint methods, statistics, and emphasized the importance of yoga; Richard Davis for a course on stochastic processes; my colleague, Ron Christensen, who bugged me for years to write a book on flowgraph models; Jay Kadane, who brought the project to Wiley; the series editorial board and anonymous reviewers; the editor, Steve Quigley, and assistant editors, Susanne Steitz and Heather Bergman; and Angioline Loredo and the production staff.

Portions of the work were completed in California during my sabbatical. I would like to thank my colleagues in the Statistics Group and Elvira Loredo in the Management Science Group at RAND for an unbelievably fun and collegial environment. I am grateful to Guillermo Marshall and Satish Garg for the diabetic retinopathy data; to Ira Longini for providing data from the San Francisco AIDS study; and to Kate Hall for discussions on cellular telephone network operations. The work was supported in part by the University of New Mexico.

My family and friends have helped in every way possible, but primarily by putting up with me through this process, especially Jayashree and Anagha Phansalkar, Ceela McElveny and Kevin Glasgow, and Noreen Gima. For helping to maintain the general health of my mind and body, I owe thanks to my yoga teachers: Herb McDonald in Albuquerque; Gail Ackerman in Santa Fe; Claudia Kuhns in Denver; and Karen Cline and Sherry Gould in Santa Monica, who took me on as a student during my sabbatical.

There are four people who helped at various stages throughout the project:

- Snehalata Huzurbazar provided much needed constructive criticism on early versions of the manuscript, early versions of S-plus saddlepoint programs, and computation for Example 3.11.
- Lillian Yau entered into many discussions and provided the retinopathy simulation of Section 8.2.
- Amit Phansalkar was always ready with a touch of humor on the all-too-frequent "bad book days" and eager to browse portions of the manuscript.
- Finally, Brian Williams graciously accepted the fact that the first year of our marriage would involve me working on this book. In addition to painstakingly reading drafts of the manuscript, he introduced me to slice sampling and convinced me that open source software was the way to go.

CHAPTER 1

Multistate Models and Flowgraph Models

Multistate models are used to describe longitudinal, time-to-event data. They model stochastic processes that progress through various stages. Flowgraph models are one type of multistate model. Flowgraphs model potential outcomes, probabilities of outcomes, and waiting times for the outcomes to occur. For example, in medicine, the process can be a disease such as cancer and the stages can be diagnosis, remission, and advanced stage of cancer leading to death. Other examples of processes include the stages of HIV leading to AIDS, the progression of diabetic retinopathy leading to eventual blindness, and the failures of organs such as kidneys. In an engineering context, processes include the internal mechanisms of cellular telephone networks or planning construction in a building project. For example, the stages of the cellular telephone network begin with a fully functioning network that proceeds through a series of degradations to a partially functioning network and eventually to a fully failed network. All of these examples involve modeling the time until the occurrence of some event or events. This is *time-to-event data*. Time-to-event data analysis focuses on modeling waiting times until events of interest occur. Since such data are obtained by observing individuals, components, or systems over time, their analysis is also called *event history analysis*. Although this book focuses on medical and engineering applications, multistate models arise in areas as diverse as demography, economics, linguistics, logistics and operations research, and sociology. One of the earliest uses of a multistate stochastic model in medical statistics is in the work of Fix and Neyman (1951), who considered relapse and recovery for cancer patients. An earlier, albeit simpler model for assays used in the standardization of insulin production can be found in Fieller (1941). In economics, Samuelson (1947) used such models for problems in macroeconomics such as demand management. The entire area of queuing theory is based on such models, beginning with Johanssen (1907) as cited by Kendall (1951). For an introduction to

Flowgraph Models for Multistate Time-to-Event Data, By Aparna V. Huzurbazar
ISBN 0-471-26514-4

multistate models applied in survival analysis, see Hougaard (1999, 2000), Klein and Moeschberger (1997), and Therneau and Grambsch (2000).

1.1 MULTISTATE MODELS

By way of introduction, we present some basics of multistate models without formal definitions. The theory underlying multistate models for event history analysis was formalized by the work of Aalen (1975, 1978), who showed that such models may be analyzed within the framework of counting processes. Aalen decomposed the *intensity process*, the process that records the uncensored observations on the events over time, into a product of the hazard function and an observable factor and used counting process methods along with techniques of product integration to convert the results to a distribution function. For an account of counting process models in survival analysis, see Andersen et al. (1993) and Fleming and Harrington (1991). While the theory of counting processes, martingales, and stochastic integration is an elegant foundation for survival analysis, application of these methods is often quite difficult in practice. In terms of data analysis, multistate models have been restricted to the realm of Markov models. In a Markov multistate model, given the current state of the process, the transition time to a future state does not depend on the past history of the process. At the initial state of a Markov process, the transition time is a minimum of the waiting time distributions corresponding to all possible transitions from the initial state. Hence, in practice, for tractability and analytical convenience, exponential distributions are assumed. Occasionally, with appropriate parametric restrictions, Weibull distributions are used (cf. Wilson and Soloman, 1994), exploiting the fact that the minimum of independent and identically distributed Weibulls is again a Weibull distribution.

A semi-Markov multistate model allows the transition time to a future state to depend on the duration of time spent in the current state. In practice, it is quite difficult to analyze data for semi-Markov multistate models. One method of analysis for multistate models consists of combining independent submodels for each transition intensity, a method that restricts the analysis to models with unidirectional or progressive flow (cf. Andersen and Keiding, 2002). In fact, Hougaard (1999) states that for nonunidirectional or nonprogressive multistate models, it is impossible to obtain general formulas for transition probabilities for models where the hazard is allowed to depend on the history in any way. Another method is the use of the popular proportional hazards model. The proportional hazards model as used in medical statistics was developed by Cox (1972); however, the idea of assuming proportional hazards to fit more parsimonious models dates back to the operations research literature (cf. Allen, 1963). Cox's model is semiparametric and assumes that the intensity of the counting process is a product of a parametric function of the covariates and an arbitrary function of time, hence its designation as *proportional hazards*. This method is also restricted to a unidirectional or progressive multistate model. The obvious restriction of the proportional hazards model is that hazards are not always proportional. In both methods, the key approach for analyzing such multistate models is based on modeling the hazard function, a quantity that is not directly observable. The end result is a hazard function model based on a set of covariates that can be converted to a survivor function if required. There are many extensions of this model; for example, the stratified Cox model divides the subjects into groups, or models that allow the use of different functional forms for the covariate. See Klein and Moeschberger (1997) and Therneau and Grambsch (2000) for application of the Cox model to survival data.

Before we discuss the differences between analysis of multistate models and flowgraph models, we present some definitions. Throughout this book, we will assume that the time to event T is a continuous random waiting time that is nonnegative.

Definition 1.1.1. Suppose that T is the random waiting time until the event of interest occurs. In survival analysis, the *survivor function* is defined as

$$S(t) = P(T > t).$$

In reliability, this is called the *reliability function* and is denoted by

$$R(t) = P(T > t).$$

We will use the terms survivor function and reliability function interchangeably.

Definition 1.1.2. Suppose that T is the random waiting time until the event of interest occurs, and let $f(t)$, $F(t)$, and $S(t)$ denote the density, cumulative distribution function (CDF), and survivor function of T. Then the *hazard function* is defined as

$$h(t) = \frac{f(t)}{1 - F(t)} = \frac{f(t)}{S(t)}.$$

We discuss these functions in detail in Chapter 4. Some common relationships between these functions are

$$\begin{aligned} f(t) &= \frac{dF(t)}{dt} = -S'(t) \\ S(t) &= \exp\left[-\int_0^t h(u)\,du\right] \\ h(t) &= -\frac{d\log S(t)}{dt}, \end{aligned} \tag{1.1}$$

where $F(t)$ is the CDF of T. Verification is left to the reader as an exercise in Chapter 4.

Figure 1.1 gives a standard three-state multistate model. This model is called an *illness–death model*. It is quite popular in the multistate models literature for

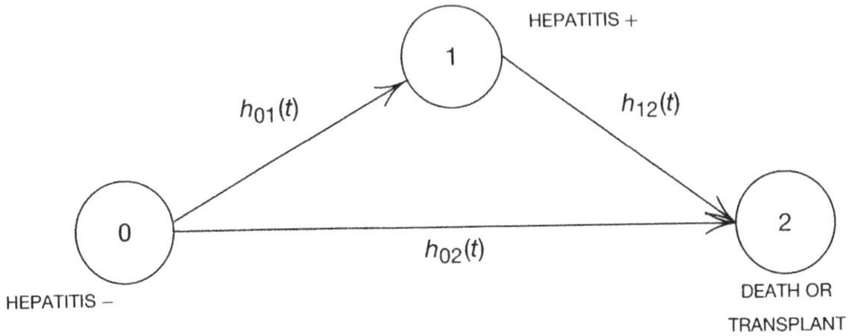

FIGURE 1.1 Three-state illness–death multistate model.

modeling the progression of a variety of diseases, including AIDS (cf. Alioum et al., 1998; Frydman, 1995; Joly and Commenges, 1999), cancer (cf. Fix and Neyman, 1951; Lagakos, 1976; Lindsay and Ryan, 1993), dementia (cf. Commenges et al., 1998), and for studies involving transplantation of organs or bone marrow with potentially more states (cf. Klein et al., 1993; Klein and Shu, 2002). A recent issue of *Statistical Methods for Medical Research* (cf. Multistate models, 2002) is devoted entirely to medical applications of multistate models. We will use the illness–death model for illustration of hepatitis, a condition marked by inflammation of the liver caused mainly by viral or toxic agents. There are five main types of hepatitis, A, B, C, D, and E, and we will assume this general model for hepatitis B. A patient in the early stages of hepatitis experiences a variety of flulike illnesses. This stage is denoted by state 0 (HEPATITIS −) in the figure. The advanced stage, denoted as state 1 (HEPATITIS +) in the figure, is associated with a wide range of liver diseases, including acute and chronic hepatitis, cirrhosis, and cancer. This may be followed by jaundice, liver enlargement, and eventual liver failure. This is represented by state 2 (DEATH OR TRANSPLANT), where death may be death due to causes other than hepatitis. The multistate model treats $0 \to 1$, $1 \to 2$, and $0 \to 2$ as separate events. In the Markov case, the transition intensities reduce to the hazard functions, and these are labeled on the branches:

- $h_{02}(t)$ is the hazard for patients who have yet to advance to HEPATITIS +.
- $h_{01}(t)$ is the hazard for patients who advance to HEPATITIS +.
- $h_{12}(t)$ is the hazard for patients who make the transition from advanced hepatitis to death or receive a transplant.

In general, one fits separate Cox or Poisson regression models to each transition (cf. Therneau and Grambsch, 2000). In a Poisson regression model, the hazard is modeled as a piecewise constant. If there is some effect due to proportionality, we can fit a stratified Cox model (cf. Klein and Moeschberger, 1997). For example, one stratification group would be patients who are alive with the mild form of hepatitis, and another would consist of patients living with advanced hepatitis. In all cases, separate models are fit to each hazard. These hazard functions or the survivor functions computed from these hazards are examined. If the form of the hazard is simple, a type of prediction probability can also be computed. The history process of a patient reflects what happens to the patient over the course of his or her lifetime in the study. Prediction of a patient's outcome based on the history of the process at time t proceeds as follows. Let $\mathcal{H}_0(t)$ be the history of a patient who is alive at time t and has a mild form of hepatitis, and let $\mathcal{H}_1(t)$ be the history of a patient who is still alive and has an advanced form of hepatitis. We can find $\pi_1(t_2|t_1)$, the probability that a patient who has an advanced form of hepatitis will die in the interval from t_1 to t_2 for $t_2 > t_1$ as

$$\begin{aligned}\pi_1(t_2|t_1) &= \pi[t_2|\mathcal{H}_1(t_1)] \\ &= \int_{t_1}^{t_2} \exp\left[-\int_{t_1}^{r} h_{12}(u)\,du\right] h_{12}(r)\,dr.\end{aligned}$$

Then we can find the probability, $\pi_0(t_2|t_1)$, that a patient with a mild form of hepatitis will die in the interval from t_1 to t_2 $(t_2 > t_1)$ given information up to t_1 as

$$\begin{aligned}\pi_0(t_2|t_1) &= \pi[t_2|\mathcal{H}_0(t_1)] \\ &= \int_{t_1}^{t_2} \exp\left[-\int_{t_1}^{r} h_{02}(u)\,du - \int_{t_1}^{r} h_{01}(u)\,du\right][h_{02}(r) + h_{01}(r)\pi_1(t_2|r)]\,dr.\end{aligned}$$

The exponential term represents the probability of not dying and not progressing to the advanced form of hepatitis in the interval from t_1 to r. One can think of the $h_{02}(r)$ term as the instantaneous probability of death or transplant at time r, and the last term, $h_{01}(r)\pi_1(t_2|r)$, as the instantaneous probability of advancing to HEPATITIS + at time r and then proceeding to state 2, DEATH OR TRANSPLANT, in the interval r to t_2. These prediction probabilities must be solved after fitting either Cox or alternative models to the hazard. As such, the form of the hazard can be quite restrictive. Work on accessing such specific response probabilities dates back to Aalen and Johansen (1978) and Temkin (1978). For the semi-Markov case, see Voelkel and Crowley (1984), and for a model-free approach, see Pepe et al. (1991). In all cases, one can see that complications arise if the form of the hazard becomes complex or if reversibility is allowed. If our interest is not a binary outcome probability but rather, a waiting time distribution for the event of interest, this approach becomes intractable without excessive simplifying assumptions.

1.2 FLOWGRAPHS AS MULTISTATE MODELS

Flowgraph models are multistate models for time-to-event data that provide an alternative, innovative approach for the analysis of such data. A flowgraph is a graphical representation of a stochastic system in which possible outcomes are connected by directed line segments. The model concerns modeling the probabilities of the outcomes, the waiting time distributions of the outcomes, and manipulating the flowgraph to access overall waiting time distributions. Flowgraphs model semi-Markov processes and allow for a variety of distributions to be used within the stages of the multistate model. They also easily handle reversibility. In a medical context, this means that in a progressive disease, a patient is allowed to improve at times. In the engineering context, this means that a failed component can be repaired. Flowgraphs model the observable waiting times rather than the hazards, and as such, they do not directly make any assumptions about the shape of the hazard. The end results from a flowgraph analysis are Bayes predictive densities, CDFs, survivor or reliability functions, and hazard functions of the waiting times of interest. If one prefers, maximum likelihood estimation is also available. Although covariates can be added to flowgraph models in much the same way as other multistate models, analysis based on time-varying covariates is beyond the scope of this book.

Figure 1.2 presents a flowgraph model for hepatitis but with an additional transition from $1 \to 0$ which adds reversibility to the multistate model. A patient with hepatitis in state 0 can proceed to state 1 or proceed directly to state 2. Once in state 1, the patient can receive treatment and improve to the milder form of the disease in state 0 or eventually, progress to state 2. This figure is also a generalization of the classical illness–death multistate model in that it allows reversibility (i.e., the patient is allowed to improve from the illness). The flowgraph is an outcomes graph, and the outcomes are the events HEPATITIS −, HEPATITIS +, and DEATH OR TRANSPLANT. The labels on the branches of the figure are transition probabilities and moment generating functions (MGFs) of waiting time distributions for transitions to occur. The flowgraph returns the MGF of the distribution of the waiting times of interest. This may be the waiting time from the early stages of hepatitis until death due to hepatitis, $0 \to 1 \to 2$, or it may be the waiting time from $0 \to 2$, regardless of which path, $0 \to 1 \to 2$ or direct transition from $0 \to 2$, is taken. It can also be the waiting time distribution for $0 \to 1$, a partial passage, given data on the entire flowgraph. Once we have the MGF, our interest is in converting it to a Bayes predictive density, CDF, survivor or reliability, or hazard function. Our data on the flowgraph can be censored. In addition, sometimes the data that we observe on a patient is incomplete for reasons other than censoring. The aim of this book is

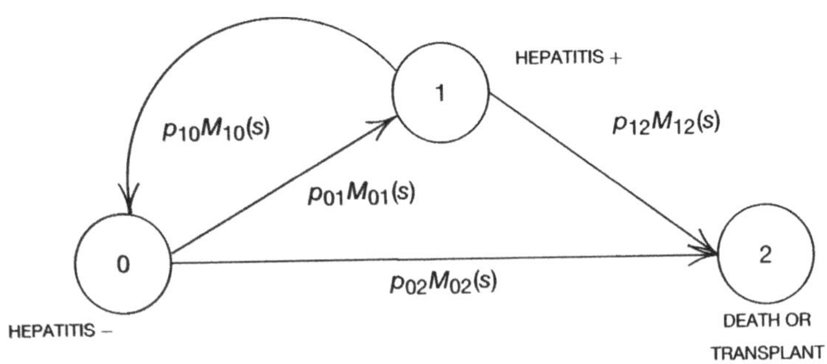

FIGURE 1.2 Flowgraph model for a reversible illness–death multistate model.

to illustrate the use of flowgraphs under all of these scenarios. The book is replete with applications of flowgraphs. Examples in the medical and engineering context are alternated throughout the book. In addition, flowgraphs have a wide range of applications from branching processes to social networks that it is hoped readers of this book will pursue.

1.3 ORGANIZATION OF THE BOOK

In Chapter 2 we present the basic flowgraph structures: series, parallel, and loop networks. Using these as building blocks, more complex structures are created. Work with flowgraphs requires flowgraph algebra, which manipulates probabilities and MGFs for accessing the waiting time distributions of interest. Most statisticians have an aversion to MGFs, and in fact, most students of statistics treat MGFs as an annoyance that is only required knowledge for an exam. Engineers, on the other hand, are quite comfortable with Laplace transforms, which are closely related to MGFs. It turns out that flowgraph algebra solves the waiting time distribution problem in the transformed world of MGFs. One should think of the MGF in the same vein as polar coordinates or as any other transformation designed to make solutions more accessible. The simplicity of working with MGFs in the context of modern computing capabilities is made possible by the availability of symbolic algebra packages. Even the person who dreads working with MGFs can have a package do the most difficult work for them. As such, the latter part of Chapter 2 illustrates how to work with flowgraphs using the symbolic algebra package MAPLE. Any symbolic algebra package (e.g., MATHEMATICA) will suffice. Detailed examples are presented for many survival analysis applications, including kidney disease progression, AIDS, and cancer. Engineering applications include failure of hydraulic pumps, parts testing, and car stereo manufacturing. Models for recurrent events are also presented for procedures such as endoscopy in survival analysis and repair of components in reliability.

In Chapter 3 we present tools that convert MGFs into densities. For the simplest of flowgraphs this can be done by hand or using MAPLE. For complicated flowgraphs we use saddlepoint methods to convert MGFs into densities. Examples originally presented in Chapter 2 are more fully developed.

Data analysis using flowgraphs requires that we model the branches of the flowgraph with parametric models. In Chapter 4 we present histograms for censored data that help to select such branch models. The chapter begins with background on survivor or reliability functions and hazard functions and then presents methods for censored data. Estimation of survivor or reliability functions using the Kaplan–Meier estimator and estimation of the hazard and the cumulative hazard using the Nelson–Aalen estimator is also presented.

Ultimately, the end results from a flowgraph analysis are a Bayes predictive density, predictive CDF, predictive survivor or reliability function, and predictive hazard function. These are derived using the Bayesian prediction methods of

Chapter 5. In this chapter we review the basics of Bayesian prediction and present four methods of sampling from the posterior: rejection sampling, Gibbs sampling, Laplace's method, and slice sampling. The use of each method with flowgraphs is illustrated. Slice sampling is a generally applicable method for use with flowgraphs, and an entire section is devoted to describing this approach. Most of the examples in subsequent chapters use slice sampling.

In Chapter 6 we unify the flowgraph computations of previous chapters with computer code. An example is presented that walks the reader through each step to enable the use of flowgraphs for his or her own applications. Readers who are not interested in the details of the computation can skip this chapter without loss of continuity.

Flowgraphs by themselves are a probabilistic tool. When used to model data, they become a data analytic tool for semi-Markov multistate models. Chapter 7 is devoted to application of flowgraphs to semi-Markov processes. In this chapter we review the basics of Markovian pure birth, pure death, and birth and death processes and then proceed with data analysis for semi-Markov processes. For readers familiar with such processes, one of the advantages of flowgraphs is that they bypass the highly mathematical characterization of semi-Markov processes and provide a computational method for handling real data. Medical examples of flowgraphs are presented using AIDS data and data for diabetic retinopathy. The last portion of the chapter is devoted to phase type distributions. These distributions arise in Markov processes, and flowgraphs generalize them to semi-Markov processes.

Chapter 8 is devoted to an advanced application of flowgraphs that is based on likelihood construction in the presence of incomplete data. Incomplete data are divided into two categories: data that are known to be incomplete and data that are incomplete but not recognized as such, the more common situation. The diabetic retinopathy data are reanalyzed incorporating incomplete observations from the study. A second example using a complex flowgraph for the degenerative stages of dementia is also presented. Engineering examples of incomplete data are relegated to Chapter 9.

In Chapter 9 we give applications of flowgraphs to queues. This chapter reviews queuing terminology and presents flowgraph equivalent models for standard queues. We use an example of hydraulic pumps to illustrate a flowgraph model for an M/M/1 queue, broaden the scope of the model to allow general waiting time distributions, and analyze the flowgraph for complete data with censoring and with partially incomplete data in addition to censoring. We present a flowgraph application for an M/M/c queue using data from a cellular telephone network and illustrate prediction of a partial waiting time from both complete and incomplete data. We also present flowgraph equivalents for an M/G/1 queue and an exotic infinite-state queue. As with semi-Markov processes, the use of flowgraphs for problems in queuing is intended to bypass the cumbersome mathematics of queues, which generally leads to limit distributions. The aim is to perform prediction for queues in the short term.

1.4 COMPUTATIONAL ASPECTS

By the nature of their application to statistical problems, flowgraph methods are computationally intensive. Computer code for the examples is given throughout the book. We urge the reader to try out the worked examples on their own. The main computational tools are MAPLE, the R system, and the C programming language with the Gnu Scientific Library (GSL).

The symbolic algebra package MAPLE is used only to access the flowgraph MGF. Chapter 3 is devoted to methods for converting flowgraph MGFs into densities. This is generally done using saddlepoint methods. MAPLE is used to generate C code for use with R.

Programming for flowgraphs can be done using any programming language or package. R program code is given for performing all the calculations. The early work on flowgraphs was done using FORTRAN and the IMSL subroutine libraries. The work then migrated to the S-plus environment. The drawbacks of S-plus are looping and the availability of S-plus to a wide audience. In writing this book, the work was migrated to the R system primarily because R is open source software. It is easy to use along with GSL, the open source library of C functions available from the Free Software Foundation, to perform most of the calculations. For readers who prefer to use S-plus, R code can also be run in S-plus, possibly with minor modifications. The R code is written with the assumption that dynamic linking of shared libraries is available on the user's system. All of the examples presented in this book were implemented on the author's laptop computer.

The C functions that are used with R are written to be transparent to the user. The code has been made as robust as possible to glitches encountered previously; nevertheless, it remains impossible to guarantee error-free performance for each new application. In Chapter 6 we give the program code and point out the necessary changes to the functions that the reader can make for a new problem. It is not necessary that the reader know MAPLE, R, or C prior to reading this book. For the interested reader, many books are available on all three topics. Kofler (1997) is an introductory book on MAPLE, and Kelley and Pohl (1995) cover the basics of programming in C. An excellent introduction to R is given by Maindonald and Braun (2003). Unfortunately, currently there is no single package to perform flowgraph computations. At present, Lillian Yau in the Department of Biostatistics at Tulane University is developing a flowgraph package that is contained entirely in MATHEMATICA. An attempt has been made to create a set of programs in MAPLE, R, and C to perform the necessary computations. The programs, given in this book, are also available from the author's web site:

`http://www.stat.unm.edu/~aparna/flowgraphmodels.html`

It is hoped that the programs will continue to improve and that interested readers will add to the contributions. R can be downloaded from

`http://cran.r-project.org/`

The GSL library can be downloaded from

`http://sources.redhat.com/gsl/`

CHAPTER 2

Flowgraph Models

Flowgraph models are useful for modeling multistate time-to-event data that result from a stochastic process. A flowgraph models potential outcomes, probabilities of outcomes, and waiting times for those outcomes to occur. For example, a patient with an initial diagnosis of cancer is considered to be in state I in Figure 2.1. The patient can have two possible outcomes: progression to an advanced stage, A, of the disease, or death from a different cause, D_2. Once in the advanced stage, the patient has three possible outcomes: her disease can be in remission, R; she can die from cancer, D_1; or she can die from other causes, D_2. If she experiences remission, she can subsequently return to an advanced stage or she can die from other causes. There is interest in the transition probabilities from one stage to other stages and in how long she remains in a given stage. In particular, interest centers on remission of cancer and on death. Flowgraphs model this entire complex stochastic network.

Block diagrams and signal flowgraphs are widely used to represent engineering systems, especially in circuit analysis. Basic flowgraph ideas were developed in engineering, but they did not incorporate probabilities, waiting times, or data analysis. The literature on flowgraph methods is vast. Introductions to flowgraph methods are contained in most circuit analysis or systems engineering textbooks, such as D'Azzo and Houpis (1981), Dorf and Bishop (1995), Gajic and Lelic (1996), Lorens (1964), and Whitehouse (1973). Statistical flowgraph models are based on flowgraph ideas, but unlike their antecedents, statistical flowgraph models can also be used to model and analyze data from complex stochastic systems. Early literature on signal flowgraphs includes Mason (1953, 1956), Sittler (1956), and Huggins (1957). For early work on statistical flowgraph models, see Butler and Huzurbazar (1997), and Huzurbazar (1999, 2000a,b). In this chapter we discuss flowgraph models and flowgraph algebra. In Sections 2.1 and 2.2 we present flowgraph basics for series and parallel structures. Combinations of series and parallel structures are discussed in Section 2.3, and the loop flowgraph model is presented in Section 2.3.1. Section 2.4 focuses on solving flowgraph models containing series,

Flowgraph Models for Multistate Time-to-Event Data, By Aparna V. Huzurbazar
ISBN 0-471-26514-4

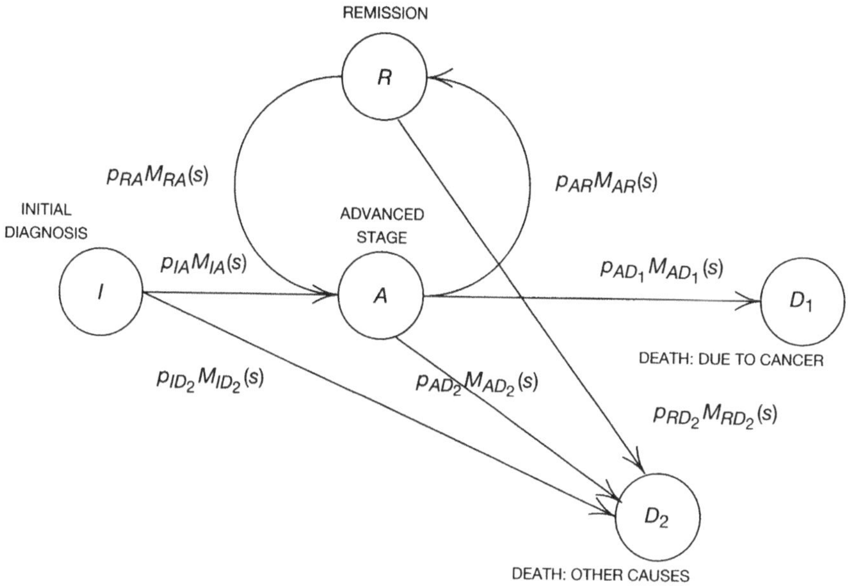

FIGURE 2.1 Flowgraph model for cancer.

parallel, and loop structures for moment generating functions (MGFs) of the distributions of the waiting times of interest. In Section 2.5 we present a systematic procedure for solving larger flowgraphs such as those of Figure 2.1.

2.1 FLOWGRAPH BASICS: MODELS FOR SERIES STRUCTURES

In a flowgraph model, the states or nodes represent outcomes. This is distinct from a graphical model where the nodes represent random variables. Flowgraphs are outcomes graphs, whereas graphical models are variables graphs. In the cancer example the nodes represent various stages of cancer or death. The nodes are connected by directed line segments called *branches*. Each branch has a transition probability and waiting time distribution associated with a transition from its beginning and ending nodes. The branches are labeled with *transmittances*, which involve probabilities and moment generating functions (MGFs).

Definition 2.1.1. The *moment generating function* of a random variable T is defined as

$$M(s) = E(e^{sT}), \tag{2.1}$$

provided that the expectation exists for s in an open neighborhood of 0.

Often, we write $M_T(s)$ to emphasize that it is the MGF of the random variable T, or equivalently, of the distribution of T.

Definition 2.1.2. A *transmittance* consists of the transition probability multiplied by the MGF of the waiting time distribution.

The Appendix contains a brief review of properties of MGFs. In Figure 2.1, probabilities and MGFs of the waiting time distributions are shown as branch transmittances. The transition from initial diagnosis to the advanced stage $I \to A$ has transmittance $p_{IA}M_{IA}(s)$. The transition probability is p_{IA}, the probability of transition to an advanced stage of cancer, A, beginning with initial diagnosis, I. The waiting time distribution in state I until transition to A is represented by its MGF, $M_{IA}(s)$. We use the branch transmittances of a flowgraph model to solve for the distribution of the waiting time of interest.

Definition 2.1.3. An *overall transmittance* is the transmittance of the entire flowgraph from the initial to the end state.

For the cancer example of Figure 2.1, the overall transmittance refers to the transmittance from initial diagnosis to death regardless of the cause of death. The overall transmittance is a quantity that must be solved for by combining the individual transmittances according to a set of rules discussed in detail in Section 2.4. We use MGFs on the branches for algebraic convenience. MGFs are useful for finding distributions of sums of independent random variables. In engineering, systems analysis uses Laplace transforms, although it does not use probabilities or statistical distributions. A MGF is a bilateral Laplace transform. See the Appendix for further details.

The most basic element of a flowgraph is the simple series system. This structure is useful for modeling progressive diseases in medicine or structures in engineering. In a series model, progression occurs from one state to the next state with certainty. No other transitions are allowed. Suppose that Figure 2.1 consisted only of states I, A, and D_1 with the transitions indicated; then the allowable transitions would be $I \to A$ and $A \to D_1$. States I, A, and D_1 would be in series.

Example 2.1: SIMPLE SERIES MODEL: KIDNEY DISEASE PROGRESSION. Figure 2.2 presents a three-state model of kidney failure. The data and original analysis are due to Gross et al. (1971). State 0 represents an initial diagnosis of kidney disease where the patient has two functioning kidneys. State 1 represents one failed kidney, and state 2 represents two failed kidneys, an absorbing state. Let Y_0 be the random waiting time in state 0 until state 1 is reached and Y_1, independent of Y_0, be the

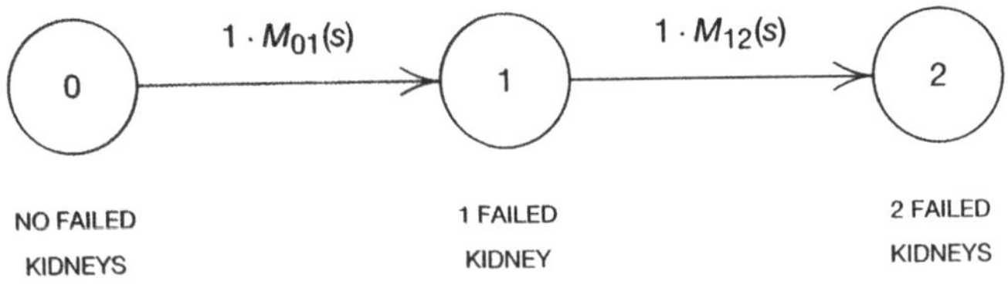

FIGURE 2.2 Series flowgraph for kidney failure.

random waiting time in state 1 until state 2 is reached. One quantity of interest is the survival time of the patient's kidneys, that is, the time spent in states 0 and 1 before reaching state 2. Let $T = Y_0 + Y_1$ be the total waiting time from $0 \rightarrow 2$. There are two branches in this flowgraph, $0 \rightarrow 1$ and $1 \rightarrow 2$, and they are labeled with transmittances. In this flowgraph, the probability that one kidney fails and the patient eventually reaches state 1 is $p_{01} = 1$, and the MGF of the waiting distribution for this outcome is $M_{01}(s) = E(e^{sY_0})$. In a flowgraph, the transmittance, $p_{01}M_{01}(s)$, is written on the branch. In state 1 we wait for the remaining kidney to fail. The probability that this occurs is $p_{12} = 1$, and the MGF of the corresponding waiting time distribution is $M_{12}(s) = E(e^{sY_1})$. □

Example 2.2: Simple Series: Hydraulic Pump System. Consider a hydraulic pump system with two pumps and appropriate valving so that the system can operate with only one pump if necessary. Limnios (1992) presents an analysis of this system with feedback. Figure 2.3 is a flowgraph of the hydraulic pump system. We assume that the pumps operate independently and that the system can operate with only one pump if necessary. State 0 represents no failed pumps, state 1 represents one failed pump, and state 2 is the absorbing state with two failed pumps. Let Y_0 be the random waiting time in state 0 until one pump fails and state 1 is reached. Let Y_1, independent of Y_0, be the random waiting time in state 1 until the second pump fails and state 2 is reached. In state 0, a pump fails with probability 1, so the transition probability from $0 \rightarrow 1$ is $p_{01} = 1$. The waiting time for one pump to fail is represented by its MGF, $M_{01}(s) = E(e^{sY_0})$. Once in state 1, the second pump eventually fails so that $p_{12} = 1$ and the waiting time MGF of the time until state 2 is reached is $M_{12}(s) = E(e^{sY_1})$. One quantity of interest is the time to total failure, $0 \rightarrow 1 \rightarrow 2$. □

2.1.1 Series Flowgraph Models versus Series Engineering Systems

For readers familiar with engineering systems, a series flowgraph model will appear to be a parallel system in engineering. This is true, and it is due to the fact that the flowgraph model is graphing outcomes of the physical system rather than the physical system itself. For instance, in Example 2.1 we know that the human body has two kidneys physically in parallel. For a kidney patient, we know that one of these will fail first, then the other. The kidneys are assumed indistinguishable (i.e., it does not matter to us which kidney fails first, only that a kidney fails). If it mattered to us which kidney failed first, we would leave this as a

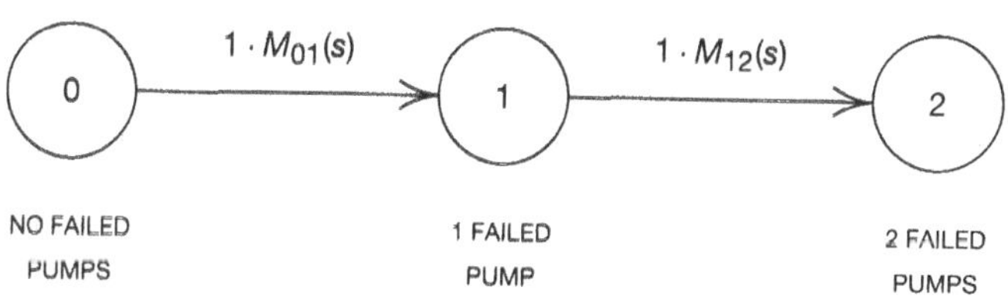

FIGURE 2.3 Flowgraph model for a hydraulic pump system.

parallel system. Since our interest is in modeling the time to failure of kidneys, we label the flowgraph with the outcomes 0, 1, and 2, representing the number of failed kidneys. Alternatively and equivalently, we could have labeled the states on the flowgraph as 2, 1, and 0, representing the number of functioning kidneys.

2.2 FLOWGRAPH MODELS FOR PARALLEL STRUCTURES

The next basic component of a flowgraph is one where branches are in parallel. In a parallel flowgraph, transition from a state is allowed to one of a set of outcomes. Figure 2.4 illustrates the connection. This represents a *competing risks* situation. In state 0 we wait for one of three possible outcomes: state 1, 2, or 3. In a medical study of time to remission of cancer, state 0 is the initial detection of a cancerous tumor, and states 1, 2, and 3 are remission, death due to cancer, and death due to other causes, respectively. In an engineering study, our quantity of interest is the time to failure of a component or equipment, say a car. State 0 is the initial state in which the car is put into operation. State 1 represents minor problems that do not interfere with the car being driven, state 2 represents partial failure, and state 3 represents total failure. In engineering, this is called an *open parallel system* because there are distinct endpoints or exits: 1, 2, and 3. Note that initially, in state 0, the risks are what the component or individual is subject to, but once an outcome has occurred, the risks become the *competing causes* of failure. The theory of competing risks is vast and we refer the reader to Crowder (2001) for a comprehensive introduction. The flowgraph models the observable parts of the competing risks distribution, that is, the actual competitive waiting times and the probabilities of each outcome.

Example 2.3: MEDICAL APPLICATION: PROGRESSION OF CANCER. Figure 2.5 is a flowgraph model for the progression of cancer. State 0 represents the initial

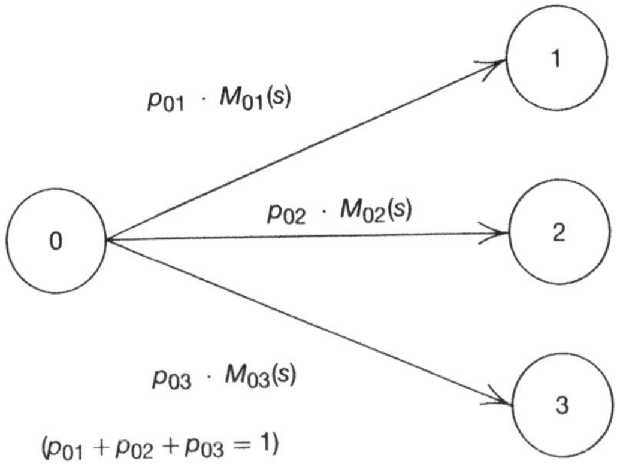

FIGURE 2.4 Flowgraph model for a parallel system.

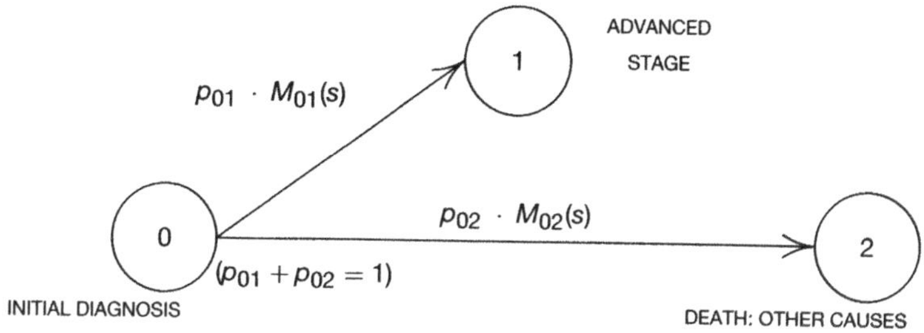

FIGURE 2.5 Parallel flowgraph model for cancer progression.

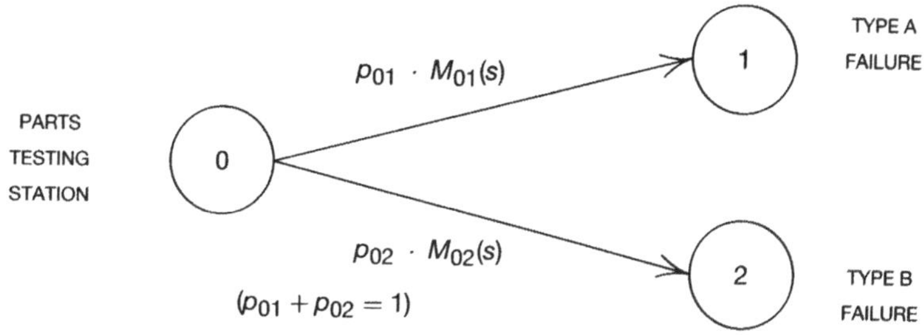

FIGURE 2.6 Flowgraph model for an open parallel system.

diagnosis of cancer, state 1 is a progressive state of the cancer, and state 2 is death. States 1 and 2 are in parallel. A patient in state 0 is in a competing risk situation in that progression to the advanced stage of cancer (passage from $0 \to 1$) is in competition with death (passage from $0 \to 2$). The MGFs of the waiting time distributions for the events to occur are $M_{01}(s)$ and $M_{02}(s)$, with transition probabilities p_{01} and p_{02}, where $p_{01} + p_{02} = 1$. □

Example 2.4: Open Parallel System: Parts Testing Station. Figure 2.6 shows the flowgraph of a parallel system. State 0 represents the testing station for a component. The component is prone to one of two types of failure which move the system to state 1 or 2. These two types of failure are the competing risks for the component. The probability of transition from $0 \to 1$ is p_{01}, and the MGF of the waiting time for passage is $M_{01}(s)$, which gives $p_{01}M_{01}(s)$ as the branch transmittance. The transition probability from $0 \to 2$ is $p_{02} = 1 - p_{01}$, with $M_{02}(s)$ as the MGF of the waiting time for passage. □

In Examples 2.3 and 2.4, suppose that Y_1 represents the waiting time to state 1 and Y_2 represents the waiting time to state 2. The flowgraph structure describes the waiting time for the occurrence of $\min\{Y_1, Y_2\}$. If distributional assumptions about Y_1 and Y_2 were made, then $p_{01} = P(Y_1 < Y_2)$. Note that $M_{01}(s)$ is not the MGF of Y_1 but rather, the MGF of the actual *competitive waiting time*, the *competing risks* distribution. Flowgraphs adopt a conditional approach to modeling competing risks. We can express the probabilities as

$$P[\min(Y_1, Y_2) \le t] = P(Y_1 \le t \mid 0 \to 1)p_{01} + P(Y_2 \le t \mid 0 \to 2)p_{02}$$
$$= P(Y_1 \le t \mid Y_1 < Y_2)P(Y_1 < Y_2)$$
$$+ P(Y_2 \le t \mid Y_2 < Y_1)P(Y_2 < Y_1). \quad (2.2)$$

In the multistate models literature, there are many applications of basic competing risks, including modeling bone marrow transplant, cancer, cardiovascular disease, drug use, and so on. See Andersen at al. (1993), Crowder (2001), or Klein and Moeschberger (1997), for details.

2.3 COMBINATIONS OF SERIES AND PARALLEL FLOWGRAPHS

We can model more complex processes by combining series and parallel flowgraphs. This creates more than one possible *path* from the initial to the end state.

Definition 2.3.1. A *path* from state i to state j is any possible sequence of nodes from i to j that does not pass through any intermediate state more than once.

For example, in the cancer example of Figure 2.1, one possible path from initial diagnosis to death is $I \to A \to R \to D_2$.

Example 2.5: Medical Application: Progression of Cancer. Figure 2.7 is a flowgraph model for the progression of lung cancer. The model is a more general form of the Markovian model with exponential waiting times due to Lagakos (1976). State 0 represents the initial diagnosis of cancer, state 1 is an advanced state of the cancer, and state 2 is death. States 1 and 2 are in parallel and states 0, 1, and 2 are in series. □

Example 2.6: Closed Parallel System: Parts Testing Station. Figure 2.8 shows a closed parallel system, a modification of Figure 2.6. State 0 once again represents a parts testing station, and states 1 and 2 represent partial and total

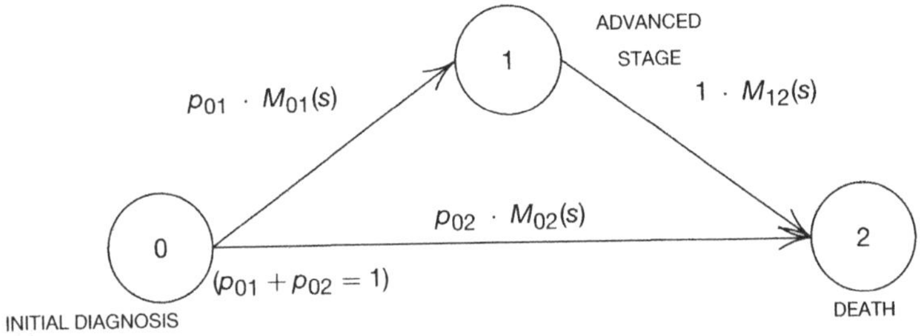

FIGURE 2.7 Flowgraph model for cancer progression.

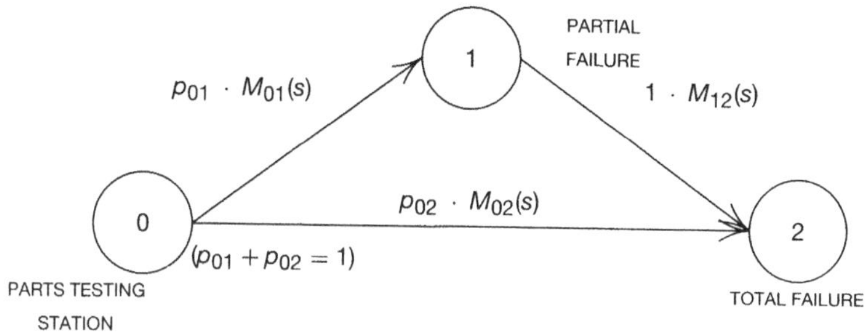

FIGURE 2.8 Flowgraph model for a closed parallel system.

failure of the part. The path $0 \to 1 \to 2$ is a series system and it is in parallel with the direct path from $0 \to 2$.

This three-state model of Examples 2.5 and 2.6 is a classic competing risks model. In Example 2.5 a patient in state 0 is in a competing risks situation in that death due to cancer (passage from $0 \to 1 \to 2$) is in competition with death due to other causes (direct passage from $0 \to 2$). Our quantities of interest are time to death due to any cause (i.e., either with or without advanced cancer, or time until death due to advanced cancer). Similarly, in Example 2.6, a part undergoing testing has competing causes for failure, either partial or total failure. In the multi-state models literature it is often called the illness–death model (cf. Andersen and Keiding, 2002). It has been used for modeling the progression of various diseases, such as dementia (cf. Commenges et al., 1998). □

2.3.1 Loop Flowgraph Model

Series and parallel elements are two basic components of flowgraphs. Series elements lead to distributions of sums of independent random variables, whereas parallel elements lead to finite mixtures of distributions of independent random variables. The feedback loop is the third basic element of a flowgraph model.

Definition 2.3.2. A *loop* is any closed path that returns to the initiating state without passing through any state more than once.

For example, in the cancer example of Figure 2.1, there is one loop, $A \to R \to A$. Theoretically, a patient can pass through this loop several times, although it is unlikely to be the case in practice for a cancer patient. Feedback loops arise naturally in both medical and engineering settings. Loops are used to model recurrent events.

Example 2.7: Recurrence of an Ulcer. Figure 2.9 presents a loop flowgraph model of ulcer recurrence. An ulcer patient who has received therapy to heal an ulcer begins in state 0. The patient undergoes an endoscopy procedure to detect ulcer recurrence. A positive endoscopy indicates that the ulcer has recurred, transiting the patient to state 1. The patient is treated and returns to state 0. The

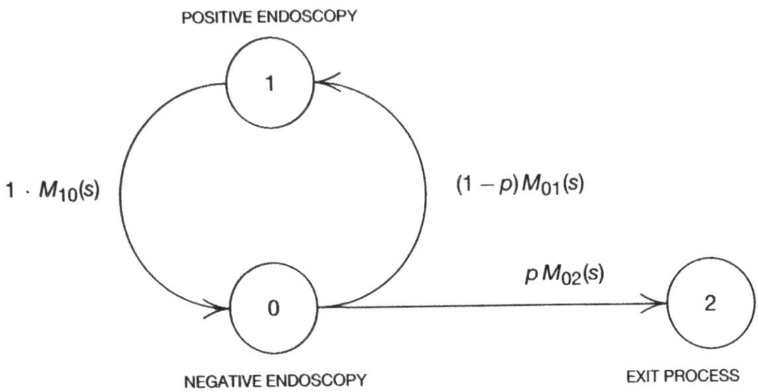

FIGURE 2.9 Flowgraph model for ulcer recurrence.

patient leaves the process when the ulcer does not recur. This may be decided by a predetermined number of negative endoscopies. A patient can have a sequence of positive and negative endoscopies before exiting the process. If the initiating state is 0, the feedback loop is $0 \to 1 \to 0$. If the initiating state is 1, the feedback loop is $1 \to 0 \to 1$. □

In the engineering setting, feedback loops can be used to represent repairs to a system or a part. Consider the hydraulic pump system of Example 2.2, but now allow for pumps to be repaired. Limnios (1992) analyzes this system for various quantities, including mean time to total failure.

Example 2.8: SERIES MODEL WITH FEEDBACK: HYDRAULIC PUMP SYSTEM. Figure 2.10 is a flowgraph of the hydraulic pump system with repair. We assume that the pumps operate independently and that the system can operate with only one pump if necessary. State 0 represents zero failed pumps, state 1 represents one failed pump, and state 2 is the absorbing state with two failed pumps. The backward transition from state 1 to state 0 represents the repair of a failed pump. This is an example of a *series system* with *feedback*. Quantities of interest are the time to total failure, $0 \to 1 \to 2$, or the number of pumps repaired before total failure of the system. □

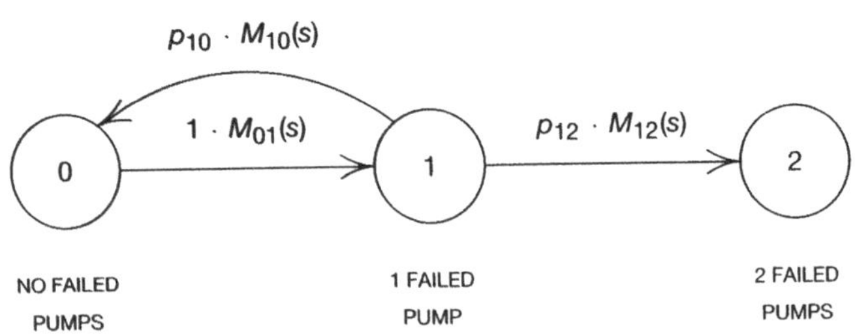

FIGURE 2.10 Flowgraph model for a hydraulic pump system.

2.4 SOLVING FLOWGRAPH MODELS

Solving a flowgraph model refers to reducing all of the branch transmittances in a flowgraph into one branch with one overall transmittance for the flowgraph.

Definition 2.4.1. The term *equivalent transmittance* refers to the transmittance that is obtained by reducing two or more branch transmittances into one transmittance. It is denoted by $T(s)$.

This is sometimes referred to as the *reduced equivalent transmittance*. When the entire flowgraph is reduced to one branch, the overall transmittance is the equivalent transmittance of the flowgraph. For finite systems, this equivalent transmittance is the overall MGF i.e., the MGF of the waiting time distribution from input to output, and in this case it is denoted using the notation for a MGF, $M(s)$. A finite system contains a finite number of states such that passage from input to output is certain to occur. The words *input* and *output* are used quite generally. This method of solving flowgraphs works for computing the MGF between any two nodes of interest, provided that there is at least one path between the two nodes. In general, the MGF of a path P is given by $M_P(s) = T_P(s)/T_P(0)$, where $T_P(0)$ represents the probability of taking path P.

2.4.1 Solving the Simple Series Flowgraph

The kidney example, Example 2.1, and the hydraulic pump system, Example 2.2, are both simple series systems with one path, $0 \to 1 \to 2$. To solve the flowgraphs of Figures 2.2 and 2.3, we compute the transmittance of the path $0 \to 1 \to 2$.

Definition 2.4.2. A *path transmittance* is the product of all of the branch transmittances for that path.

In terms of random variables, the total waiting time distribution is the distribution of the sum of two independent random waiting times from $0 \to 1$ and $1 \to 2$. If Y_0 and Y_1 are the independent random waiting times $0 \to 1$ and $1 \to 2$ with MGFs $M_{Y_0}(s) = M_{01}(s)$ and $M_{Y_1}(s) = M_{12}(s)$, respectively, we can find the MGF of $T = Y_0 + Y_1$ as

$$\begin{aligned} M_T(s) &= M_{Y_0+Y_1}(s) \\ &= E(e^{s(Y_0+Y_1)}) \\ &= E(e^{sY_0})E(e^{sY_1}) \\ &= M_{Y_0}(s)M_{Y_1}(s) \\ &= M_{01}(s)M_{12}(s) \end{aligned}$$

Here, the transmittance of the path, which is the MGF of the sum T, is $M_{01}(s)M_{12}(s)$. In this example, the path transmittance is also the equivalent transmittance of

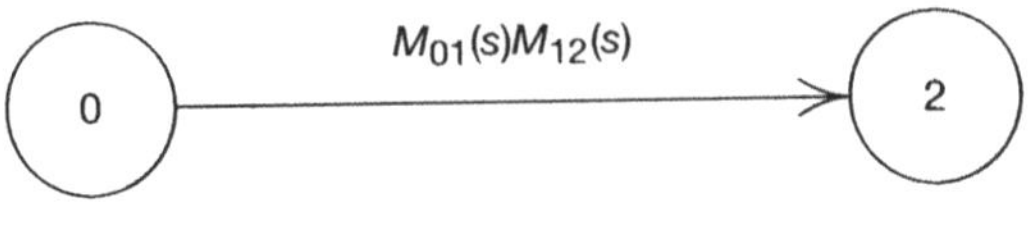

FIGURE 2.11 Solved flowgraph for a series system.

the flowgraph. We can replace Figures 2.2 and 2.3 with an equivalent flowgraph, Figure 2.11, in which node 1 is removed and passage is directly from node 0 to node 2. This equivalent flowgraph is labeled with the overall transmittance $M_{01}(s)M_{12}(s)$. This procedure of replacing a more complicated flowgraph with an equivalent one consisting only of nodes 0 and 2, for example, is called *solving* the flowgraph from 0 to 2. Note that this solved flowgraph is appropriate for both Examples 2.1 and 2.2, even though the states represent different outcomes, the number of failed kidneys or the number of failed pumps. This equivalent transmittance is the MGF of the distribution of the overall waiting time T from $0 \to 2$,

$$M_T(s) = M_{01}(s)M_{12}(s).$$

Example 2.9: KIDNEY DISEASE PROGRESSION. Let us return to Example 2.1 but add some distributional assumptions to the model. Suppose that kidneys fail independently and according to an exponential distribution, Exp(λ_0), with mean $1/\lambda_0$. In state 0 we observe the minimum of two independent exponentials. Let W_1 be the waiting time for the first kidney to fail and let W_2 be the waiting time for the second kidney to fail. Then W_1 and W_2 are independent and identically distributed (i.i.d.) with an Exp(λ_0) distribution. The density and CDF of W_1 are

$$f(w) = \lambda_0 e^{-\lambda_0 w}, \qquad w > 0,\ \lambda_0 > 0 \tag{2.3}$$

$$F(w) = 1 - e^{-\lambda_0 w}, \qquad w > 0,\ \lambda_0 > 0. \tag{2.4}$$

Let $W_{(1)} = \min(W_1, W_2)$ be the minimum waiting time for the first kidney to fail. Then the distribution of $W_{(1)}$ is computed as

$$\begin{aligned} P(W_{(1)} \le t) &= 1 - P(W_1 > t, W_2 > t) \\ &= 1 - P(W_1 > t)P(W_2 > t) \\ &= 1 - [1 - F(t)]^2 \\ &= 1 - [1 - (1 - e^{-\lambda_0 t})]^2 \\ &= 1 - e^{-2\lambda_0 t}. \end{aligned} \tag{2.5}$$

We see that the minimum, $W_{(1)}$, has an Exp($2\lambda_0$) distribution. This is the waiting time in state 0 until one kidney fails and the patient is in state 1. Since the kidney patient will eventually make the transition from state 0 to state 1, the probability that this occurs is 1. Once in state 1, we assume that the remaining kidney has

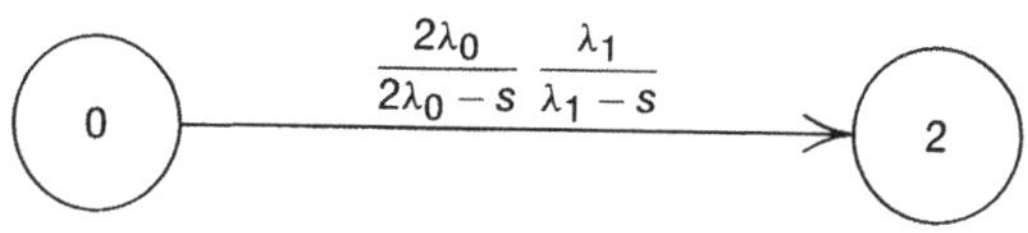

FIGURE 2.12 Solved flowgraph for kidney disease progression.

a failure time distributed according to an Exp(λ_1) distribution such that $\lambda_1 > \lambda_0$, and this transition occurs with probability 1. This constraint on the rate parameter λ_1 is a way to account for additional stress on the remaining kidney, once one kidney has failed. The waiting time distribution for passage from state 0 to 2 is the sum of the two independent waiting times, $0 \to 1$ and $1 \to 2$ (i.e., the convolution of two independent exponential distributions). If we let $Y_0 = W_{(1)}$ be the waiting time from $0 \to 1$ and Y_1 be the waiting time from $1 \to 2$, the total waiting time from $0 \to 2$ is $T = Y_0 + Y_1$. Since Y_0 and Y_1 are independent, the MGF of the total waiting time T, $M_T(s)$, is the product of the MGFs of the waiting times from $0 \to 1$ and $1 \to 2$, which are $M_{01}(s)$ and $M_{12}(s)$, respectively. The MGF of an Exp(λ) distribution is $M(s) = \lambda/(\lambda - s)$ for $s < \lambda$. The MGF of T is

$$\begin{aligned} M_T(s) &= M_{Y_0+Y_1}(s) \\ &= M_{01}(s)M_{12}(s) \\ &= \frac{2\lambda_0}{2\lambda_0 - s}\frac{\lambda_1}{\lambda_1 - s} \qquad \text{for } s < \min(2\lambda_0, \lambda_1). \end{aligned} \tag{2.6}$$

Note that we still have not identified the distribution of T, but we have its MGF, which uniquely determines its distribution (cf. Casella and Berger, 2002, p. 65). We can replace Figure 2.2 with an equivalent flowgraph, Figure 2.12, in which node 1 is removed and passage is directly from node 0 to node 2. This equivalent flowgraph is the *solved* flowgraph and is labeled with the overall transmittance $M_{01}(s)M_{12}(s)$. □

Example 2.10: Human Immunodeficiency Virus (HIV) Blood Transfusion Model. Figure 2.13 is a flowgraph model for patients infected with HIV through blood transfusions. The model is motivated by data from the San Francisco Men's Health Study (SFMHS), which is analyzed in Section 7.2. State 0 is HIV infection, antibody negative, state 1 is antibody positive but without acquired immune

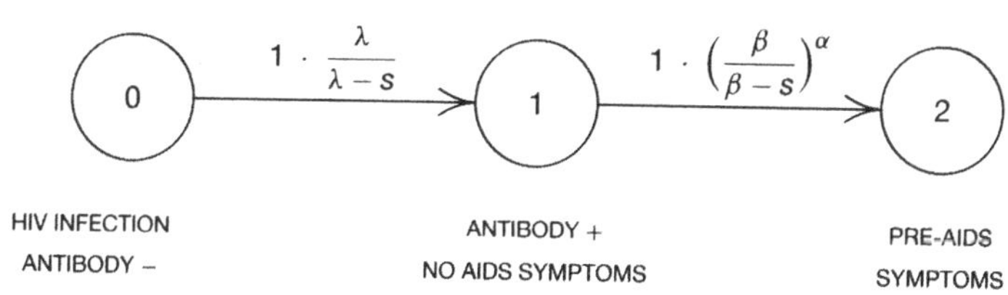

FIGURE 2.13 Flowgraph model for transfusion data.

deficiency syndrome (AIDS) symptoms, and state 2 is pre-AIDS symptoms. Let T_0 be the waiting time in state 0 until transition to state 1 and let T_1 be the waiting time in state 1 until transition to state 2. The total waiting time for passage from $0 \to 2$ is $T = T_0 + T_1$. Model $T_0 \sim \text{Exp}(\lambda)$ with mean $1/\lambda$ and $T_1 \sim \text{Gamma}(\alpha, \beta)$ with mean α/β. The branches of the flowgraph are labeled with their respective transmittances. The MGFs, $M_{01}(s)$ and $M_{12}(s)$, are the MGFs of exponential and gamma distributions, given by

$$M_{01}(s) = \frac{\lambda}{\lambda - s} \quad \text{for } s < \lambda \quad \text{and} \quad M_{12}(s) = \left(\frac{\beta}{\beta - s}\right)^{\alpha} \qquad \text{for } s < \beta.$$

Solving this flowgraph results in the flowgraph given in Figure 2.11 but with an overall waiting time distribution MGF given by

$$M(s) = M_{01}(s)M_{12}(s) = \left(\frac{\lambda}{\lambda - s}\right)\left(\frac{\beta}{\beta - s}\right)^{\alpha} \qquad \text{for } s < \min(\lambda, \beta). \tag{2.7}$$

□

Example 2.11: Hydraulic Pump Series System. Consider the hydraulic pump series system of Example 2.2 with distributional assumptions such that the waiting time for $0 \to 1$ and the waiting time for $1 \to 2$ are independent Weibull random variables. The Weibull distribution is widely used in both engineering and biostatistics. However, the distribution of a convolution of Weibull distributions is rarely used since it is analytically intractable, although it is numerically tractable. This example serves as an illustration of how to use the methods when the MGF is tractable numerically but not analytically. A Weibull random variable $T \sim \text{Weibull}(\omega, \nu)$ has probability density function given by

$$f(t) = \nu\omega^{\nu}t^{\nu-1}\exp[-(\omega t)^{\nu}] \qquad \text{for } \nu > 0,\ \omega > 0,\ t > 0.$$

Let the waiting time for $0 \to 1$ be $T_0 \sim \text{Weibull}(\omega_0, \nu_0)$ with mean $\omega_0\Gamma(1 + 2\nu_0)$, and let the waiting time for $1 \to 2$ be $T_1 \sim \text{Weibull}(\omega_1, \nu_1)$ with mean $\omega_1\Gamma(1 + 2\nu_1)$. The flowgraph model is that of a series system with transmittances given by the MGFs, $M_{01}(s)$ and $M_{12}(s)$, as

$$\begin{aligned} M_{01}(s) &= \int_0^{\infty} \nu_0{\omega_0}^{\nu_0}t^{\nu_0-1}\exp[st - (\omega_0 t)^{\nu_0}]\,dt \\ M_{12}(s) &= \int_0^{\infty} \nu_1\omega_1^{\nu_1}t^{\nu_1-1}\exp[st - (\omega_1 t)^{\nu_1}]\,dt. \end{aligned} \tag{2.8}$$

The equivalent flowgraph is still that of Figure 2.11; however, the overall MGF is the product of these two integrals. □

As seen here, when the distribution of the sum of two independent random variables is desired, we multiply the individual MGFs to get the MGF of the distribution of the sum. This is the fundamental reason why flowgraph algebra uses MGFs.

Remark 2.4.1. The result from a flowgraph model is the MGF of the distribution of the waiting time of interest. This MGF can then be converted to a density, survivor or reliability, or hazard function.

2.4.2 General Results for Convolution

In series models, the total waiting time is a sum of independent random waiting times. The distribution of this total waiting time is a convolution of the distributions of these independent random waiting times. In a series flowgraph, this translates into a product of the MGFs of the independent random waiting time distributions that comprise the total waiting time. Hence, series flowgraphs are quite straightforward to analyze. For example, a series model with $n+1$ states has an equivalent transmittance given by the product of the MGFs of the n independent waiting times comprising the convolution. This property has been amply exploited in reliability analysis for independent and often identically distributed random variables where the sum is tractable and leads to tractable functional forms for expressions of system reliability. With a flowgraph model, the random variables need only be independent, not identically distributed. For example, the distribution of CD4+ T-cell counts, a measure of HIV disease in AIDS, is often modeled using a series model with exponential waiting times as shown in Figure 2.14 (cf. Longini et al., 1991; Satten and Longini, 1994). This nine-state model represents decreasing CD4 counts in patients with HIV and is a convolution of eight waiting time distributions, or equivalently, a series flowgraph model with nine states.

Suppose that $Y_0, \ldots, Y_{n-1}$ are independent random variables such that Y_0 is the waiting time from $0 \to 1$, Y_1 is the waiting time from $1 \to 2$, etc., and so on, and Y_{n-1} is the waiting time from $(n-1) \to n$. Then the MGF of the total waiting time $T = \sum_{i=0}^{n-1} Y_i$ is

$$M_T(s) = \prod_{i=0}^{n-1} M_{Y_i}(s), \tag{2.9}$$

where $M_{Y_i}(s)$ is the MGF of Y_i.

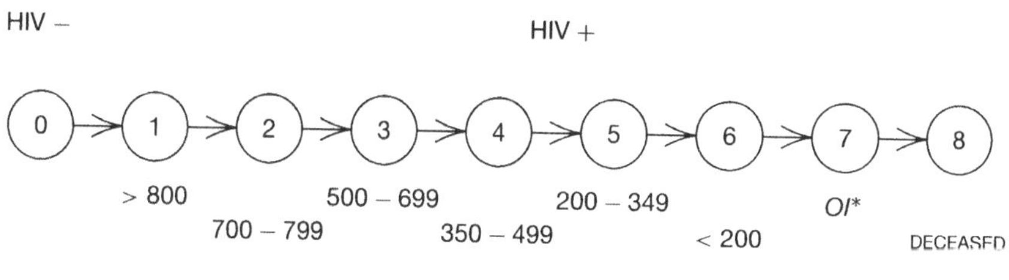

FIGURE 2.14 CD4+ T4-cell count. *OI, opportunistic infection.

2.4.3 Solving Parallel Flowgraphs

In parallel models, the total waiting time is the minimum waiting time of multiple paths from input to output, leading to flowgraphs with finite mixture distributions. For example, consider solving the open parallel flowgraph of Figure 2.6. The probability of transition from $0 \to 1$ is p_{01} and the MGF of the waiting time for passage is $M_{01}(s)$, which gives $p_{01}M_{01}(s)$ as the branch transmittance. The transition probability from $0 \to 2$ is $p_{02} = 1 - p_{01}$, with $M_{02}(s)$ as the MGF of the waiting time for passage, giving $p_{02}M_{02}(s)$ as the branch transmittance. The MGF of the overall waiting time is a finite mixture distribution: With probability p_{01} it is $M_{01}(s)$, and with probability p_{02} it is $M_{02}(s)$. Therefore, the MGF of the waiting time for first passage to either state 1 or state 2 follows from (2.2) and is given by

$$M(s) = M_{\min(Y_1,Y_2)}(s) = p_{01}M_{01}(s) + p_{02}M_{02}(s). \tag{2.10}$$

If the competitive waiting time from $0 \to 1$ were inverse Gaussian and from $0 \to 2$ were gamma, the overall waiting time distribution would be the mixture of these with the mixing parameters determined by the transition probabilities. For a general $(n + 1)$-state parallel system where transition is possible from state $0 \to 1, 0 \to 2, \ldots,$ or $0 \to n$, the MGF of the waiting time distribution for first passage to state $1,\ 2,\ \ldots,$ or n, is

$$M(s) = \sum_{j=1}^{n} p_{0j}M_{0j}(s), \qquad \text{where} \quad \sum_{j=1}^{n} p_{0j} = 1. \tag{2.11}$$

2.4.4 Solving Combinations of Series and Parallel Flowgraphs

Examples 2.5 and 2.6 dealt with combinations of series and parallel flowgraphs. To solve these flowgraphs, we consider the upper path first. Note that the path $0 \to 1 \to 2$ is a series flowgraph. Thus the transmittance of this path is

$$p_{01}M_{01}(s)M_{12}(s),$$

where $p_{12} = 1$. We can replace the original flowgraphs of Figures 2.7 and 2.8 with the one in Figure 2.15, a flowgraph that has two paths in parallel. We reduce this further by treating the flowgraph as having just two branches in parallel, as in Figures 2.5 and 2.6. If transition to state 1 occurs prior to transition to state 2, the $0 \to 1$ branch is taken with probability p_{01}, leading to the series path through the flowgraph. If transition to state 2 occurs prior to state 1, the transition is direct passage from $0 \to 2$. Solving this using (2.10) gives the overall MGF,

$$M(s) = p_{01}M_{01}(s)M_{12}(s) + p_{02}M_{02}(s). \tag{2.12}$$

Expression (2.12) is the equivalent transmittance labeling the solved equivalent flowgraph shown in Figure 2.16. This flowgraph is the reduced equivalent flowgraph of Figures 2.7 and 2.8.

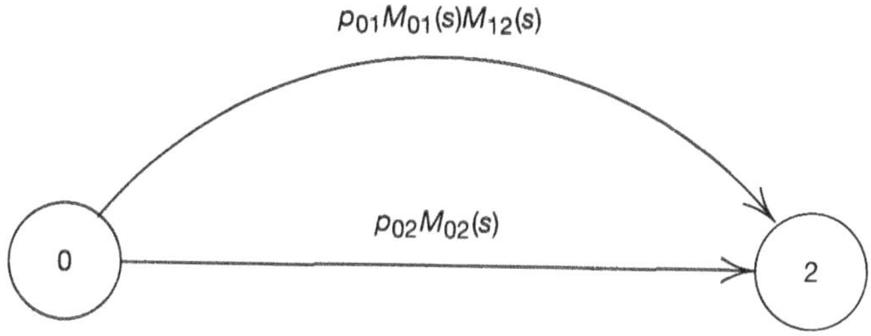

FIGURE 2.15 Reduced flowgraph model for a closed parallel system.

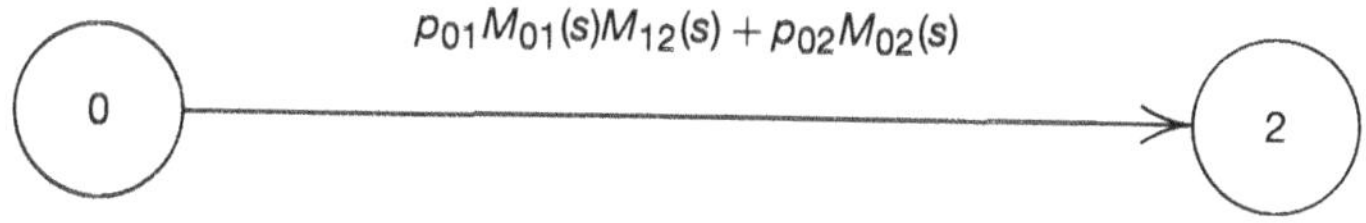

FIGURE 2.16 Solved flowgraph model for a closed parallel system.

Example 2.12: MEDICAL APPLICATION: PROGRESSION OF CANCER. Example 2.5 presented a flowgraph model for cancer progression. Recall that a patient in state 0 is in a competing risks situation in that death due to cancer (passage from $0 \to 1 \to 2$) is in competition with death due to other causes (direct passage from $0 \to 2$). We solve this flowgraph with some distributional assumptions. Let $W_1 \sim \text{Exp}(\lambda_1)$, the waiting time for passage to state 1, be independent of $W_2 \sim \text{Exp}(\lambda_2)$, the waiting time for direct passage to state 2. Then $P(W_1 < W_2)$ is the probability that the patient progresses to state 1 and is given by

$$\begin{aligned}
P(W_1 < W_2) &= \int_0^\infty \int_{w_1}^\infty f_{W_1,W_2}(w_1, w_2)\, dw_2\, dw_1 \\
&= \int_0^\infty \int_{w_1}^\infty \lambda_1 \lambda_2 e^{-\lambda_1 w_1} e^{-\lambda_2 w_2}\, dw_2\, dw_1 \\
&= \int_0^\infty \lambda_1 e^{-\lambda_1 w_1} (-e^{-\lambda_2 w_2})\Big|_{w_1}^\infty dw_1 \\
&= \int_0^\infty \lambda_1 e^{-(\lambda_1+\lambda_2)w_1}\, dw_1 \\
&= -\frac{\lambda_1}{\lambda_1+\lambda_2} e^{-(\lambda_1+\lambda_2)w_1}\Big|_0^\infty \\
&= \frac{\lambda_1}{\lambda_1+\lambda_2}. \qquad (2.13)
\end{aligned}$$

The complementary probability for direct passage to state 2 is

$$P(W_2 < W_1) = 1 - P(W_1 < W_2) = \frac{\lambda_2}{\lambda_1+\lambda_2}.$$

In state 0 we observe the minimum of W_1 and W_2. If passage to state 1 occurs before passage to state 2, the waiting time distribution is W_1 conditioned on the event $\{W_1 < W_2\}$ as in (2.2). If passage to state 2 occurs first, the waiting time distribution is W_2 conditioned on the event $\{W_2 < W_1\}$. These are *competitive waiting times*. The competing risks distribution is computed as follows:

$$\begin{aligned}
P(W_1 \le t \mid W_1 < W_2) &= \frac{P[W_1 \le \min(t, W_2)]}{P(W_1 < W_2)} \\
&= \frac{\lambda_1 + \lambda_2}{\lambda_1} \int_0^t \int_{w_1}^{\infty} \lambda_1 \lambda_2 e^{-\lambda_1 w_1} e^{-\lambda_2 w_2}\, dw_2\, dw_1 \\
&= \frac{\lambda_1 + \lambda_2}{\lambda_1} \int_0^t \lambda_1 e^{-\lambda_1 w_1} (-e^{-\lambda_2 w_2})|_{w_1}^{\infty}\, dw_1 \\
&= (\lambda_1 + \lambda_2) \int_0^t e^{-(\lambda_1+\lambda_2) w_1}\, dw_1 \\
&= -e^{-(\lambda_1+\lambda_2) w_1}|_0^t \\
&= 1 - e^{-(\lambda_1+\lambda_2)t} \qquad \text{for } t > 0.
\end{aligned} \tag{2.14}$$

The corresponding density function and MGF are

$$\begin{aligned}
f_{W_1|W_1<W_2}(t) &= (\lambda_1 + \lambda_2) e^{-(\lambda_1+\lambda_2)t} \qquad \text{for } t > 0 \\
M(s) &= \frac{\lambda_1 + \lambda_2}{\lambda_1 + \lambda_2 - s}.
\end{aligned}$$

The probability that W_2 occurs first is derived analogously (see Exercise 2.4). In this case, due to the independent exponential distributional assumptions, we have

$$W_1 \mid W_1 < W_2 \overset{d}{=} W_2 \mid W_2 < W_1 \overset{d}{=} \min(W_1, W_2),$$

where $\overset{d}{=}$ denotes *equal in distribution*, but in general, this is not the case. We can calculate the competing risks distribution as the distribution of $W_{(1)} = \min(W_1, W_2)$, plugging (2.13) and (2.14) into (2.2) to obtain

$$P(W_{(1)} \le t) = 1 - e^{-(\lambda_1+\lambda_2)t}, \tag{2.15}$$

which is again an Exp$(\lambda_1 + \lambda_2)$ distribution. The waiting time from $1 \to 2$, W_3, is assumed to be independent of W_1 and W_2 and has an Exp(λ_3) distribution. This flowgraph has two paths, $0 \to 1 \to 2$ and $0 \to 2$. To solve this flowgraph, we consider the upper path first. Note that the path $0 \to 1 \to 2$ is a series flowgraph. Thus, the transmittance of the upper path is

$$p_{01} M_{01}(s) p_{12} M_{12}(s) = \left(\frac{\lambda_1}{\lambda_1 + \lambda_2}\right) \left(\frac{\lambda_1 + \lambda_2}{\lambda_1 + \lambda_2 - s}\right) \left(\frac{\lambda_3}{\lambda_3 - s}\right)$$

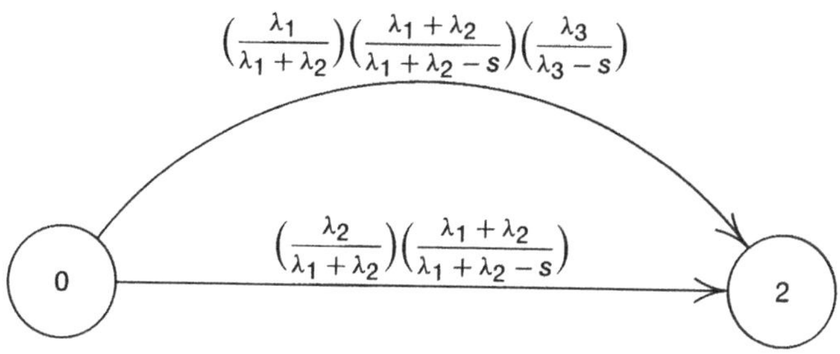

FIGURE 2.17 Reduced flowgraph model for cancer progression.

where $p_{12} = 1$. The transmittance of the lower path is

$$p_{02}M_{02} = \left(\frac{\lambda_2}{\lambda_1+\lambda_2}\right)\left(\frac{\lambda_1+\lambda_2}{\lambda_1+\lambda_2-s}\right).$$

We can replace the original flowgraph with the one shown in Figure 2.17, a flowgraph that has two paths in parallel. The survival time of a patient in state 0 is now a finite mixture distribution: With probability $\lambda_2/(\lambda_1+\lambda_2)$ it is Exp$(\lambda_1+\lambda_2)$, and with probability $\lambda_1/(\lambda_1+\lambda_2)$ it is the convolution of an Exp$(\lambda_1+\lambda_2)$ with an independent Exp(λ_3), the passage time from state 1 to 2. The MGF of the time to death is

$$M(s) = \left(\frac{\lambda_2}{\lambda_1+\lambda_2}\right)\left(\frac{\lambda_1+\lambda_2}{\lambda_1+\lambda_2-s}\right) + \left(\frac{\lambda_1}{\lambda_1+\lambda_2}\right)\left(\frac{\lambda_1+\lambda_2}{\lambda_1+\lambda_2-s}\right)\left(\frac{\lambda_3}{\lambda_3-s}\right)$$
$$\text{for } s < \min(\lambda_1+\lambda_2, \lambda_3). \qquad (2.16)$$

Expression (2.16) is the equivalent transmittance labeling the solved equivalent flowgraph shown in Figure 2.18. □

In this analysis, we assumed unconditional independent exponential waiting time distributions for the failure times of the kidneys and used these to drive the competing risks distributions. With this approach, one has to be comfortable assuming a distribution for a waiting time that is never observed. We observe $W_1 \mid \{W_1 < W_2\}$ or $W_2 \mid \{W_2 < W_1\}$. We do not observe W_1 or W_2 directly. Furthermore, this analytical derivation of the competing risks distribution is possible only because we have exponential distributions, and these calculations will be difficult either analytically

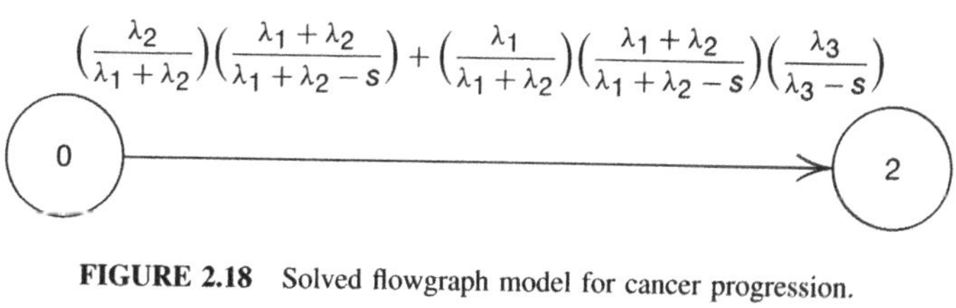

FIGURE 2.18 Solved flowgraph model for cancer progression.

or numerically for richer families of distributions. Flowgraphs use the observed data to model the conditional distributions directly. For example, if the observed data indicated that the waiting time from $0 \rightarrow 1$ would be best modeled by a gamma distribution and the waiting time from $0 \rightarrow 2$ would be best modeled by a Weibull distribution, the MGFs corresponding to these distributions would be used in (2.12).

2.4.5 Solving Flowgraphs with Feedback Loops

Series and parallel elements are two basic components of flowgraphs. Series elements lead to distributions of sums of independent random variables, whereas parallel elements lead to finite mixtures of distributions of independent random variables. The feedback loop is the third basic element of a flowgraph model. For example, in the cancer example of Figure 2.1, there is one loop, $A \rightarrow R \rightarrow A$. A patient can pass through this loop more than once. Feedback loops arise naturally in both medical and engineering settings. The feedback loop has a nice probabilistic relationship to the geometric distribution. Figure 2.19 shows a flowgraph model representation of the geometric distribution. The states are the number of successes. The feedback loop is taken each time a failure is observed and it is taken with the probability of failure $(1 - p)$. When a success is observed, the process is complete and transition occurs to state 1 with probability p. Since we are not interested in the waiting times or tracking data on them, they are set to constant and the flowgraph has the MGF of the constant 1 on the branches. The MGF of a geometric is

$$M(s) = \frac{pe^s}{1 - (1 - p)e^s}. \tag{2.17}$$

The numerator is the $0 \rightarrow 1$ transmittance and the denominator is 1 minus the $0 \rightarrow 0$ transmittance, the feedback portion. There are three different methods for deriving the overall MGF of a feedback loop. The first method is direct and is derived using a medical example, Example 2.13. The second method is engineering based and related to balance equations much like those used to derive Kirchhoff's laws (cf. Thomas and Rosa, 2003) in circuit analysis. This method is illustrated in Example 2.14. The section concludes with a technical derivation of the feedback loop MGF using random variables and conditional probability arguments.

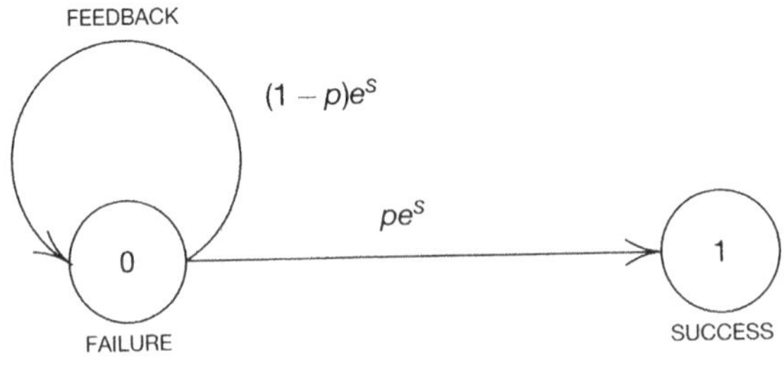

FIGURE 2.19 Flowgraph model for a geometric distribution.

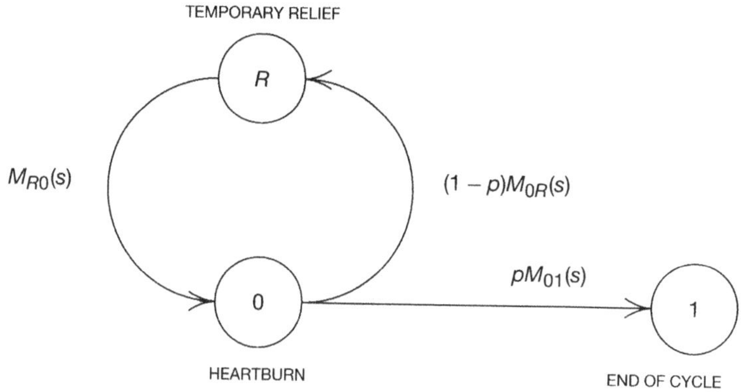

FIGURE 2.20 Flowgraph model for heartburn.

Example 2.13: MEDICAL EXAMPLE: HEARTBURN. Figure 2.20 shows a flowgraph model for a cycle of heartburn. A patient is in state 0 at the first sign of experiencing heartburn. The patient can take medication that will provide temporary relief, R, putting the patient in a loop around state 0. The relief will wear off, returning the patient to state 0, at which time he would again take medication for temporary relief. This could take several cycles, but eventually the patient's heartburn cycle will end and he will move to state 1, end of cycle. Let T be the total time of the heartburn cycle. Note that the states $0 \to R \to 0$ are in series and we can reduce that portion of the feedback loop. For simplicity, let

$$M_{00}(s) = M_{0R}(s)M_{R0}(s)$$

so that the reduced equivalent transmittance of the feedback loop is

$$(1-p)M_{00}(s) = (1-p)M_{0R}(s)M_{R0}(s).$$

The reduced flowgraph model is shown in Figure 2.21.

Beginning in state 0, a patient's heartburn cycle could end, giving direct passage to state 1. This gives an overall waiting time MGF of $pM_{01}(s)$. If the patient takes the feedback loop once, his contribution would be magnified by the transmittance

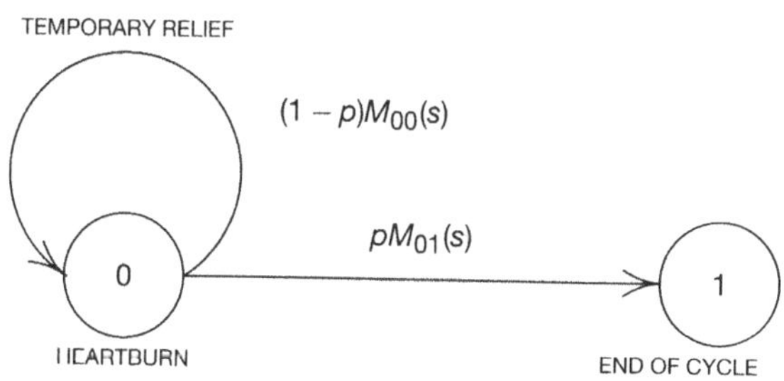

FIGURE 2.21 Partially reduced flowgraph model for heartburn.

of the loop, so the total waiting time MGF would be $pM_{01}(s)(1-p)M_{00}(s)$. If he takes the feedback loop twice, this becomes $pM_{01}(s)[(1-p)M_{00}(s)]^2$, and three times gives $pM_{01}(s)[(1-p)M_{00}(s)]^3$. Iterating in this way gives the overall MGF,

$$\begin{aligned}
M_T(s) &= pM_{01}(s) + pM_{01}(s)(1-p)M_{00}(s) + pM_{01}(s)[(1-p)M_{00}(s)]^2 \\
&\quad + pM_{01}(s)[(1-p)M_{00}(s)]^3 + \cdots \\
&= pM_{01}(s)\{1 + (1-p)M_{00}(s) + [(1-p)M_{00}(s)]^2 + \cdots\} \\
&= pM_{01}(s)\sum_{j=0}^{\infty}[(1-p)M_{00}(s)]^j \\
&= pM_{01}(s)\frac{1}{1-(1-p)M_{00}(s)} \qquad \text{for } |(1-p)M_{00}(s)| < 1, \qquad (2.18)
\end{aligned}$$

where the last condition is true for an open neighborhood containing $s = 0$. An alternative way to conceptualize this loop is to consider the two branches, $0 \to R$ and $0 \to 1$, exiting state 0 as branches in parallel. Recall that parallel flowgraphs lead to finite mixture distributions. In this case, each time the feedback loop is taken, another term in the mixture is generated so that we have infinite parallel branches. This is an infinite mixture with weights p, $p(1-p)$, $p(1-p)^2, \ldots$. This direct approach of (2.18) is the least elegant of the three feedback loop derivations. □

Example 2.14: Engineering Application: Hydraulic Pump with Feedback. We return to the full hydraulic pump system of Example 2.8. To solve this flowgraph (Figure 2.10) for the MGF of T, $M_T(s)$, we set up a balance equation at state 0. A balance equation equates the inputs into a state with the outputs of the state. The idea is based on conservation laws applied to systems and originates from Kirchhoff's laws governing circuit behavior, which are derived from conservation laws applied to circuits. In our case, the input into state 0 is $p_{10}M_{10}(s)$. The output from state 0 is $M_{01}(s)$. This is magnified by the factor $M_T(s)$, which is the MGF of the loop that is fed back into itself. In systems analysis, this factor is called the *gain* of the system. The balance equation for state 0 in Figure 2.10 is

$$M_T(s) = M_{01}(s)p_{10}M_{10}(s)M_T(s) + M_{01}(s)p_{12}M_{12}(s). \qquad (2.19)$$

The right-hand side of this equation contains a mixture of two parallel components: taking the loop and returning to the initial state and immediately exiting the system. Note that the term $M_{01}(s)$ appears in both terms on the right-hand side since it is a transmittance passed through the system. The first term on the right-hand side is the feedback term from the loop. This is multiplied by $M_T(s)$ since that MGF is fed back into itself through the loop; hence $M_T(s)$ appears on both sides of the equation. Solving for $M_T(s)$ gives

$$M_T(s) = \frac{p_{12}M_{01}(s)M_{12}(s)}{1 - p_{10}M_{01}(s)M_{10}(s)}, \qquad \text{where} \quad p_{12} + p_{10} = 1. \qquad (2.20)$$

This is the overall MGF of the distribution of T.

Alternatively, we can solve the feedback loop using a more formal argument. Let N be a geometric random variable that represents the feedback loop of Figure 2.19. Let U be the random waiting time spent in the loop so that U has density $f_U(u)$ and MGF $M_{00}(s)$. Let V, independent of U, be the random waiting time until state 1 is reached so that V has density $f_V(v)$ and MGF $M_{01}(s)$. The distribution of waiting time (first passage time) from $0 \to 1$ is the convolution of N distributions $f_U(u)$ and a single distribution $f_V(v)$ so that the total waiting time, W, for $0 \to 1$ can be written as

$$W = \sum_{i=1}^{N} U_i \; + \; V.$$

The density of W is

$$f_W(w) = \sum_{j=0}^{\infty} f_W(w|N = j)P(N = j).$$

The MGF of W is

$$\begin{aligned} M_W(s) &= E(e^{sW}) \\ &= \sum_{j=0}^{\infty} \int e^{sw} f_W(w|N = j)P(N = j)\,dw \\ &= \sum_{j=0}^{\infty} (1 - p)^j p \int e^{sw} f_W(w|N = j)\,dw \end{aligned} \tag{2.21}$$

where integration and summation can be interchanged as a consequence of the monotone convergence theorem (cf. Billingsley, 1986, Theorem 16.6). Noting that the integral in (2.21) is the MGF of $W|N = j$, we substitute

$$M_{W|N=j}(s) = [M_U(s)]^j M_V(s),$$

giving

$$\begin{aligned} M_W(s) &= \sum_{j=0}^{\infty} (1 - p)^j p[M_U(s)]^j M_V(s) \\ &= pM_V(s) \sum_{j=0}^{\infty} [(1 - p)M_U(s)]^j \end{aligned}$$

$$= \frac{pM_V(s)}{1-(1-p)M_U(s)} \quad \text{for } |(1-p)M_U(s)| < 1$$

$$= \frac{pM_{01}(s)}{1-(1-p)M_{00}(s)}. \tag{2.22}$$

Note that $|(1-p)M_U(s)| < 1$ holds in an open neighborhood of 0 as $M_U(s)$ is a MGF. Therefore, $M_W(s)$ is a MGF. □

2.4.6 Combining Series, Parallel, and Loop Flowgraphs

In this section we present a detailed solution of a flowgraph with combined series, parallel, and loop structures. We use a general reversible illness–death model as shown in Figure 2.22. Note that without the transition from state 1 to state 0, this is identical to the flowgraphs of Figures 2.7 and 2.8. State 0 represents the healthy state. From this state, a patient will transition to a diseased state with probability p_{01} or die from other causes with probability $p_{02} = 1 - p_{01}$. The waiting time distributions for these events to occur have MGFs $M_{01}(s)$ and $M_{02}(s)$. Once in a diseased state, a patient can recover to a healthy state with probability p_{10}, or die from the disease with probability $p_{12} = 1 - p_{10}$. The waiting time distributions for these events to occur have MGFs $M_{10}(s)$ and $M_{12}(s)$. Suppose that our interest is in the waiting time from $0 \to 2$ regardless of the transitions made. We illustrate the solution of this flowgraph model for the MGF of the $0 \to 2$ waiting time distribution.

To solve this flowgraph, consider the lower path, $0 \to 2$, and the upper path, $0 \to 1 \to 2$, separately. The lower path is shown in the subflowgraph of Figure 2.23. This flowgraph has the same structure as the heartburn flowgraph of Figure 2.20 and is redrawn in Figure 2.24. Transition from $0 \to 2$ can occur directly or involve the feedback loop, $0 \to 1 \to 0$. We reduce the lower path by reducing the feedback loop in Figure 2.24. The transmittance of the lower path is derived using (2.22) to give

$$p_{02}\left[\frac{M_{02}(s)}{1 - p_{01}p_{10}M_{01}(s)M_{10}(s)}\right]. \tag{2.23}$$

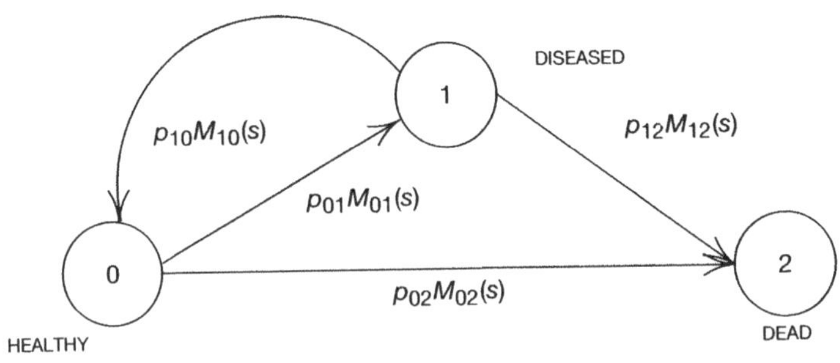

FIGURE 2.22 Flowgraph model for a reversible illness–death multistate model.

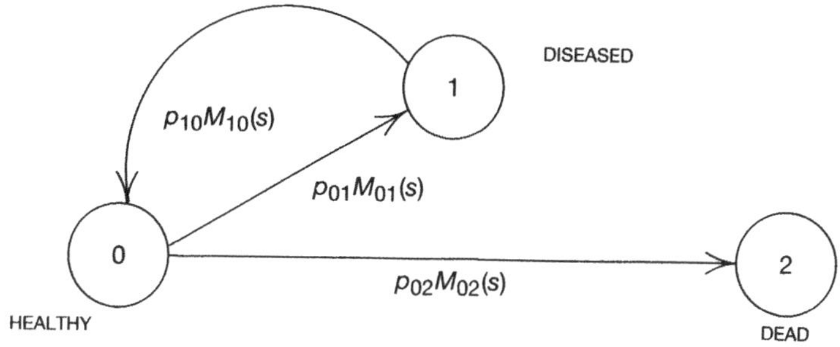

FIGURE 2.23 Subflowgraph for the lower path of Figure 2.22.

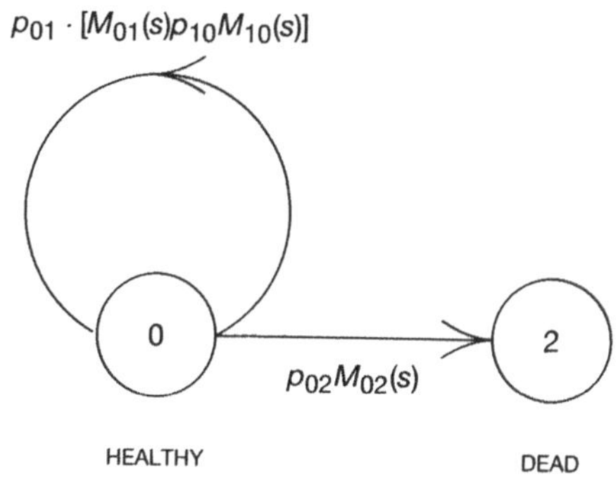

FIGURE 2.24 Reducing the lower path to solve Figure 2.22.

We have written this loosely in the format of a transmittance (i.e., as a probability multiplied by the MGF of the waiting time distribution). Note that this transmittance is not a MGF because passage from $0 \to 2$ via the lower path is not certain to occur. If p_{10} were equal to 1, this transmittance would be a MGF. In (2.23), we are treating p_{02} loosely as the probability of taking the lower path. In fact, the probability of taking the lower path is always greater than p_{02}; it is $p_{02}/(1 - p_{10}p_{01})$. Similarly, the expression within the brackets is not quite "the MGF of the lower path." We can divide this quantity by its value at $s = 0$ to convert it to a proper MGF:

$$p_{02}\left[\frac{M_{02}(s)}{1 - p_{01}p_{10}M_{01}(s)M_{10}(s)}\right]$$
$$= \left(\frac{p_{02}}{1 - p_{01}p_{10}}\right)\left[\frac{(1 - p_{01}p_{10})M_{02}(s)}{1 - p_{01}p_{10}M_{01}(s)M_{10}(s)}\right]. \qquad (2.24)$$

The right-hand side of (2.24) is now a "probability" multiplied by a "MGF."

The upper path is shown in the subflowgraph of Figure 2.25. This has the same structure as the flowgraph for the hydraulic pump system with feedback of Figure 2.10. To reduce the upper path, we must reduce the feedback loop, $1 \to 0 \to 1$. This loop can be reduced to state 0 or to state 1. For illustration, we

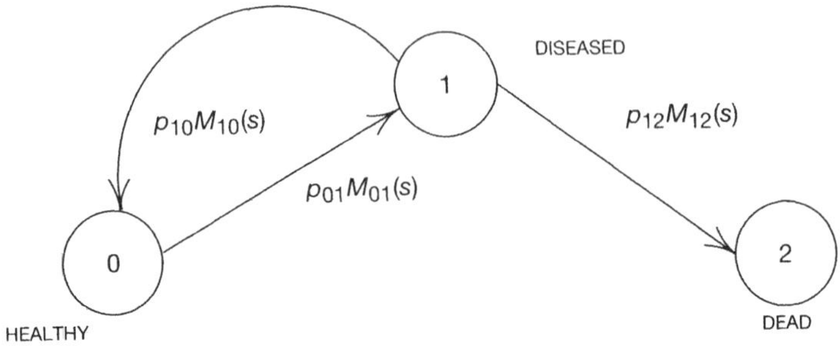

FIGURE 2.25 Subflowgraph for the upper path of Figure 2.22.

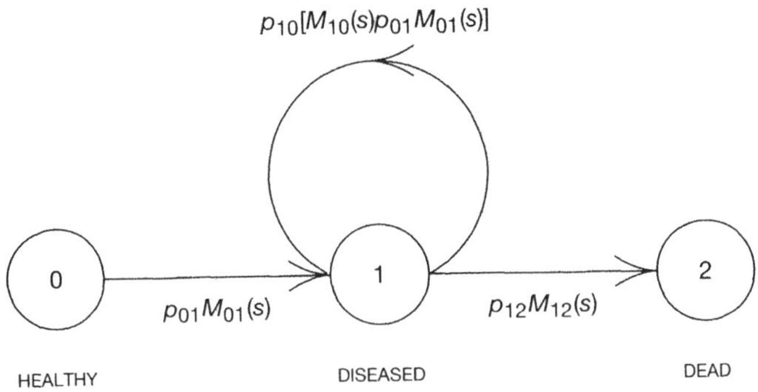

FIGURE 2.26 Reducing the feedback loop for the upper path.

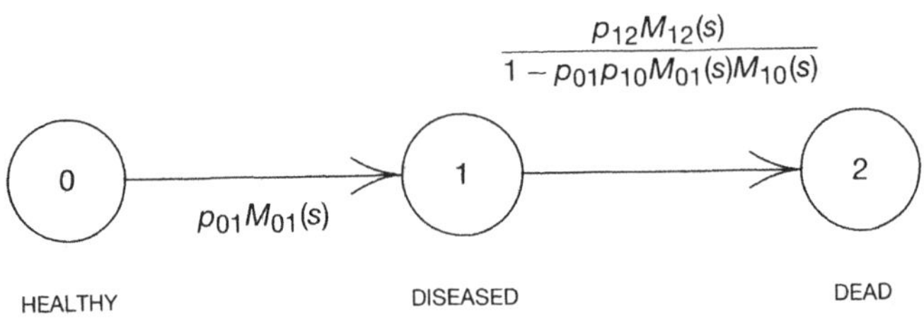

FIGURE 2.27 Reduced upper path for the flowgraph of Figure 2.22.

reduce it to state 1 as shown in Figure 2.26. Figure 2.27 shows the fully reduced loop for the upper path. The transmittance of the upper path is

$$p_{01}\left[\frac{p_{12}M_{01}(s)M_{12}(s)}{1-p_{01}p_{10}M_{01}(s)M_{10}(s)}\right].$$

As with the lower path, this is loosely written in the form of a probability times a MGF. We can write this as a probability times a MGF if we divide the term in

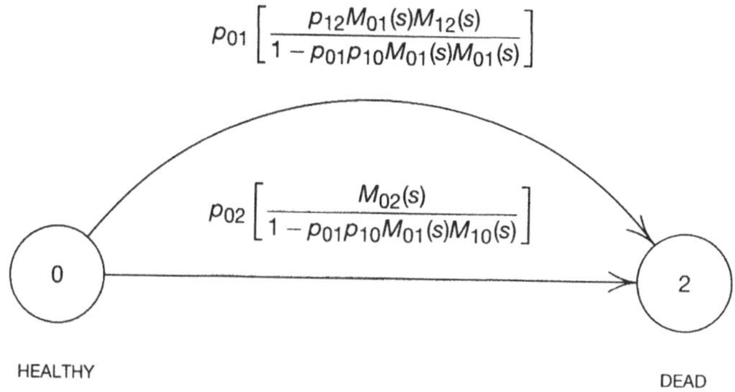

FIGURE 2.28 Reduced flowgraph model for Figure 2.22.

brackets by its value at $s = 0$ to give

$$\left(\frac{p_{01}p_{12}}{1-p_{10}p_{01}}\right)\left[\frac{(1-p_{10}p_{01})M_{01}(s)M_{12}(s)}{1-p_{01}p_{10}M_{01}(s)M_{10}(s)}\right].$$

The reduced lower and upper paths can be combined to give the flowgraph of Figure 2.28. This flowgraph is now a parallel flowgraph and we can reduce it to give the overall waiting time MGF for the reversible illness death model:

$$M(s) = \frac{p_{01}p_{12}M_{01}(s)M_{12}(s) + p_{02}M_{02}(s)}{1-p_{01}p_{10}M_{01}(s)M_{10}(s)}. \tag{2.25}$$

2.5 SYSTEMATIC PROCEDURE FOR SOLVING FLOWGRAPHS

To solve a complicated flowgraph such as Figure 2.1 requires several steps for reducing the series, parallel, and loop components. We illustrated this in Section 2.4 for the reversible illness–death model of Figure 2.22. For such complex structures we use a procedure based on Mason's rule (cf. Mason, 1953) to solve flowgraphs. Mason's rule was developed in the context of graph theory for solving systems of linear equations. In its original implementation, Mason's rule did not involve probabilities or MGFs. However, flowgraph models can be solved by applying Mason's rule to the branch transmittances. This gives a systematic procedure for computing the transmittance of the flowgraph from any state A to any state B for any finite system. When the system is certain to pass eventually from state A to state B, the transmittance is the MGF of the waiting time distribution. Practical use of Mason's rule entails identifying all of the paths from A to B and the loops involved in those paths. It requires computing the transmittance for every distinct path from the initial state to the end state and adjusting for the transmittances of various loops. We illustrate using the cancer example of Figure 2.1. Suppose that our initial state is I and that our waiting time of interest is simply death, D_1 or D_2.

Path transmittances are simply products. There are four paths from $I \to D_1$ and $I \to D_2$: $I \to A \to D_1$, $I \to A \to D_2$, $I \to D_2$, and $I \to A \to R \to D_2$.

TABLE 2.1 Summary of Paths and Loops for Solving the Cancer Flowgraph

Path	Transmittance
Path 1: $I \to A \to D_1$	$p_{IA}p_{AD_1}M_{IA}(s)M_{AD_1}(s)$
Path 2: $I \to A \to D_2$	$p_{IA}p_{AD_2}M_{IA}(s)M_{AD_2}(s)$
Path 3: $I \to D_2$	$p_{ID_2}M_{ID_2}(s)$
Path 4: $I \to A \to R \to D_2$	$p_{IA}p_{AR}p_{RD_2}M_{IA}(s)M_{AR}(s)M_{RD_2}(s)$
First-order loop:	
$A \to R \to A$	$p_{AR}p_{RA}M_{AR}(s)M_{RA}(s)$
Loop not touching path:	
Path 3: $A \to R \to A$	$p_{AR}p_{RA}M_{AR}(s)M_{RA}(s)$

The transmittances of these paths and all other transmittances required for the application of Mason's rule are given in Table 2.1. Mason's rule differentiates among different types of loops. These are defined as follows.

Definition 2.5.1. A *first-order loop* is any closed path that returns to the initiating node without passing through any node more than once.

Definition 2.5.2. A *second-order loop* consists of two nontouching first-order loops.

Definition 2.5.3. A *jth-order loop* consists of j nontouching first-order loops.

Figure 2.1 contains one *first-order* loop, $A \to R \to A$. The transmittance of a first-order loop is again the product of the individual branch transmittances involved in the path. Table 2.1 gives the transmittance for this loop. The transmittance of a higher-order loop is the product of the transmittances of the first-order loops it contains. There are no higher-order loops in this flowgraph.

The general form of Mason's rule gives the overall transmittance i.e., the MGF from input to output, as

$$M(s) = \frac{\sum_i P_i(s)[1 + \sum_j (-1)^j L_j^i(s)]}{1 + \sum_j (-1)^j L_j(s)}, \tag{2.26}$$

where:

- $P_i(s)$ is the transmittance for the ith path.
- $L_j(s)$ in the denominator is the sum of the transmittances over the jth-order loops.
- $L_j^i(s)$ is the sum of the transmittances over jth-order loops sharing no common nodes with the ith path (i.e., loops not touching that path).

There is one first-order loop that does not touch path 3, so that

$$L_1^3(s) = p_{AR}p_{RA}M_{AR}(s)M_{RA}(s).$$

For this cancer model, (2.26) is

$$M(s) = \frac{P_1(s) + P_2(s) + P_3(s)[1 - L_1^3(s)] + P_4(s)}{1 - L_1(s)}, \tag{2.27}$$

where

$$P_1(s) = p_{IA}p_{AD_1}M_{IA}(s)M_{AD_1}(s)$$
$$P_2(s) = p_{IA}p_{AD_2}M_{IA}(s)M_{AD_2}(s)$$
$$P_3(s) = p_{ID_2}M_{ID_2}(s)$$
$$P_4(s) = p_{IA}p_{AR}p_{RD_2}M_{IA}(s)M_{AR}(s)M_{RD_2}(s)$$
$$L_1(s) = p_{AR}p_{RA}M_{AR}(s)M_{RA}(s).$$

Example 2.15: ENGINEERING EXAMPLE: CAR STEREO MANUFACTURING. Figure 2.29 is a flowgraph model of a complex system consisting of outcomes in series and cascaded in parallel with feedback loops. This flowgraph has higher-order loops. The system is an assembly line for a manufacturing process for car stereos. State 0 represents an initial detection of a problem with a stereo. The problem is categorized into either severity type A or severity type B. Type A problems consist of problems with wiring, sound, and so on. Type B problems are considered less severe and are cosmetic problems with materials finish, fittings of buttons and switches, and so on. If the severity is of type A, the item is repaired at station A, represented by state 1. Once the problem is fixed, the item is inspected at inspection station A, represented by state 3. If the item does not pass inspection, it is returned to repair

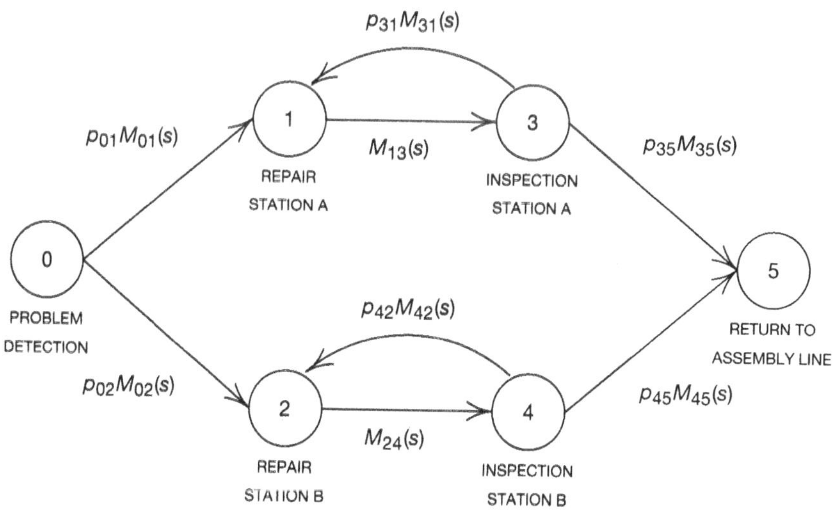

FIGURE 2.29 Flowgraph model for a manufacturing system.

TABLE 2.2 Summary of Paths and Loops for Solving the Manufacturing System Flowgraph

Path	Transmittance
Path 1: $0 \to 1 \to 3 \to 5$	$p_{01}p_{35}M_{01}(s)M_{13}(s)M_{35}(s)$
Path 2: $0 \to 2 \to 4 \to 5$	$p_{02}p_{45}M_{02}(s)M_{24}(s)M_{45}(s)$
First-order loops:	
$1 \to 3 \to 1$	$p_{31}M_{13}(s)M_{31}(s)$
$2 \to 4 \to 2$	$p_{42}M_{24}(s)M_{42}(s)$
Second-order loops:	
$1 \to 3 \to 1$ and $2 \to 4 \to 2$	$p_{31}p_{42}M_{13}(s)M_{31}(s)M_{24}(s)M_{42}(s)$
Loop not touching path:	
Path 1: $2 \to 4 \to 2$	$p_{42}M_{24}(s)M_{42}(s)$
Path 2: $1 \to 3 \to 1$	$p_{31}M_{13}(s)M_{31}(s)$

state A as represented by the $3 \to 1$ transition. Eventually, the item is permanently fixed and returned to the assembly line as represented by state 5. The lower half of the flowgraph represents an analogous process for the less severe type B problems.

This flowgraph has two paths: $0 \to 1 \to 3 \to 5$ and $0 \to 2 \to 4 \to 5$. There are two first-order loops and they do not touch each other. There is one second-order loop, made up of the two nontouching first-order loops. Table 2.2 gives a summary of the paths and loops along with their transmittances. For this flowgraph, the MGF of the overall waiting time distribution i.e., the waiting time from $0 \to 5$, is found using (2.26):

$$M(s) = \frac{P_1(s)[1 - L_1^1(s)] + P_2(s)[1 - L_1^2(s)]}{1 - L_1(s) + L_2(s)}, \tag{2.28}$$

where

$$P_1(s) = p_{01}p_{35}M_{01}(s)M_{13}(s)M_{35}(s)$$
$$P_2(s) = p_{02}p_{45}M_{02}(s)M_{24}(s)M_{45}(s)$$
$$L_1(s) = p_{31}M_{13}(s)M_{31}(s) + p_{42}M_{24}(s)M_{42}(s)$$
$$L_2(s) = p_{31}p_{42}M_{13}(s)M_{31}(s)M_{24}(s)M_{42}(s)$$
$$L_1^1(s) = p_{42}M_{24}(s)M_{42}(s)$$
$$L_1^2(s) = p_{31}M_{13}(s)M_{31}(s). \qquad \square$$

Although expressions (2.27) and (2.28) may appear complicated, in general, we do not calculate these by hand. Having a systematic procedure such as (2.26) allows us to program equations (2.27) and (2.28) using symbolic algebra. In Chapter 3 we learn how to do this using the symbolic algebra package MAPLE. Solving flowgraph models gives us the MGF of the waiting time distribution of interest; however, we still have not obtained the distribution itself. In Chapter 3 we describe how these flowgraph MGFs are converted into density, survivor or reliability, and hazard functions.

2.6 FLOWGRAPHS FOR COUNTS

Sometimes our quantity of interest is not the time to an event but rather the number of occurrences of an event. In this section we present a brief introduction to the use of flowgraphs to access counts. The example is based on a reliability repair problem due to Lloyd and Lipow (1964) and considered in Whitehouse (1970, 1983). Suppose that a device is being developed for a given application. When the device is put into operation, it either succeeds or fails to perform as designed. The device is simple enough that only one thing can go wrong and the device will eventually fail due to this fault. The purpose of the development effort is to discover the cause of failure and attempt to redesign the device so that it will not fail at all. The device is tested repeatedly as follows. If the device works on a given trial, it proceeds to its next trial. The probability that the device fails on any given trial is p_{12}. If the device fails, it is worked on and there is a probability p_{23} that it is fixed. The outcomes on a given trial are:

1. The trial is successful given that the device is faulty.
2. The trial is a failure given that the device is faulty.
3. The trial is successful given that the device is fixed.

These outcomes represent the states of the system in the flowgraph of Figure 2.30. In this context, $M_{11}(s)$ and $M_{12}(s)$ represent MGFs of testing times, and $M_{21}(s)$, $M_{22}(s)$, and $M_{23}(s)$ are MGFs of repair times. State 0 is the start of the process and we assume that every device reached state 1 instantly i.e., we assume a constant probability 1 and a constant waiting time of 0 to reach state 1. The model assumes

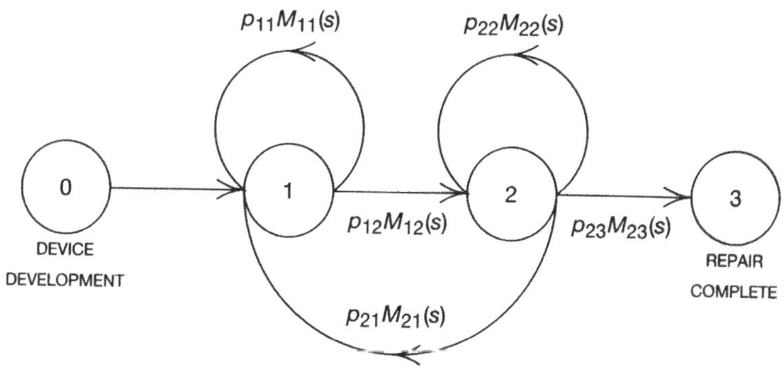

FIGURE 2.30 Flowgraph model for a reliability repair problem.

that the probabilities are related as follows:

$$p_{12} = 1 - p_{11}$$
$$p_{22} = (1 - p_{11})(1 - p_{23})$$
$$p_{21} = (1 - p_{23})p_{11}.$$

The flowgraph MGF is

$$M(s) = p_{12}p_{23}M_{12}(s)M_{23}(s)[1 - p_{11}M_{11}(s) - p_{22}M_{22}(s) - p_{21}p_{12}M_{21}(s)M_{12}(s) + p_{11}p_{22}M_{11}(s)M_{22}(s)]^{-1}. \tag{2.29}$$

The derivation of this MGF is left as an exercise for the reader (Exercise 2.5). Notice that this flowgraph has a second-order loop because the self-loops at states 1 and 2 do not touch.

Suppose that our interest is in how many times the loop around state 1 is taken. Let N_{11} be the random variable that counts the number of times the $1 \to 1$ loop is taken. One approach to computing the distribution of N_{11} with flowgraphs is to create an auxiliary constant 1 and attach its MGF, e^u, to the branch in question. This is similar to the idea of tagging for delayed signals from systems analysis (cf. Sittler, 1956). This means that the transmittance about the $1 \to 1$ loop is

$$p_{11}e^u M_{11}(s).$$

We can compute the overall waiting time MGF as in (2.29). This is the joint MGF of the distribution of the overall waiting time T and N_{11}, $M(s, u)$,

$$M(s,u) = p_{12}p_{23}M_{12}(s)M_{23}(s)[1 - p_{11}e^u M_{11}(s) - p_{22}M_{22}(s) - p_{21}p_{12}M_{21}(s)M_{12}(s) + p_{11}p_{22}e^u M_{11}(s)M_{22}(s)]^{-1}. \tag{2.30}$$

We can get the marginal MGF of the distribution of the count N_{11} as

$$M(s,u)|_{s=0} = \frac{p_{12}p_{23}}{1 - p_{22} - p_{21}p_{12} - e^u p_{11}(1 - p_{22})}$$
$$= \frac{p_{12}p_{23}/(1 - p_{22} - p_{21}p_{12})}{1 - p_{11}(1 - p_{22})e^u/(1 - p_{22} - p_{21}p_{12})} \tag{2.31}$$

This is the MGF of a geometric distribution with success probability

$$p = \frac{p_{12}p_{23}}{1 - p_{22} - p_{21}p_{12}}. \tag{2.32}$$

Note that (2.31) differs from (2.17) as the count N_{11} starts at 0 here whereas the counting starts at 1 in (2.17). It takes some algebra to verify these results. Similarly, we can derive the distribution of the count N_{22}, the number of times the loop around state 2 is taken. We derive the MGF of N_{21}, the count of the number of times the $2 \to 1$ branch is taken. This time we attach an e^u term to the $2 \to 1$ branch and proceed to find the joint MGF of T and N_{21},

$$M(s,u) = p_{12}p_{23}M_{12}(s)M_{23}(s)[1 - p_{11}M_{11}(s) - p_{22}M_{22}(s) - p_{21}p_{12}e^u M_{21}(s)M_{12}(s) + p_{11}p_{22}M_{11}(s)M_{22}(s)]^{-1}. \tag{2.33}$$

The marginal MGF of N_{21} is

$$\begin{aligned} M(s,u)|_{s=0} &= \frac{p_{12}p_{23}}{1 - p_{11} - p_{22} - p_{21}p_{12}e^u + p_{11}p_{22}} \\ &= \frac{p_{23}}{p_{23} + p_{11}(1 - p_{23}) - p_{21}e^u} \\ &= \frac{p_{23}/[p_{23} + p_{11}(1 - p_{23})]}{1 - p_{21}e^u/[p_{23} + p_{11}(1 - p_{23})]} \end{aligned} \tag{2.34}$$

The idea of attaching MGFs of auxiliary constants to branches extends to deriving conditional distributions such as $N_{11} \mid N_{21}$; however, this is beyond the scope of this book. The distributions of N_{11} and N_{21} can be derived using probability theory, which is left as an exercise for the reader.

EXERCISES

2.1 Consider the flowgraph of Figure 2.22.

(a) Reduce this flowgraph by reducing the loop to state 0 rather than state 1 for the upper path. Verify that the $0 \to 2$ transmittance is the same as (2.25).

(b) Use Mason's rule to derive the overall transmittance from $0 \to 2$.

2.2 Consider the cancer progression flowgraph of Figure 2.1. Modify this flowgraph by removing the branches from $R \to D_2$ and $I \to D_2$. Find the MGF of the distribution of the overall survival time of a cancer patient, $I \to (D_1 \text{ or } D_2)$.

2.3 Consider the *prisoner's dilemma problem*, given under various guises in many introductory probability texts (cf. Ross, 1988, p. 286). A prisoner is locked in a cell containing three doors. The first door leads to a tunnel that will take him to freedom after three days. The second door leads to a tunnel that will return him to the cell after five days. The third door leads to a tunnel that will return him to the cell after seven days. Assume that the prisoner is at all times equally likely to choose any one of the doors to escape.

(a) Using your knowledge of conditional probabilities, find the expected length of time until the prisoner reaches freedom.

(b) Construct a flowgraph model for this problem. Recall that the MGF of a constant k is just e^{ks}.

(c) Find the MGF of the distribution of the overall waiting time until the prisoner is free.

(d) How would your analysis change if each waiting time were a continuous distribution rather than a constant?

2.4 In Example 2.12, the cancer progression example, derive the conditional probability that the patient dies from other causes without progression to an advanced stage of cancer [i.e., $P(W_2 \leq t \mid W_2 < W_1)$].

2.5 Consider the reliability repair flowgraph of Figure 2.30.

(a) Derive the MGF of the distribution of the overall waiting time T using Mason's rule and verify that it is the same as (2.29).

(b) Derive the marginal MGF of the distribution of N_{22}, the number of times the branch from $2 \to 2$ is taken.

(c) Derive the distributions of N_{11} and N_{21} using principles of probability theory.

CHAPTER 3

Inversion of Flowgraph Moment Generating Functions

The flowgraph model gives us the MGF of the waiting time distribution of interest. This is usually the time to total failure, whether it be of kidneys or of hydraulic pumps. MGFs by themselves are not useful or intuitive for inference. Our interest is in computing the Bayes predictive density of the total waiting time, and this requires converting the MGF into a density function. In Section 3.1 we introduce the ideas with special cases where exact inversion of the MGF using symbolic algebra is possible. Code from the package MAPLE is provided. In Section 3.2 we discuss approximate inversion of MGFs using saddlepoint approximations. In Section 3.3 we discuss the inversion of flowgraph MGFs using saddlepoint approximations for one value of the branch waiting time parameter vector $\boldsymbol{\theta}$. Code from the package R is provided. In Section 3.4 we illustrate the inversion of a complex flowgraph that uses Mason's rule (2.26) and MAPLE to solve for the overall MGF. In Section 3.5 we present a general program to convert MGFs to densities, CDFs, survivor, or hazard functions.

3.1 EXACT INVERSION OF FLOWGRAPH MGFs

Example 3.1: MEDICAL APPLICATION: KIDNEY DISEASE PROGRESSION. We consider the MGF of the total time to kidney failure as given in (2.6) in Example 2.9. Because each waiting time is exponential, we can solve this analytically for the density of T. We substitute the appropriate exponential MGFs in (2.6) and use

Flowgraph Models for Multistate Time-to-Event Data, By Aparna V. Huzurbazar
ISBN 0-471-26514-4

partial fraction expansion to simplify and convert $M_T(s)$ to a density as follows:

$$\begin{aligned} M_T(s) &= M_{2\lambda_0}(s)M_{\lambda_1}(s) \\ &= \Big(\frac{2\lambda_0}{2\lambda_0 - s}\Big)\Big(\frac{\lambda_1}{\lambda_1 - s}\Big) \qquad \text{for } s < \min(2\lambda_0, \lambda_1) \qquad (3.1) \\ &= \Big(\frac{2\lambda_0\lambda_1}{\lambda_1 - 2\lambda_0}\Big)\Big(\frac{1}{2\lambda_0 - s}\Big) + \Big(\frac{2\lambda_0\lambda_1}{2\lambda_0 - \lambda_1}\Big)\Big(\frac{1}{\lambda_1 - s}\Big). \qquad (3.2) \end{aligned}$$

We rearrange the terms to get known MGFs, giving

$$M_T(s) = \Big(\frac{2\lambda_0}{2\lambda_0 - \lambda_1}\Big)\Big(\frac{\lambda_1}{\lambda_1 - s}\Big) - \Big(\frac{\lambda_1}{2\lambda_0 - \lambda_1}\Big)\Big(\frac{2\lambda_0}{2\lambda_0 - s}\Big). \qquad (3.3)$$

We convert each MGF on the right-hand side of (3.3) to a density, considering the cases $2\lambda_0 > \lambda_1$, $2\lambda_0 < \lambda_1$, and $2\lambda_0 = \lambda_1$ separately. For the case $2\lambda_0 > \lambda_1$, we have

$$\begin{aligned} f_T(t) &= \Big(\frac{2\lambda_0}{2\lambda_0 - \lambda_1}\Big)\lambda_1 e^{-\lambda_1 t} - \Big(\frac{\lambda_1}{2\lambda_0 - \lambda_1}\Big)(2\lambda_0)e^{-2\lambda_0 t} \\ &= \Big(\frac{2\lambda_0\lambda_1}{2\lambda_0 - \lambda_1}\Big)(e^{-\lambda_1 t} - e^{-2\lambda_0 t}) \qquad (3.4) \\ &\qquad \text{for } t > 0, \lambda_0 > 0, \lambda_1 > 0, 2\lambda_0 > \lambda_1. \end{aligned}$$

Similarly, for the case $2\lambda_0 < \lambda_1$, we have

$$\begin{aligned} f_T(t) &= \Big(\frac{\lambda_1}{\lambda_1 - 2\lambda_0}\Big)(2\lambda_0)e^{-2\lambda_0 t} - \Big(\frac{2\lambda_0}{\lambda_1 - 2\lambda_0}\Big)\lambda_1 e^{-\lambda_1 t} \\ &= \Big(\frac{2\lambda_0\lambda_1}{\lambda_1 - 2\lambda_0}\Big)(e^{-2\lambda_0 t} - e^{-\lambda_1 t}) \qquad (3.5) \\ &\qquad \text{for } t > 0, \lambda_0 > 0, \lambda_1 > 0, 2\lambda_0 < \lambda_1. \end{aligned}$$

Note that (3.4) and (3.5) are identical, so that we can write

$$f_T(t) = \Big(\frac{2\lambda_0\lambda_1}{2\lambda_0 - \lambda_1}\Big)(e^{-\lambda_1 t} - e^{-2\lambda_0 t}) \qquad \text{for } t > 0, \lambda_0 > 0, \lambda_1 > 0, 2\lambda_0 \neq \lambda_1. \qquad (3.6)$$

For the case where $2\lambda_0 = \lambda_1$, the MGF reduces to

$$M_T(s) = \Big(\frac{\lambda_1}{\lambda_1 - s}\Big)\Big(\frac{\lambda_1}{\lambda_1 - s}\Big) = \Big(\frac{\lambda_1}{\lambda_1 - s}\Big)^2, \qquad s < \lambda_1, \qquad (3.7)$$

so that

$$f_T(s) = \lambda_1^2 t e^{-\lambda_1 t}, \qquad t > 0, \lambda_1 > 0, \tag{3.8}$$

which is a gamma density with mean $2/\lambda_1$. The key step is in recognizing that each term in (3.3) contains the MGF of an exponential distribution. MGFs of distributions not given here can be found in handbooks of distributions such as Johnson et al. (1994). In more complicated problems, particularly those without exponentials, we cannot simply look up corresponding distributions on a table, and we must convert $M_T(s)$ numerically to get the density $f_T(t)$ in (3.6). □

Example 3.2: ENGINEERING APPLICATION: HYDRAULIC PUMP SERIES SYSTEM. Consider the hydraulic pump series system of Example 2.2. Suppose that each pump fails according to an Exp(λ) distribution. Then the waiting time distribution for $0 \to 1$ is the minimum of two exponential random variables, which is Exp(2λ), computed as in (2.5). The waiting time distribution from 1 to 2 is just Exp(λ). Then the MGF, $M_T(s)$, of the total waiting time, T, can be computed using partial fraction expansion to convert the MGF to a density, $f_T(t)$, as follows:

$$\begin{aligned} M_T(s) &= M_{01}(s)M_{12}(s) && (3.9)\\ &= \left(\frac{2\lambda}{2\lambda - s}\right)\left(\frac{\lambda}{\lambda - s}\right) && (3.10)\\ &= \left(\frac{2\lambda}{\lambda - s}\right) - \left(\frac{2\lambda}{2\lambda - s}\right). && (3.11) \end{aligned}$$

Noting that each term corresponds to a known MGF, and then converting these MGFs to densities, gives

$$f_T(t) = 2\lambda(e^{-\lambda t} - e^{-2\lambda t}), \qquad t > 0, \lambda > 0. \tag{3.12}$$

□

Example 3.3: SURVIVAL ANALYSIS APPLICATION: A FEEDFORWARD MODEL FOR HIV. A type of feedforward model is useful in modeling the incubation time for AIDS. A Markov model for this problem was suggested by Brookmeyer and Liao (1992). Aalen (1995) proposed a phase type (PH) distribution model for this problem. A PH distribution is used to model the time until absorption in a finite-state continuous-time Markov chain and is studied further in Section 7.4. Huzurbazar (1999) developed flowgraph models for PH distributions.

Figure 3.1 shows a general flowgraph model for this problem. This model is similar to a series model but with an additional feedforward to account for treatment. This model allows for nonexponential waiting times and is more general than the models of Aalen (1995) and Brookmeyer and Liao (1992). The MGF of the time to incubation, $1 \to 5$, is the MGF of the overall waiting time distribution,

$$M(s) = p_{35}M_{12}(s)M_{23}(s)M_{35}(s) + (1 - p_{35})M_{12}(s)M_{23}(s)M_{34}(s)M_{45}(s). \tag{3.13}$$

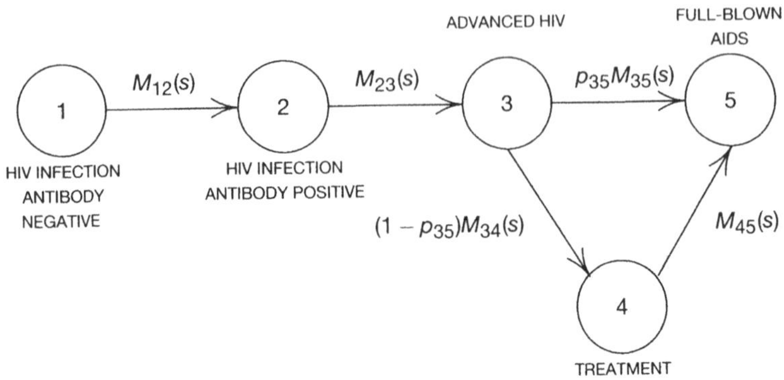

FIGURE 3.1 Feedforward flowgraph model for incubation time of HIV.

For illustration, we use a Markov model with exponential waiting time distributions Exp(α) for $1 \to 2$, Exp(β) for $2 \to 3$, Exp(λ) for $3 \to 5$, Exp(γ) for $3 \to 4$, and Exp($\theta\lambda$) for $4 \to 5$. We use these exponential distributions only to retain comparability with models suggested previously. We can use any other distribution with a tractable MGF in the flowgraph model of (3.13). The parameter γ represents the rate at which treatment can be offered. The parameter θ represents the factor with which the progression of AIDS is slowed down as an effect of treatment. The MGF of the survival time distribution of the time to incubation is

$$M(s) = \left(\frac{\lambda}{\lambda+\gamma}\right)\left(\frac{\lambda+\gamma}{\lambda+\gamma-s}\right)\left(\frac{\alpha}{\alpha-s}\right)\left(\frac{\beta}{\beta-s}\right) + \left(\frac{\gamma}{\lambda+\gamma}\right)\left(\frac{\alpha}{\alpha-s}\right)\left(\frac{\beta}{\beta-s}\right)\left(\frac{\lambda+\gamma}{\lambda+\gamma-s}\right)\left(\frac{\theta\lambda}{\theta\lambda-s}\right). \tag{3.14}$$

Using a partial fraction expansion, the overall waiting time density is

$$\begin{aligned} f(t) = {} & \frac{\lambda^2\beta\theta - \alpha\beta\lambda + \gamma\beta\theta\lambda}{(\lambda+\gamma-\alpha)(\beta-\alpha)(\theta\lambda-\alpha)}\alpha e^{-\alpha t} \\ & + \frac{\lambda^2\alpha\theta - \alpha\beta\lambda + \gamma\alpha\theta\lambda}{(\lambda+\gamma-\beta)(\alpha-\beta)(\theta\lambda-\beta)}\beta e^{-\beta t} \\ & + \frac{\alpha\beta\lambda(\theta-1)}{(\alpha-\lambda-\gamma)(\beta-\lambda-\gamma)(\lambda\theta-\gamma-\lambda)}(\lambda+\gamma)e^{-(\lambda+\gamma)t} \\ & + \frac{\alpha\beta\gamma}{(\alpha-\theta\lambda)(\beta-\theta\lambda)(\lambda+\gamma-\theta\lambda)}\theta\lambda e^{-\theta\lambda t}, \end{aligned} \tag{3.15}$$

a finite mixture of exponentials. Note that even though it is possible, partial fraction expansion is a tedious process as the number of transitions increases. The form in (3.15) is equivalent to the overall survival function for this model given in Aalen (1995). More generally, for a model with some nonexponential waiting times, analytic inversion of the MGF without using symbolic algebra can be cumbersome or even impossible. □

3.1.1 Comments on Exact Inversion

The key step in the applications of Section 3.1 is in recognizing that the MGFs in (3.3), (3.10), and (3.14) are MGFs of exponential distributions. Generally, we must convert $M_T(s)$ numerically to get the density $f_T(t)$ in (3.6), (3.12), or (3.15). As long as we restrict ourselves to convolutions of exponentials or gamma distributions with integer-valued shape parameters, we can also use symbolic algebra packages such as MAPLE to convert the MGFs to densities. This is done by using commands for inverting Laplace transforms. Let the nonnegative random variable T have density $f(t)$. We can write the Laplace transform of $f(t)$ as

$$\mathcal{L}(f(t)) = \int_0^\infty e^{-st} f(t)\,dt = E(e^{-sT}) = M_T(-s), \tag{3.16}$$

where $M_T(s)$ is the MGF of T. This is just the MGF with argument $-s$ instead of s. Hence we can use the inverse Laplace functions in MAPLE but with the argument changed. Further background on Laplace transforms is given in the Appendix. The code in the examples below is annotated with comments indicated by #. Note that in MAPLE a colon at the end of a line of code suppresses the output to the screen, whereas a semicolon at the end of a line of code echoes the output to the screen.

Example 3.4: MAPLE CODE FOR KIDNEY DISEASE PROGRESSION. This example illustrates the use of MAPLE with the MGF found in Example 3.1. The MAPLE program code is given in Figure 3.2.

The MAPLE output is

$$f(t) = 2\frac{\lambda 0\,\lambda 1\left(e^{-\lambda 1\,t} - e^{-2\lambda 0\,t}\right)}{2\lambda 0 - \lambda 1}. \tag{3.17}$$

This is identical to (3.6) except that MAPLE does not create subscripts automatically. □

Example 3.5: MAPLE CODE FOR HYDRAULIC PUMP SYSTEM. This example illustrates the use of the symbolic algebra package MAPLE with the MGF found in Example 3.2. The MAPLE program code is given in Figure 3.3. MAPLE output is

$$f(t) = 4\lambda e^{-3/2\,\lambda t}\sinh\left(1/2\,\lambda t\right). \tag{3.18}$$

```
with(inttrans):                       # include the inverse transforms
m01(s) := (2*lambda0/(2*lambda0+s)):  # define MGF for 0 to 1
m12(s) := (lambda1/(lambda1+s)):      # define MGF for 1 to 2
m(s) := m01(s)*m12(s):                # define overall MGF for series
f(t) := invlaplace(m(s),s,t);                     # get the density
latex(f(t),ftkidney):                 # save density as LATEX code
```

FIGURE 3.2 MAPLE program code for Example 3.4.

```
with(inttrans):                    # include the inverse transforms
m01(s) := (2*lambda/(2*lambda+s)):    # define MGF for 0 to 1
m12(s) := (lambda/(lambda+s)):        # define MGF for 1 to 2
m(s) := m01(s)*m12(s):             # define overall MGF for series
f(t) := invlaplace(m(s),s,t);                 # get the density
latex(f(t),fthydro):                # save density as LATEX code
```

FIGURE 3.3 MAPLE program code for Example 3.5.

We see that this is identical to (3.12) after substituting for $\sinh(u) = \frac{1}{2}(e^u - e^{-u})$:

$$
\begin{aligned}
f(t) &= 4\lambda e^{-3/2\,\lambda t}\left(\frac{1}{2}\right)(e^{\lambda t/2} - e^{-\lambda t/2}) \\
&= 2\lambda(e^{-\lambda t} - e^{-2\lambda t}), \qquad t > 0, \lambda > 0.
\end{aligned}
$$

The presence of other waiting time distributions requires alternative means of inversion. For example, the sum of n i.i.d. exponential distributions is gamma. However, the distribution of the sum of an exponential, a Weibull, and a gamma random variable does not have a well-known form. Most symbolic algebra packages allow distributions such as the exponential or gamma (with integer-valued shape parameter) but the inverse Gaussian, Weibull, Gompertz, and most other survival and reliability analysis distributions are not allowed. Complicated mixtures of exponentials and gammas, and gammas with non-integer-valued shape parameters, all run into trouble in the standard packages. □

3.2 APPROXIMATE INVERSION OF FLOWGRAPH MGFs: SADDLEPOINT APPROXIMATION

One method for numerically converting a MGF to a density is to use a saddlepoint approximation. In this section we discuss the univariate saddlepoint approximation and conclude with a simple example that is intended to walk the reader through its implementation. The literature on saddlepoint methods is vast, beginning with Daniels (1954), Skovgaard (1987), Reid (1988), Barndorff-Nielsen and Cox (1989), Jensen (1995), and Kolassa (1997). Practical applications of saddlepoint methods to problems in survival analysis are given in Huzurbazar and Huzurbazar (1999), while more general applications of the methods are given in Huzurbazar, S. (1999). For a recent review of asymptotic analysis, including saddlepoint methods for deriving distributions in statistics, see Reid (2003).

The basic saddlepoint approximations are analytical formulas that provide a density or cumulative distribution function (CDF) by performing an approximate inversion of the MGF. As such, they are restricted to distributions having tractable MGFs, but this is a rich class of useful distributions in survival and reliability analysis, including distributions such as the exponential, gamma, inverse Gaussian, compound exponential, Weibull, and Gompertz. One notable exception is the lognormal. However, the inverse Gaussian mimics the shape of the lognormal and is more appropriate for use in practice because the hazard function of the lognormal goes to zero as $t \to \infty$, whereas the hazard function of the inverse Gaussian approaches a limit that is a function of the parameters of the distribution (cf. Hoyland and Rausand, 1994). In addition to handling these distributions individually, the saddlepoint approximation can be used on MGFs resulting from their finite mixtures or convolutions. Let T be the random waiting time of interest.

Theorem 3.2.1: SADDLEPOINT APPROXIMATION. Let $K(s) = \log[M(s)]$ be the cumulant generating function (CGF) of T. Let c_1 and c_2 be constants such that $c_1 < 0 < c_2$. Suppose that $M(s)$ exists for $s \in (c_1, c_2)$, an open neighborhood of zero. Then the saddlepoint approximation for the density of T is

$$\tilde{f}_T(t) = [2\pi K''(\hat{s})]^{-1/2} \exp[K(\hat{s}) - \hat{s}t], \tag{3.19}$$

where $K''(s) = d^2K(s)/ds^2$ and

$$K'(\hat{s}) = t. \tag{3.20}$$

The saddlepoint approximation is typically used to find the distribution of the mean of independent and identically distributed random variables. The saddlepoint approximation presented in (3.19) is for one random variable T. With flowgraphs, we are using the usual saddlepoint approximation for the mean but with one random variable. For a proof of Theorem 3.2.1, we refer the reader to three main derivations of the saddlepoint approximation. The original derivation is by Daniels (1954) who introduced saddlepoint methods to statistics. This derivation begins with the characteristic function of the desired density and computes a contour integral in the complex plane using the method of steepest descent to derive the saddlepoint approximation. A simpler derivation, due to Barndorff-Nielsen and Cox (1979), is based on using Edgeworth expansions and the method of exponential tilting. This can be traced back to Esscher (1932). This derivation shows that the saddlepoint is the best Edgeworth expansion at any given point and does not suffer from the problem of Edgeworth expansions, namely that of giving negative values in the tails of the distribution. The third main derivation is due to Barndorff-Nielsen (1988), presented in the discussion of Reid (1988), which is based on Laplace's method. This derivation shows that the saddlepoint approximation is the best Laplace approximation at every point. For most problems involving flowgraphs, $\hat{s}$ is a complicated implicit function of both t and the parameters of the distribution. The constants c_1 and c_2 are found numerically. These quantities are used to bound the solution to the saddlepoint equation (3.20). Practically, the upper bound c_2 is the smallest positive root of (3.20). The saddlepoint is unique and it is a monotonically increasing function of t, a property that can be exploited in its computation. The computational details for the general saddlepoint program are presented in Section 3.5. The following example illustrates practical computation of a saddlepoint approximation.

Example 3.6: EXPLICIT SADDLEPOINT. Gamma distributions arise naturally as waiting time distributions, especially as the family includes sums of identically and independently distributed exponential variables. Unfortunately, the survival and hazard function calculations involve the incomplete gamma integral, which limits the use of this family of distributions in practice (Cox and Oakes, 1984). The use of the gamma distribution for these quantities is possible with saddlepoint approximations. The saddlepoint approximation is exact (up to a constant of proportionality) for the density of the gamma distribution and computation of the density approximation is relatively straightforward.

A gamma random variable $T \sim \text{Gamma}(\alpha, \beta)$ with mean α/β has MGF

$$M(s) = \left(\frac{\beta}{\beta - s}\right)^{\alpha}, \qquad s < \beta.$$

The CGF, $K(s) = \log[M(s)]$, is

$$K(s) = \alpha\ [\log(\beta) - \log(\beta - s)] \qquad \text{for } s < \beta \tag{3.21}$$

and

$$K'(s) = \frac{\alpha}{\beta - s} \tag{3.22}$$

$$K''(s) = \frac{\alpha}{(\beta - s)^2}. \tag{3.23}$$

The saddlepoint equation,

$$\frac{\alpha}{\beta - s} = t,$$

is solved by

$$\hat{s} = \beta - \frac{\alpha}{t}.$$

Evaluating the quantities in equations (3.21) and (3.23) at $\hat{s}$ and substituting into (3.19) gives

$$\tilde{f}_T(t) = \left(\frac{2\pi t^2}{\alpha}\right)^{-1/2} \exp\left\{\alpha \log\left(\frac{\beta t}{\alpha}\right) - t\beta + \alpha\right\} \qquad \text{for } t > 0. \tag{3.24}$$

We can see that this approximation is exact up to a constant of proportionality by rewriting (3.24) to give

$$\tilde{f}_T(t) = \left(\frac{1}{\sqrt{2\pi}}\,\frac{e^{\alpha}}{\alpha^{\alpha - 1/2}}\right)\beta^{\alpha} t^{\alpha-1} e^{-\beta t}.$$

The constant of proportionality is Stirling's approximation to the gamma function,

$$\Gamma(\alpha) \approx \sqrt{2\pi}\,\alpha^{\alpha-1/2}\,e^{-\alpha}.$$

Stirling's approximation improves as $\alpha \to \infty$; that is,

$$\frac{\Gamma(\alpha)}{\sqrt{2\pi}e^{-\alpha}\alpha^{\alpha-1/2}} \to 1 \qquad \text{as } \alpha \to \infty.$$

This gives the saddlepoint approximated density as

$$\tilde{f}_T(t) \approx \frac{\beta^\alpha}{\Gamma(\alpha)} t^{\alpha-1} e^{-\beta t} \qquad \text{for } t > 0.$$

Thus we say that the saddlepoint approximation of a gamma distribution is exact up to a normalizing constant. This is also true for two other distributions, the normal and the inverse Gaussian, as illustrated by Daniels (1954) and proven by Blaesild and Jensen (1985). The saddlepoint density is generally normalized to integrate to 1, so that in the case of time-to-event data, we have

$$\tilde{f}_T(t) \leftarrow \frac{\tilde{f}_T(t)}{\int_0^\infty \tilde{f}_T(t)\,dt}. \tag{3.25}$$

The denominator of (3.25) is the normalizing constant. In practice, we integrate the saddlepoint density approximation numerically to calculate the normalizing constant. □

There is a saddlepoint CDF approximation due to Lugannani and Rice (1980); however, it is unstable near the mean. An alternative to the Lugannani and Rice approximation is to integrate the saddlepoint density approximation numerically as suggested by Daniels (1954, 1987). The saddlepoint approximation to the CDF due to Lugannani and Rice (1980) is stated as follows.

Theorem 3.2.2: Saddlepoint Approximation to the CDF. Let $K(s) = \log[M(s)]$ be the cumulant generating function (CGF) of T. Let c_1 and c_2 be constants such that $c_1 < 0 < c_2$. Suppose that $M(s)$ exists for $s \in (c_1, c_2)$, an open neighborhood of zero. Let $\hat{s}$ solve $K'(\hat{s}) = t$, the saddlepoint equation (3.20). Let $K''(s) = d^2K(s)/ds^2$ and $K'''(s) = d^3K(s)/ds^3$. Then the saddlepoint approximation for the CDF of T is

$$\hat{F}(t) = \begin{cases} \Phi(\hat{w}) + \phi(\hat{w})[\hat{w}^{-1} - \hat{u}^{-1}], & t \neq E(T) \\ 1/2 + K'''(0)\left[6\sqrt{2\pi K''(0)^3}\,\right]^{-1}, & t = E(T), \end{cases} \tag{3.26}$$

where Φ and ϕ are the standard normal CDF and density

$$\hat{w} = \text{sgn}(\hat{s})\sqrt{2[\hat{s}t - K(\hat{s})]} \quad \text{and} \quad \hat{u} = \hat{s}\sqrt{K''(\hat{s})}.$$

We refer the reader to Lugannani and Rice (1980) for a proof. Note that $\hat{w}$ and $\hat{u}$ are also complicated implicit functions of t and $\boldsymbol{\theta}$ through $\hat{s}$. While there is a separate expression for the formula at the mean, the problems with the approximation occur in a neighborhood of the mean. The Lugannani and Rice CDF approximation was intended to be an approximation to tail probabilities. Huzurbazar and Huzurbazar (1999) show that the numerically summed density performs as well as the Lugannani and Rice approximation in the tails of the distribution and does not suffer from the instability near the mean of the distribution. It is the preferred saddlepoint CDF approximation for use with flowgraphs. In addition to the density and the CDF, we are also interested in computing the survivor or reliability function and the hazard function.

Figures 3.4, 3.5, and 3.6 give plots of the exact and saddlepoint approximated density, survival, and hazard functions for a Gamma(5.4, 2) distribution. We can see that the density and survivor function approximations are graphically indistinguishable and the hazard function approximation deviates only slightly from the exact as t increases. R code for the computations involved in the plots is given following this example. Although we have an explicit saddlepoint in this example, the R code is written to solve the saddlepoint equation numerically, as that will generally be the method of solution in practice. The saddlepoint density normalizing constant for this example is included in the code below and it is 1.015534, quite close to 1, for the saddlepoint density computed to $t = 20$. This is important for computing the hazard function since the normalized integrated density is forced to 1.0 at the last endpoint. If we had only computed the saddlepoint density to $t = 10$, we would have missed some mass and the effect would have been apparent in the hazard function approximation.

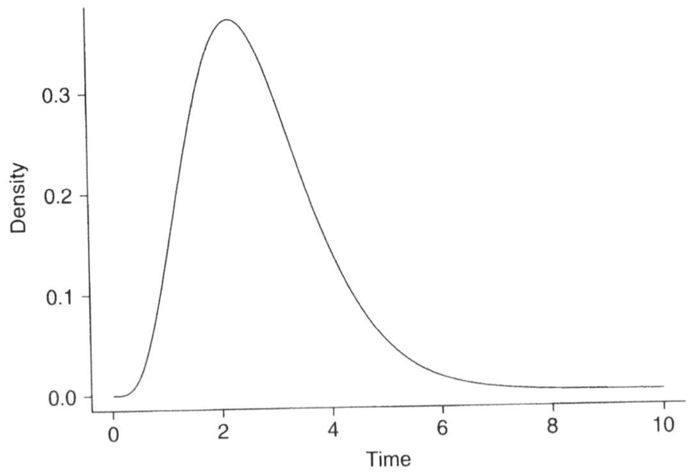

FIGURE 3.4 Exact and saddlepoint approximated densities for Gamma(5.4, 2). The two lines overlap perfectly.

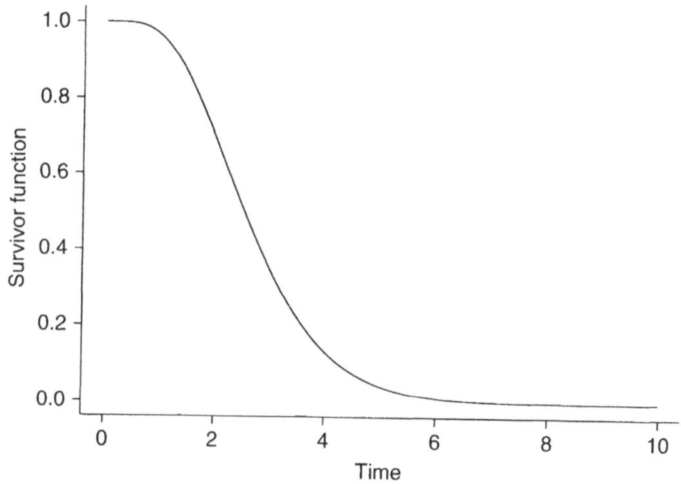

FIGURE 3.5 Exact and saddlepoint approximated survivor functions for Gamma(5.4, 2). The two lines overlap perfectly.

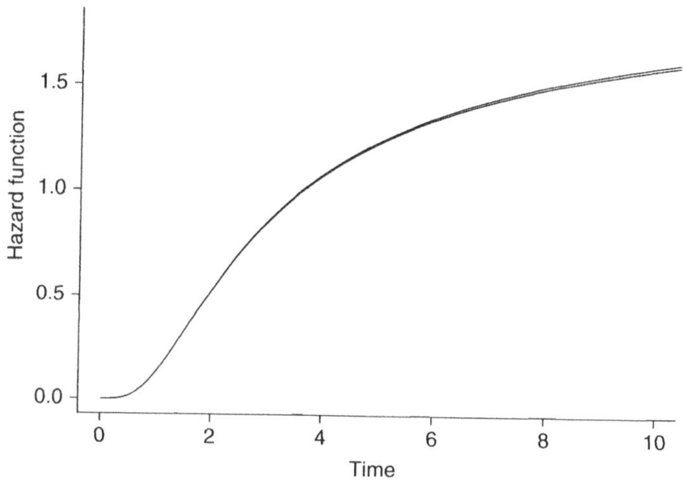

FIGURE 3.6 Exact and saddlepoint approximated hazard functions for Gamma(5.4, 2). The two lines deviate slightly as t increases.

Let T_M be the largest value of T at which the density is computed. Figure 3.7 shows the effect on the density when the normalizing constant is not estimated accurately. The values of T_M and corresponding normalizing constants are given in Table 3.1. For $T_M \geq 10$, the exact and saddlepoint densities are indistinguishable from one another and the normalizing constant has stabilized.

In general, when dealing with complicated flowgraph problems, we do not know the exact density. Figure 3.7 and Table 3.1 provide a way to decide when the normalizing constant has stabilized and one has captured all of the density.

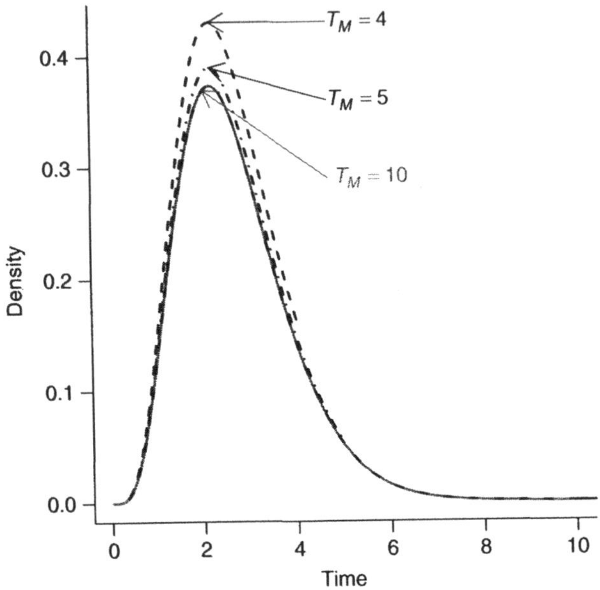

FIGURE 3.7 Exact and saddlepoint approximated densities for Gamma(5.4, 2) and various values of T_M. The solid line for the exact and the dashed line for $T_M = 10$ overlap perfectly.

TABLE 3.1 Effect of T_M on Values of the Normalizing Constant for Example 3.6

T_M	Normalizing Constant
4	0.8820708
5	0.9734477
10	1.015503
15	1.015534
20	1.015534

In the right tail of the hazard function, generally both the numerator and the denominator are approaching 0. To avoid problems with the denominator approaching 0, we must compute the saddlepoint approximation much farther out in the tail than the largest value of t desired for accuracy. Figure 3.8 gives the hazard function for various values of T_M, along with the exact hazard for comparison. For example, although $T_M = 10$ is adequate for the density plot, it is clear from Figure 3.8 that a larger value of T_M is required for the hazard computation.

Example 3.7: R CODE FOR SADDLEPOINT APPROXIMATION. This example presents the R code for computing the saddlepoint approximation to the density, survivor, and hazard functions in Example 3.6. Two functions are needed: a function to

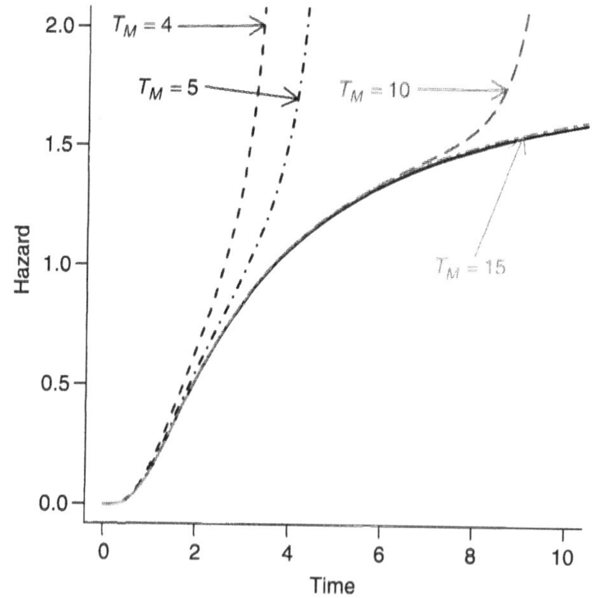

FIGURE 3.8 Exact and saddlepoint approximated hazard functions for Gamma(5.4, 2) and various values of T_M. The solid line is the exact, matched closely by $T_M = 15$.

solve the saddlepoint equation (3.20), called `saeqn` in the code; and a function to compute the saddlepoint density approximation (3.19), called `fhat` in the code. This program is a rudimentary version of the general program (see Section 3.5) and it is presented here to illustrate the steps of the computations with a simple example.

```
# R code for computing the saddlepoint density,
# survivor, and hazard function approximations.
# The functions that are needed:  saeqn and fhat.
#
# saeqn:  defines the saddlepoint equation
# fhat:  defines the saddlepoint density approximation

# Define the saddlepoint equation

saeqn <- function(s, l)
{
   k1 <- l$gp$alpha/(l$gp$beta - s)
   k1 - l$t
}

# Define the saddlepoint approximation

fhat <- function(t, s, gp)
```

```
{
   k <- gp$alpha*(log(gp$beta) - log(gp$beta - s))
   k2 <- gp$alpha/(gp$beta - s)^2
   fh <- (2 * pi * k2)^(-0.5) * exp(k - t * s)
   fh
}

gp <- list(alpha=5.4, beta=2)  # gamma parameters
delta <- 0.01  # interval length for density
w <- seq(0.01, 20, delta)  # support of the density

# s:  saddlepoint
# fh:  vector of saddlepoint density values
# nsd:  normalizing constant

# Numerically solve for the saddlepoint

s <- sapply(1:length(w), function(i)
            {
               uniroot(saeqn, lower=-1999, upper=1.999,
                       l=list(t=w[i], gp=gp))$root
            })

# Compute the saddlepoint density approximation

fh <- sapply(1:length(w), function(i)
             {
                fhat(w[i], s[i], gp)
             })

# Compute the normalizing constant

nsd <- sum(fh[2:length(w)]) * delta

nsd

# Normalize saddlepoint density approximation
# Compute saddlepoint survivor function approximation
# Compute saddlepoint hazard function approximation

fh <- fh / nsd
sasurv <- 1 - cumsum(fh) * delta
sahaz <- fh / sasurv

# Compute and plot density functions

eden <- dgamma(w, gp$alpha, rate=gp$beta)
postscript("density.ps", horizontal=F)
par(pty="s")
```

```
plot(w, eden, type="l", xlim=c(0, 10), xlab="t", ylab="density")
lines(w, fh, lty=2, col="red")
graphics.off()

# Compute and plot survivor functions

esurv <- pgamma(w, gp$alpha, rate=gp$beta, lower.tail=FALSE)
postscript("survivor.ps", horizontal=F)
par(pty="s")
plot(w, esurv, type="l", xlim=c(0, 10), ylim=c(0, 1), xlab="t",
     ylab="survivor function")
lines(w, sasurv, lty=2, col="red")
graphics.off()

# Compute and plot hazard functions

ehaz <- eden / esurv
postscript("hazard.ps", horizontal=F)
par(pty="s")
plot(w, ehaz, type="l", xlim=c(0, 10), xlab="t",
     ylab="hazard function")
lines(w, sahaz, lty=2, col="red")
graphics.off()
```
□

When the MGF is not in closed form, the main portion of the saddlepoint computation does not change. For example, the MGF may exist as an integral as in the case of the Weibull density. In this case, we compute it numerically and compute the cumulant generating function, $K(s)$, and its derivatives, $K'(s)$ and $K''(s)$, numerically as well. We will see when working with flowgraph models that the MGF is often a complicated form that must be coded as a function; this also applies to the CGF and its derivatives. We return to the Weibull MGF in Example 3.11.

3.3 USING SADDLEPOINT APPROXIMATIONS WITH FLOWGRAPH MODELS

In a flowgraph model, the random variable T is usually a total or partial waiting time. T is not necessarily just the sum of independent and identically distributed random variables, which is the assumption in the usual saddlepoint approximation literature. T can be the result of convolutions, finite mixtures, and feedback loops. We have the MGF of T from the flowgraph and we can use the saddlepoint approximation to convert this to a density. We will compute saddlepoint approximations to the density for the kidney disease progression and hydraulic pump examples. Let T be our waiting time of interest, the total waiting time. The saddlepoint density approximation given in (3.19) to the density of T, $f(t|\boldsymbol{\theta})$, must be computed for a given set of parametric model parameters $\boldsymbol{\theta}$. In this section, for convenience, we will illustrate the computation using $\hat{\boldsymbol{\theta}}$, the MLE of $\boldsymbol{\theta}$, as one value of $\boldsymbol{\theta}$ for which we compute the approximation and defer computation of the Bayes predictive density to Chapter 5.

Example 3.8: KIDNEY DISEASE PROGRESSION: SADDLEPOINT APPROXIMATION. We demonstrate the inversion of the MGF in Example 3.1 using a saddlepoint approximation. We will use the MLEs for $\boldsymbol{\theta} = (\lambda_0, \lambda_1)$ as reported in Gross et al. (1971). These are $\hat{\lambda}_0 = \frac{1}{15}$ and $\hat{\lambda}_1 = \frac{1}{3}$. To compute the saddlepoint density in equation (3.19), we need the CGF, $K(s)$, and its first and second derivatives. The MGF for the flowgraph model is given in (3.1). The inputs to the saddlepoint are

$$\begin{aligned} K(s) &= \log[M(s)] \\ &= \log(2\lambda_0\lambda_1) - \log(2\lambda_0 - s) - \log(\lambda_1 - s) \qquad (3.27) \\ K'(s) &= \frac{1}{2\lambda_0 - s} + \frac{1}{\lambda_1 - s} \qquad (3.28) \\ K''(s) &= \frac{1}{(2\lambda_0 - s)^2} + \frac{1}{(\lambda_1 - s)^2} \qquad (3.29) \end{aligned}$$

for $s < \min(2\lambda_0, \lambda_1)$. We solve the saddlepoint equation by setting $K'(s) = t$ and solving for $\hat{s}$. Solving this equation at the MLEs gives two roots,

$$\hat{s} = \left(\frac{1}{90} \frac{-90 + 21t + 9\sqrt{100 + t^2}}{t}, \frac{1}{90} \frac{-90 + 21t - 9\sqrt{100 + t^2}}{t} \right). \qquad (3.30)$$

The roots are exactly the output from MAPLE and are not simplified or formatted in any way. The general solution is

$$\hat{s} = \frac{[(\lambda_1 + 2\lambda_0)t - 2] \pm \sqrt{4 + (\lambda_1 - 2\lambda_0)^2 t^2}}{2t}$$

and the condition $s < \min(2\lambda_0, \lambda_1)$ implies that the smaller root must always be taken for $t \geq 0$. This is seen by noting that $\hat{s}_-(t)$ and $\hat{s}_+(t)$ are continuous in t for $t > 0$ and

$$\lim_{t\to\infty} \hat{s}_-(t) = \min(2\lambda_0, \lambda_1) \quad \text{while} \quad \lim_{t\to\infty} \hat{s}_+(t) = \max(2\lambda_0, \lambda_1).$$

The smaller root $\hat{s}_-(t)$ is increasing in t for $t > 0$ and $\lim_{t\to 0} \hat{s}_-(t) = -\infty$. The larger root $\hat{s}_+(t)$ is increasing in t for $t > 0$ and $\lim_{t\to 0} \hat{s}_+(t) = (\lambda_1 + 2\lambda_0)/2$. This implies that $\hat{s}_+(t) > \min(2\lambda_0, \lambda_1)$ for all $t > 0$. The MAPLE code for this example is given in Figure 3.9. □

Example 3.9: HYDRAULIC PUMP SYSTEM: SADDLEPOINT APPROXIMATION. We show how to convert the flowgraph MGF of Example 3.2 using a saddlepoint

```
with(linalg):                     # include linear algebra package
m(s) := 2*lambda0*lambda1/((2*lambda0-s)*(lambda1-s));
k(s) := ln(m(s));                 # define CGF, ln=natural log
k1(s) := diff(k(s),s);            # take the first derivative
k1(s) := simplify(k1(s));         # simplify (not necessary)
lambda0 := 1/15;
lambda1 := 1/3;
sp(s) := k1(s)-t;                 # define the saddlepoint equation

# Solve saddlepoint equation and put roots into
# a 2 x 1 vector called sproot
sproot := matrix(2,1,[solve(sp(s)=0,s)]);
latex(sproot[1,1],kidroot1);
latex(sproot[2,1],kidroot2);
```

FIGURE 3.9 MAPLE program code for Example 3.8.

approximation. For the hydraulic pump system, Limnios (1992) reports $\frac{1}{1000}$ as the MLE for $\theta = \lambda$. We will use this to illustrate the saddlepoint density approximation in (3.19). The inputs needed to compute the saddlepoint density are the CGF, $K(s)$, and its first and second derivatives, $K'(s)$ and $K''(s)$. These are

$$K(s) = \log(2\lambda^2) - \log(2\lambda - s) - \log(\lambda - s) \tag{3.31}$$

$$K'(s) = \frac{1}{2\lambda - s} + \frac{1}{\lambda - s} \tag{3.32}$$

$$K''(s) = \frac{1}{(2\lambda - s)^2} + \frac{1}{(\lambda - s)^2} \tag{3.33}$$

for $s < \lambda$. Using MAPLE to solve the saddlepoint equation (3.20) for $\hat{s}$ and evaluating it at the MLE gives two roots,

$$\hat{s} = \frac{1}{10^6} \frac{-10^6 + 1500\,t \pm 500\sqrt{(4 \times 10^6) + t^2}}{t}. \tag{3.34}$$

The MAPLE output is not simplified or formatted in any way. We discard the larger root since it violates our constraint that $s < \lambda$ for all $t > 0$. The corresponding MAPLE code is given in Figure 3.10. □

In both examples, the last two LaTeX commands can be avoided on some personal computers by cutting and pasting the displayed equation from MAPLE directly into LaTeX. The conversion is automatic. Note that the result of using the saddlepoint approximation on the flowgraph MGF is to give us an approximation to the entire density, $f(t|\boldsymbol{\theta})$. When we use $\hat{\boldsymbol{\theta}}$, the MLE of $\boldsymbol{\theta}$, the result is the maximum likelihood estimated density.

```
with(linalg):                     # include linear algebra package
m(s) := 2*lambda^2/( (2*lambda-s)*(lambda-s));
k(s) := ln(m(s));                 # define CGF, ln=natural log
k1(s) := diff(k(s),s);            # take the first derivative
k1(s) := simplify(k1(s));         # simplify (not necessary)
lambda := 1/1000;
sp(s) := k1(s)-t;                 # define the saddlepoint equation

# solve saddlepoint equation and put roots into
# a 2 x 1 vector called sproot
sproot := matrix(2,1,[solve(sp(s)=0,s)]);
latex(sproot[1,1],root1);
latex(sproot[2,1],root2);
```

FIGURE 3.10 MAPLE program code for Example 3.9.

3.4 INVERSION OF COMPLEX FLOWGRAPHS

When dealing with more complicated flowgraphs such as those of Figures 2.1, 2.22, or 2.29, we can use MAPLE to derive the MGF using Mason's rule and to compute $K(s)$, $K'(s)$, and $K''(s)$. These functions can be written as C or FORTRAN functions by MAPLE and called by R, S-plus, or any relevant program to solve for the saddlepoint numerically as we did in Example 3.6.

TABLE 3.2 Summary of Waiting Time Distributions for the Manufacturing System Flowgraph of Figure 2.29

Transmittance	Distribution	MGF
$0 \to 1$	Inverse Gaussian(λ_1, λ_2)	$M_{01}(s) = \exp\left[\lambda_1\lambda_2 - \sqrt{\lambda_1^2(\lambda_2^2 - 2s)}\right]$
$1 \to 3$	Gamma(α_1, β_1)	$M_{13}(s) = \left(\frac{\beta_1}{\beta_1 - s}\right)^{\alpha_1}$
$3 \to 1$	Exponential(μ_1)	$M_{31}(s) = \frac{\mu_1}{\mu_1 - s}$
$3 \to 5$	Exponential(μ_2)	$M_{35}(s) = \frac{\mu_2}{\mu_2 - s}$
$0 \to 2$	Exponential(μ_3)	$M_{02}(s) = \frac{\mu_3}{\mu_3 - s}$
$2 \to 4$	Inverse Gaussian(δ_1, δ_2)	$M_{24}(s) = \exp\left[\delta_1\delta_2 - \sqrt{\delta_1^2(\delta_2^2 - 2s)}\right]$
$4 \to 2$	Gamma(α_2, β_2)	$M_{42}(s) = \left(\frac{\beta_2}{\beta_2 - s}\right)^{\alpha_2}$
$4 \to 5$	Exponential(μ_4)	$M_{45}(s) = \frac{\mu_4}{\mu_4 - s}$

```
# Include package to generate C code
with(codegen,makeproc,C):

# Define the MGF as a procedure function of s
ms := proc(s)

# Use w's for transmittances
   local w01,w13,w31,w35,w02,w24,w42,w45;

# Enter each transmittance

    w01 := p01*exp(l1*l2-((l1^2)*((l2^2)-2*s))^(.5));
    w13 := (b1/(b1-s))^a1;
    w31 := p31*mu1/(mu1-s);
    w35 := (1-p31)*mu2/(mu2-s);
    w02 := (1-p01)*mu3/(mu3-s);
    w24 := exp(d1*d2-((d1^2)*((d2^2)-2*s))^(.5));
    w42 := p42*((b2/(b2-s))^a2);
    w45 := (1-p42)*mu4/(mu4-s);

# The next line defines M(s)
    (w01*w13*w35*(1-w24*w42)+w02*w24*w45*(1-w13*w31))/
         (1-w13*w31-w24*w42+w13*w31*w24*w42);

end proc:

# Define the CGF, K(s), and write it as a C function
k := makeproc(ln(ms(s)),[s]):
C(k,optimized,ansi,filename="c:/k.c"):

# Define K'(s) and write it as a C function
dk := makeproc(diff(k(s),s),[s]):
C(dk,optimized,ansi,filename="c:/dk.c"):

# Define K''(s) and write it as a C function
ddk := makeproc(diff(dk(s),s),[s]):
C(ddk,optimized,ansi,filename="c:/ddk.c"):
```

FIGURE 3.11 MAPLE program code for Example 3.10.

Example 3.10: CAR STEREO MANUFACTURING. We return to the car stereo manufacturing problem of Example 2.15. The MGF of the distribution of the waiting time until an item completes inspection and repair and is returned to the assembly line was found in (2.28). For illustration, we assume the variety of waiting time distributions given in Table 3.2. The inverse Gaussian density (cf. Johnson et al., 1994) is parameterized as

$$f(t) = \frac{\lambda_1}{\sqrt{2\pi}\ t^{3/2}} \exp\left(-\frac{\lambda_1^2/t + \lambda_2^2 t}{2} + \lambda_1\lambda_2\right) \qquad \text{for } t > 0, \lambda_1 > 0, \text{ and } \lambda_2 \geq 0.$$

```
#include <math.h>
double k(double s)
{
   double t14;
   double t2;
   double t21;
   double t22;
   double t25;
   double t27;
   double t3;
   double t32;
   double t33;
   double t4;
   double t48;
   double t50;
   double t52;
   double t7;
   double t9;
   {
      t2 = l1*l1;
      t3 = l2*l2;
      t4 = 2.0*s;
      t7 = pow(t2*(t3-t4),0.5);
      t9 = exp(l1*l2-t7);
      t14 = pow(b1/(b1-s),1.0*a1);
      t21 = d1*d1;
      t22 = d2*d2;
      t25 = pow(t21*(t22-t4),0.5);
      t27 = exp(d1*d2-t25);
      t32 = pow(b2/(b2-s),1.0*a2);
      t33 = t27*p42*t32;
      t48 = t14*p31;
      t50 = 1/(mu1-s);
      t52 = t48*mu1*t50;
      return(log((p01*t9*t14*(1.0-p31)*mu2/(mu2-s)*(1.0-t33)
         +(1.0-p01)*mu3/(mu3-s)*t27*(1.0-p42)*mu4/(mu4-s)
         *(1.0-t52))/(1.0-t52-t33+t48*mu1*t50*t27*p42*t32)));
   }
}
```

FIGURE 3.12 C function k.c for Example 3.10.

It is denoted by $IG(\lambda_1, \lambda_2)$. The exponential distribution is parameterized so that an Exp(μ) has mean $1/\mu$ and the gamma is parameterized as

$$f(t) = \frac{\beta^\alpha}{\Gamma(\alpha)} t^{\alpha-1} e^{-\beta t} \qquad \text{for } \alpha > 0, \beta > 0, t > 0$$

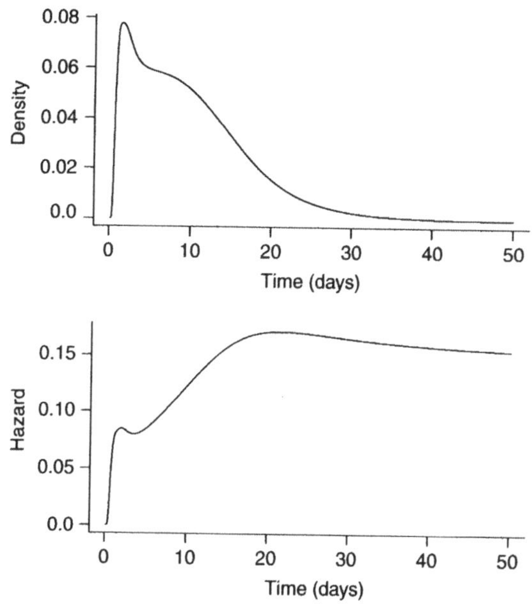

FIGURE 3.13 Density and hazard functions for the car stereo manufacturing system.

so that a Gamma(α, β) has mean α/β. Note that when $\alpha = 1$, this specializes to the exponential density. The MAPLE program code to compute the MGF of (2.28) using the distributions given in Table 3.2 is given in Figure 3.11. The resulting C functions for $K(s)$, $K'(s)$, and $K''(s)$ are k.c, dk.c, and ddk.c. These can contain many lines of C code. However, their complexity does not matter, as we only need our R program to be able to evaluate them as necessary. For illustration, the function k.c is given in Figure 3.12. This C function is output directly from MAPLE. Two minor additions to this code must be made before it can be integrated with the general saddlepoint program: A vector containing the current values of the transition density parameters must be added to the call list, and the names of these parameters in the main body of the function must be declared and assigned to their current values. Code with these changes added is presented in Chapter 6.

For illustration we assume the following parameter values: $p_{01} = 1 - p_{02} = 0.6$, $p_{31} = 1 - p_{35} = 0.3$, $p_{42} = 1 - p_{45} = 0.5$, $\lambda_1 = 25$, $\lambda_2 = 3$, $\alpha_1 = 5$, $\beta_1 = 0.8$, $\mu_1 = 0.8$, $\mu_2 = 0.5$, $\mu_3 = 0.3$, $\delta_1 = 2$, $\delta_2 = 4$, $\alpha_2 = 7$, $\beta_2 = 3$, $\mu_4 = 4$. We have deliberately chosen parameter values so that the transition $0 \to 1 \to 3 \to 5$ has a much longer waiting time (heavier-tailed distribution) than the competing path $0 \to 2 \to 4 \to 5$.

The resulting density and hazard functions are given in Figure 3.13. The general saddlepoint function to compute these is discussed further in Section 3.5. The bump in the density is reflective of the manner in which our finite mixture was chosen. This type of modeling allows us to model the shape of the hazard realistically, to include portions that are increasing, decreasing, or flat. For a given system, there is a wide scope for fine tuning the underlying distributions to reflect the actual system behavior. □

In this example, we picked some candidate parametric models for illustration. In Chapter 4 we discuss how to formulate the parametric models that define the branch waiting time distributions of the flowgraph model based on data. Once the models are chosen, we must either estimate the parameters or generate samples from their joint posterior distribution. This is discussed in Chapter 5. However, first we present a general function to perform saddlepoint inversion.

3.5 GENERAL SADDLEPOINT PROGRAM

Complex flowgraphs such as those of Example 3.10 do not have explicit analytical solutions for their saddlepoints such as those of Examples 3.8 and 3.9. We must solve for the saddlepoint numerically. Recall that the equation to be solved for the saddlepoint, $\hat{s}$, is

$$K'(\hat{s}) = t,$$

where

$$K'(s) = \frac{dK(s)}{ds} = \frac{d\{\log[M(s)]\}}{ds}.$$

In equation (3.20), although the dependence is suppressed, $\hat{s}$ is a function of both t and $\boldsymbol{\theta}$. The interval in which we search for $\hat{s}$ is determined by the interval in which the MGF exists, namely, for $s \in (c_1 < 0 < c_2)$, an open neighborhood of zero. The search algorithm involves finding $c_1(t_{\min}, \boldsymbol{\theta})$ and $c_2(t_{\max}, \boldsymbol{\theta})$ based on the value of the parameter vector, $\boldsymbol{\theta}$, and the smallest and largest value of t, denoted by $t_{\min}$ and $t_{\max}$, for which the resulting density is to be computed. For notational simplicity, we write $c_1 = c_1(t_{\min}, \boldsymbol{\theta})$ and $c_2 = c_2(t_{\max}, \boldsymbol{\theta})$. The quantities c_1 and c_2 must satisfy

$$K'(c_1) < t_{\min} < t_{\max} < K'(c_2-), \tag{3.35}$$

where c_2- represents a point to the left of c_2 satisfying $K'(c_2-) < \infty$. $K'(\cdot)$ has a singularity at c_2 such that $K'(c_2)$ approaches infinity as s increases to c_2. The algorithm uses properties of $K(s)$, $K'(s)$, and the property that the saddlepoint is monotonically increasing. The main steps are as follows:

1. Begin at $s = 0$. $K(0) = \log[M(0)] = 0$ and $K'(0)$ is the mean of the distribution of T for a given $\boldsymbol{\theta}$.
2. Increment $s \leftarrow s + \epsilon$.
3. Check that the following conditions on $K(s)$ and $K'(s)$ are satisfied:
   (a) $K'(s) < t_{\max}$.
   (b) $K'(s - \epsilon) < K'(s)$.
   (c) $K(s - \epsilon) < K(s)$.

The algorithm stops at the first violation of any of these conditions. A maximum of 1000 iterations are allowed in the following iterative process. If only the first condition fails, c_2 is the current value of s. If any of the other conditions fail, the current value of s has overshot the singularity and returns to $s \leftarrow s - \epsilon$. The steps are repeated with a smaller value of ϵ. The algorithm is terminated when the first condition is the only one that is violated. Generally, starting with a value of $\epsilon = 10^{-3}$ followed by 10^{-5}, 10^{-7}, and 10^{-9} is sufficient. If the program is unable to find an upper bound c_2 corresponding to a user-specified $t_{\max}$, the program sets $c_2 = s_{\max}$, the largest upper bound it was able to find, and returns a new $t_{\max} = K'(s_{\max})$. This situation can occur if $t_{\max}$ is chosen too far in the right tail of the density. Because of this feature, there is no danger in setting $t_{\max}$ too large. For example, suppose that the desired (unknown) density is concentrated between t-values of 0 and 10 and $t_{\max}$ is chosen to be 10^6. The program will truncate automatically at the largest numerically sustained value, set c_2, and return the corresponding $t_{\max}$.

To find the lower bound c_1 we work on the log scale. For time-to-event data, $t_{\min}$ is generally 0. We calculate c_1 as follows:

1. Set $\epsilon = 2$ and begin with $\log s = 0$.
2. If there is an error in evaluating $K'[-\exp(\log s)]$,
   (a) Set $\log s \leftarrow \log s - \epsilon$ and evaluate $K'[-\exp(\log s)]$.
   (b) Stop when $K'[-\exp(\log s)]$ evaluates without error.
3. If $K'[-\exp(\log s)] < t_{\min}$ then $c_1 \leftarrow -\exp(\log s)$.
4. Set $\epsilon = 1, \log s \leftarrow \log s + \epsilon$.
5. Evaluate $K'[-\exp(\log s)]$ and check (6). If (6) does not occur, go to (7) if $K'[-\exp(\log s)] < t_{\min}$. Otherwise repeat step (5) with $\log s \leftarrow \log s + \epsilon$.
6. In the event that there is an error in evaluating $K'[-\exp(\log s)]$, or $K'[-\exp(\log s)] \geq K'[-\exp(\log s - \epsilon)]$,
   (a) Set $\log s \leftarrow \log s - \epsilon$.
   (b) Set $\epsilon \leftarrow \epsilon/2$.
   (c) Set $\log s \leftarrow \log s + \epsilon$ and repeat (5).
7. $c_1 = -\exp(\log s)$ and stop.

The situation in step 2 can arise if the $t_{\min}$ chosen is too small. For example, suppose that the desired density is shifted so that it has mass beginning at $t = 4$ but $t_{\min}$ is chosen to be 0. Then the program will return an appropriate value of $t_{\min}$, corresponding to the smallest feasible value of s, $s_{\min}$. The program will set $c_1 = -\exp(\log s_{\min})$ and return a value of $t_{\min} = K'[-\exp(\log s_{\min})] + 10^{-8}$.

The general saddlepoint function given in Figure 3.14 passes the required input to C programs that first calculate $s_{\min}$ and $s_{\max}$, and then the saddlepoint density approximation at all desired times t for a given value of $\boldsymbol{\theta}$. In the saddlepoint function, the smallest and largest values of t desired are denoted by m and M, respectively. The default values are set to $m = 10^{-6}$ and $M = 10^6$.

```
saddle <- function (N, Theta, t, m=1e-6, M=1e6,
          maxiter=1000, epsrel=1e-7, epsabs=0,
          max.abserr=1e-4)
{
   if (!is.matrix(Theta)) { Theta <- matrix(Theta, nrow=1) }

   system("cd ..; make calc_spb.so; make saddle.so")
   dyn.load("calc_spb.so")

   List <- .C("calc_spb", as.double(rep(m, N)),
       as.double(rep(M, N)), as.double(as.vector(t(Theta))),
       as.integer(N), as.integer(dim(Theta)[2]),
       sp.lb=as.double(numeric(N)),
       sp.ub=as.double(numeric(N)))

   dyn.unload("calc_spb.so")

   sp.lb <- List$sp.lb; sp.ub <- List$sp.ub;

   dyn.load("saddle.so")

   List <- .C("saddle", as.double(sp.lb), as.double(sp.ub),
       as.integer(length(t)), as.double(t),
       as.double(as.vector(t(Theta))), as.integer(N),
       as.integer(dim(Theta)[2]), as.integer(maxiter),
       as.double(epsrel), as.double(epsabs),
       as.double(max.abserr),
       Sp=as.double(numeric(length(t))),
       nc=as.double(numeric(N)),
       err.sd=as.integer(0))

   dyn.unload("saddle.so")
   system("cd ..; make clean; make clean.so")

   err.sd <- List$err.sd
   if( err.sd ) {
      stop("Error in calculating normalizing constant.")
   }

   list(pd=List$Sp/N, nc=List$nc, sp.lb=sp.lb, sp.ub=sp.ub)
}
```

FIGURE 3.14 Saddlepoint function.

Example 3.11: HYDRAULIC PUMP SYSTEM: CONVOLUTION OF TWO WEIBULL DISTRIBUTIONS. We reconsider the hydraulic pump series system of Example 2.11 with Weibull failure times for the pumps. This example is due to S. Huzurbazar. Let T_0 be the random failure time from $0 \rightarrow 1$ with MGF $M_{01}(s)$, and let T_1 be the random failure time from $1 \rightarrow 2$ with MGF $M_{12}(s)$. Let $T_0 \sim \text{Weibull}(0.2, 1.5)$ and let $T_1 \sim \text{Weibull}(0.9, 1.5)$. The MGF of the distribution of the waiting time to total failure is

$$M(s) = M_{01}(s)M_{12}(s).$$

Approximation to the distribution of the Weibull presents additional complexities since the cumulant generating function is not available in closed form. The CGF for the total failure time $T = T_1 + T_2$ is found using (2.8):

$$K(s) = \log\left\{\int_0^\infty \nu_0 \omega_0{}^{\nu_0} t^{\nu_0 - 1} \exp[st - (\omega_0 t)^{\nu_0}]\, dt\right\} + \log\left\{\int_0^\infty \nu_1 \omega_1^{\nu_1} t^{\nu_1 - 1} \exp[st - (\omega_1 t)^{\nu_1}]\, dt\right\}$$

for $s \in (c_1, c_2)$. For a given set of parameters, c_1 and c_2 are found numerically using the saddlepoint algorithm. The saddlepoint is found by solving (3.20) for $\hat{s}$ in the equation

$$K'(\hat{s}) = \frac{\int_0^\infty \nu_0\, \omega_0^{\nu_0}\, t^{\nu_0} \exp[\hat{s}t - (\omega_0 t)^{\nu_0}]\, dt}{\int_0^\infty \nu_0\, \omega_0^{\nu_0}\, t^{\nu_0 - 1} \exp[\hat{s}t - (\omega_0 t)^{\nu_0}]\, dt} + \frac{\int_0^\infty \nu_1\, \omega_1^{\nu_1}\, t^{\nu_1} \exp[\hat{s}t - (\omega_1 t)^{\nu_1}]\, dt}{\int_0^\infty \nu_1\, \omega_1^{\nu_1}\, t^{\nu_1 - 1} \exp[\hat{s}t - (\omega_1 t)^{\nu_1}]\, dt} = t. \tag{3.36}$$

Note that the saddlepoint can be calculated as long as the MGF, and alternatively, the CGF is numerically tractable even though neither is available in closed form. The above, together with $K''(\hat{s})$ and $K'''(\hat{s})$, provide the necessary inputs for calculating the desired approximations.

Figure 3.15 presents a saddlepoint density approximation along with the exact density for the Weibull(0.2, 1.5), and Figure 3.16 presents a saddlepoint density

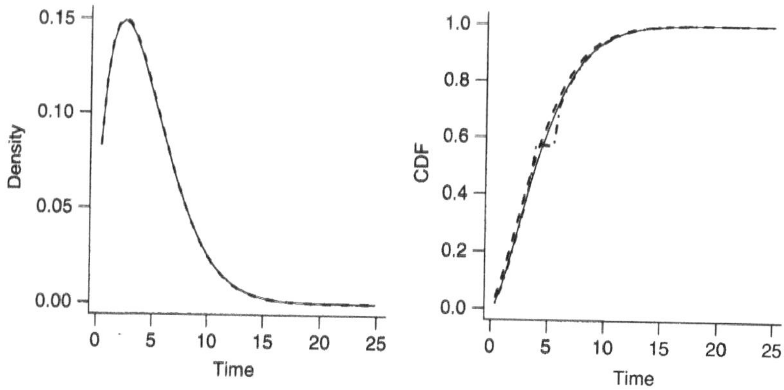

FIGURE 3.15 Exact (solid lines) and saddlepoint (dashed lines) approximated densities and CDFs for Weibull(0.2, 1.5). On the CDF plot, the dotted-dashed line that breaks down is the Lugannani and Rice saddlepoint CDF approximation.

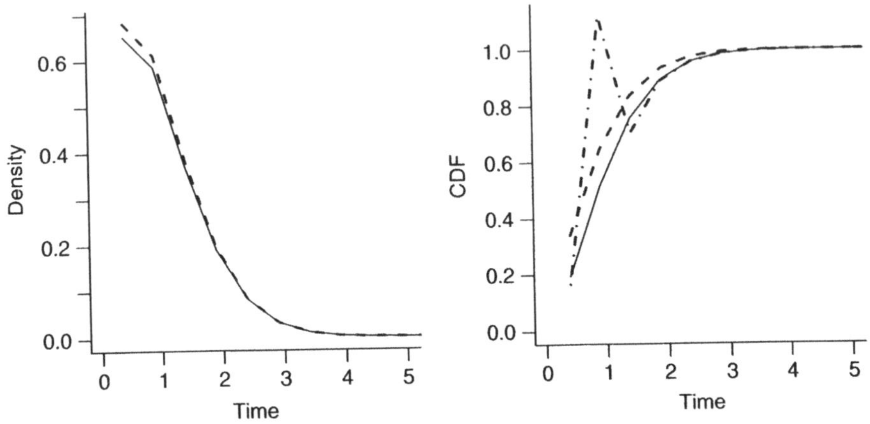

FIGURE 3.16 Exact (solid lines) and saddlepoint (dashed lines) approximated densities and CDFs for Weibull(0.9, 1.5). On the CDF plot, the dotted-dashed line that breaks down is the Lugannani and Rice saddlepoint CDF approximation.

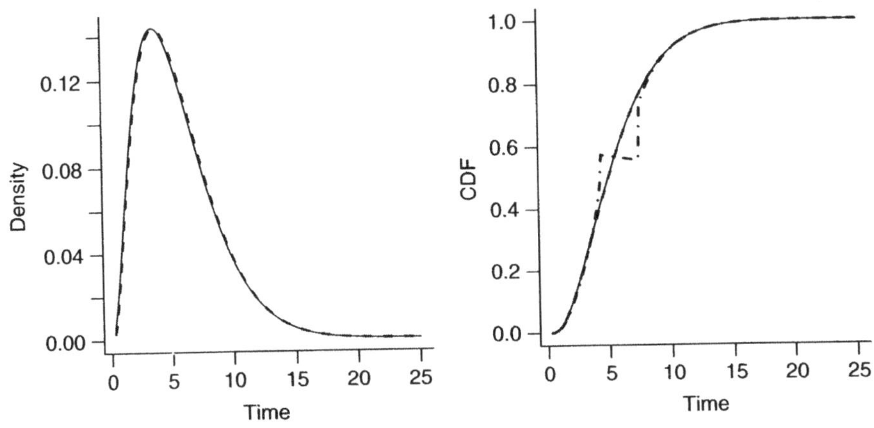

FIGURE 3.17 Exact (solid lines) and saddlepoint (dashed lines) approximated densities and CDFs for convolution of two Weibull distributions. On the CDF plot, the dotted-dashed line that breaks down is the Lugannani and Rice saddlepoint CDF approximation.

approximation along with the exact density for the Weibull(0.9, 1.5). Figure 3.17 presents the exact convolution density along with its saddlepoint approximation for purposes of comparison. On all of the plots, we can see that the saddlepoint density approximation does very well. The three figures also plot the CDFs corresponding to the displayed densities. Each CDF plot has three curves: the exact CDF, the Lugannani and Rice (1980) saddlepoint CDF of (3.26), and the approximate CDF obtained by numerically integrating the saddlepoint density. The line that breaks down on each plot is the Lugannani and Rice (1980) saddlepoint CDF approximation. This approximation is unstable near the mean of the distribution. In general, with flowgraphs, we use the more stable CDF computed by numerically integrating the saddlepoint density approximation. The exact and saddlepoint approximated convolution densities of Figure 3.17 are virtually indistinguishable. A convolution of more than two Weibull distributions can be handled by the saddlepoint approximations through the addition of more terms to the CGF of T, while approximation by numerical integration necessitates evaluation of higher-dimensional integrals. In parametric modeling, there are a variety of distributions from which one can choose. This example illustrates that while the Weibull can be used in flowgraph modeling, from the point of view of flowgraph algebra, it is more convenient to use other distributions that can mimic the shape of the Weibull, such as the gamma or inverse Gaussian. □

EXERCISES

3.1 Using partial fraction expansion, derive the density of T in Example 3.1 for the case where $\lambda_1 > 2\lambda_0$ and verify that it is equivalent to (3.6).

3.2 Derive the density of the exact survival time, T, from state 0 to state 2 for a cancer patient of Example 2.12. Use partial fraction expansion to invert the MGF given in (2.16). Verify that the density of T is

$$f(t) = \frac{(\lambda_2 - \lambda_3)(\lambda_1 + \lambda_2)\exp[-(\lambda_1 + \lambda_2)t] + \lambda_1\lambda_3 \exp(-\lambda_3 t)}{\lambda_1 + \lambda_2 - \lambda_3}, \tag{3.37}$$

for $t \geq 0$ and $\lambda_1 + \lambda_2 \neq \lambda_3$, and for $\lambda_1 + \lambda_2 = \lambda_3$ it is

$$f(t) = (\lambda_2 + \lambda_1\lambda_3 t)\exp(-\lambda_3 t). \tag{3.38}$$

Confirm your results using your favorite symbolic algebra package.

3.3 Compute the saddlepoint approximation to the Gamma(8.4, 4) density. Experiment with normalizing the density. Compute the approximation by varying time intervals and grid sizes. Compare your approximations and normalizing constants. Compute the hazard function approximation.

3.4 Suppose that T_1 and T_2 are independent with $T_1 \sim \text{Exp}(4)$ and $T_2 \sim \text{Exp}(7)$. Compute the saddlepoint density approximation to the distribution of $T_1 + T_2$. Compare your results to an exact computation via either numerical integration or analytical methods.

3.5 Verify that the cumulant generating function output from MAPLE is identical to the CGF found in (2.25).

3.6 Consider the feedforward model for HIV given in Example 3.3.

(a) Derive the MGF of the time to incubation using MAPLE or any symbolic algebra package.

(b) With the exponential parametric assumptions made in the example, convert this MGF to a density using MAPLE. Remember to use the Laplace transform inversion option with the proper change of variable.

CHAPTER 4

Censored Data Histograms

Flowgraph models involve parametric modeling of the state-to-state waiting time distributions. In the examples presented thus far, the parametric models were assumed to be known. In general, we may have a rationale for using a specific parametric model, or we may want to select one from the data. In the latter case, we can use histograms of the data to suggest parametric models. Histograms provide an important tool for an exploratory first look at data. For example, histograms can suggest appropriate parametric models, whether the data arise from more than one group, or whether the data appear to be truncated. Although it is often worthwhile to plot a histogram of the data, it is not clear how to do this when the data are censored. A histogram of just the observed (uncensored) survival times is clearly not appropriate, however, it is sometimes used as a tool with the hope that it will behave similar to a histogram of all (unobservable) uncensored survival times.

4.1 CENSORED DATA

Censored data occur commonly in survival analysis and reliability studies. An observation is censored if the time-until-event occurrence is only partially observed. For instance, in survival data, the lifetime of a person is censored if the event of interest has not been observed. This may be because at the time of the analysis, the person has not experienced the event or the person is lost to follow-up. The latter occurs when the person drops out of the study and the only information on the person's survival experience is based on the last time that he or she was observed. This is the most common type of censoring in survival data and is termed *right censoring*. A right-censored survival time is less than the actual (unobserved) survival time. Other types of censoring that can occur are *left censoring* and *interval censoring*. Left censoring occurs when the time origin for the event of interest is not observed. For example, if the study focus were on estimating the incubation time of West Nile virus, the exact time of infection must be known. A patient

Flowgraph Models for Multistate Time-to-Event Data, By Aparna V. Huzurbazar
ISBN 0-471-26514-4

who begins showing characteristic symptoms of infection may not remember the exact time of being bitten by mosquitoes. For that matter, it would not be known which mosquito was the culprit, assuming that is how the patient contracted the virus. In this case the actual survival time, the time to incubation, is less than the true survival time. Interval censoring occurs when a person is known to have experienced the event of interest within a time interval. For example, suppose that a patient is believed to have been bitten by a mosquito carrying the West Nile virus. The patient might be examined once every morning for any symptoms that may be indicative of this virus. Suppose that up to the fifth morning the patient is free of symptoms but has a fever on the sixth morning. Then we know the time to incubation up to this 24-hour period (perhaps we can narrow it down to a smaller interval based on interviewing the patient), and this survival time is interval censored. Some will argue that all survival data are interval censored, due to the rounding accuracy of measurements. We restrict ourselves to the most common case of right-censored data.

In reliability analysis, there are four types of censoring—Types I, II, III, and IV—all of which are variations on right censoring. The most common form of censoring in reliability is Type I. This occurs when a study design requires observing all n units up to a specified time t. All the units that have not failed by time t are right censored. Type II censoring occurs when the study design has a stopping rule that requires stopping after a specified number, r, of failures has been observed. In this case, only the r smallest lifetimes are observed and we do not know how long the experiment will last. Type III censoring combines Types I and II censoring. Here the design requires a specified stopping time, t, and also a specified number of failures, r, at which to stop. The stopping rule requires stopping at the first occurrence of one of these criteria. Type IV censoring is a type of right censoring that is most closely related to typical survival analysis clinical trial studies. In this case, units are activated at different time points, but the test is terminated at a fixed time point. This is analogous to a clinical trial where patients enter the study at random points in time but the trial ends at a prespecified point in time. This is the least common form of censoring in reliability studies but the most common form of censoring in survival analysis.

4.2 SURVIVOR AND RELIABILITY FUNCTIONS

In the analysis of time-to-event data, one quantity of interest is the probability that a person or component survives beyond some time t starting from the initial time $t = 0$. The word *survives* is used quite generally to mean that the event of interest has not yet occurred. Let T be the random waiting time until the event of interest occurs. The quantity $P(T > t)$ represents the probability that the event of interest has not occurred. In survival analysis, this is called the *survivor function* and is denoted by $S(t) = P(T > t)$, defined in (1.1.1). For instance, the survivor function gives the probability that a patient survives beyond some specified time t. Certain times are of specific interest, for example, the median survival time t_M for which

$S(t_M) = 0.5$. In engineering applications, this is called the *reliability function* and is denoted by $R(t) = P(T > t)$. The word *reliability* has different meanings under different contexts. A technical definition of *reliability* is "the ability of an item to perform a required function, under given environmental and operational conditions and for a stated period of time" (cf. ISO 8402). The reliability function is only one measure of reliability. In practice, these functions must be estimated from the time-to-event data.

Example 4.1: RELIABILITY/SURVIVOR FUNCTION FOR A WEIBULL DISTRIBUTION. The Weibull distribution was derived by Weibull (1951) for the propagation of cracks in aircraft wings. Since then it has been used extensively for modeling time-to-event data in both reliability and survival analysis. For example, in survival analysis applications it has been useful for modeling the time to incubation of the AIDS virus (cf. Medley et al., 1987). The Weibull density, CDF, and reliability or survivor function are given by

$$f(t|\omega, \nu) = \nu\omega^{\nu}t^{\nu-1}\exp[-(\omega t)^{\nu}] \tag{4.1}$$

$$F(t|\omega, \nu) = 1 - \exp[-(\omega t)^{\nu}] \tag{4.2}$$

$$R(t) = \exp[-(\omega t)^{\nu}] \tag{4.3}$$

for $\nu > 0$, $\omega > 0$, $t > 0$. The shape parameter of the Weibull is denoted by ν. The exponential distribution is a special case of the Weibull with $\nu = 1$. Figures 4.1

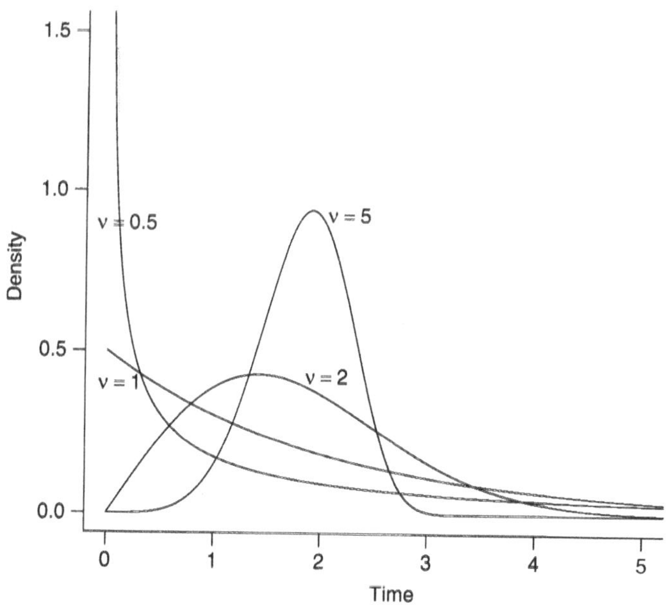

FIGURE 4.1 Weibull densities for $\omega = .5$ with varying shape parameter ν for Example 4.1.

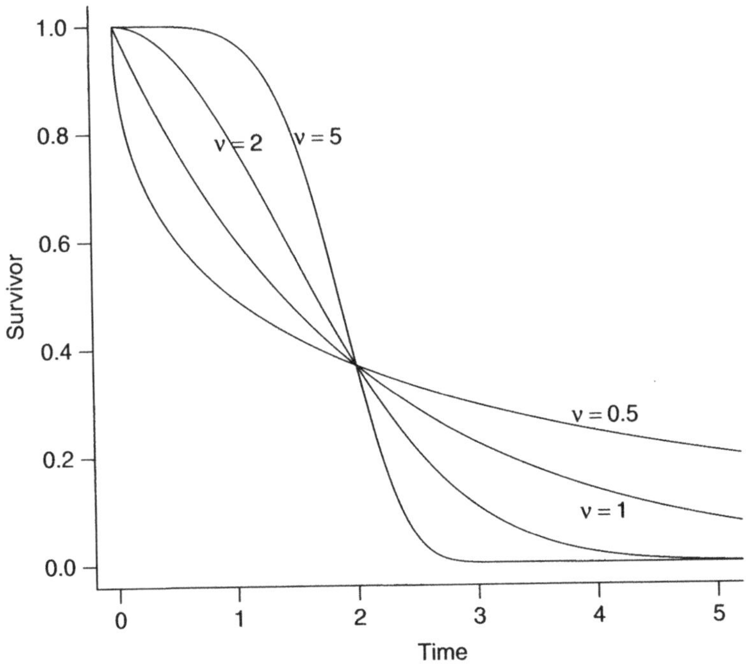

FIGURE 4.2 Weibull survivor functions for $\omega = .5$ with varying shape parameter ν for Example 4.1.

and 4.2 give examples of some different Weibull densities and survivor functions. Figure 4.3 gives R program code for plotting the density, survivor, and hazard functions. Note that *R* parameterizes the Weibull with the scale parameter as $1/\omega$, so we must invert ours in the code. □

4.3 HAZARD AND CUMULATIVE HAZARD FUNCTIONS

The hazard function was defined in Definition 1.1.2 as

$$h(t) = \frac{f(t)}{1 - F(t)} = \frac{f(t)}{S(t)}. \tag{4.4}$$

The hazard function arises from the failure rate, the rate at which failures occur in an infinitesimally small time interval. For a given time interval $(t, t + \delta t]$, the failure rate for the interval is expressed by

$$\frac{1}{\delta t} P(T \in (t, t + \delta t] \mid T > t) = \left(\frac{F(t + \delta t) - F(t)}{1 - F(t)} \right) \left(\frac{1}{\delta t} \right). \tag{4.5}$$

We can think of the failure rate as the probability that a failure per unit time occurs in the interval given that a failure has not occurred prior to the beginning of the

```
# Define the time axis
  t<-seq(0,10,.01)

# Create densities
  den<-rep(0,length(t))
  den3<-den2<-den1<-den

# Invert scale parameter for program
  den<-dweibull(t,shape=.5,scale=2)
  den1<-dweibull(t,shape=1,scale=2)
  den2<-dweibull(t,shape=2,scale=2)
  den3<-dweibull(t,shape=5,scale=2)

 plot(t,den,type="l",xlab="t",ylab="density",
      xlim=c(0,5), ylim=c(0,1.5))
      lines(t,den1)
      lines(t,den2)
      lines(t,den3)

# Define survivor function
  surv1<-surv2<-surv3<-surv<-den
  surv<-pweibull(t,shape=.5,scale=2,lowertail=FALSE)
  surv1<-pweibull(t,shape=1,scale=2,lowertail=FALSE)
  surv2<-pweibull(t,shape=2,scale=2,lowertail=FALSE)
  surv3<-pweibull(t,shape=5,scale=2,lowertail=FALSE)

 plot(t,surv,type="l",xlab="t",ylab="survivor",
      xlim=c(0,5),ylim=c(0,1))
      lines(t,surv1)
      lines(t,surv2)
      lines(t,surv3)

# Define hazard
  haz1<-haz2<-haz3<-haz<-den
  haz<-den/surv
  haz1<-den1/surv1
  haz2<-den2/surv2
  haz3<-den3/surv3

 plot(t,haz,type="l",xlab="t",ylab="hazard",
      xlim=c(0,5),ylim=c(0,2))
      lines(t,haz1)
      lines(t,haz2)
      lines(t,haz3)
```

FIGURE 4.3 R program code to generate the plots for Example 4.1.

interval. Considered in this manner, the hazard function is the limit of the failure rate as $\delta t \to 0$,

$$\begin{aligned} h(t) &= \lim_{\delta t \to 0} \left(\frac{F(t+\delta t) - F(t)}{1 - F(t)}\right)\left(\frac{1}{\delta t}\right) \\ &= \lim_{\delta t \to 0} \left(\frac{F(t+\delta t) - F(t)}{\delta t}\right)\left(\frac{1}{1 - F(t)}\right) \\ &= \frac{f(t)}{R(t)} \end{aligned} \tag{4.6}$$

This is also called the *hazard rate function, hazard rate, instantaneous failure rate, age-specific failure rate*, or *force of mortality*, and to add confusion with (4.5), the *failure rate*. The hazard gives the instantaneous rate of death or failure at time t given that the person survives up to time t. Although the hazard is technically a rate, it is often spoken of in probabilistic terms as the chance that a patient or component "fails" at time t conditional on having survived up to that time. In fact, $h(t)$ is *not* a probability density function (see Exercise 4.2) and it can take on any value greater than or equal to zero. Given survival up to time t, the approximate probability of death in a small interval δt immediately after t is $h(t)\delta t$. Viewed in this manner, the density $f(t)$ is the rate of change of the probability of failure. Figure 4.4 displays some hazard shapes for the Weibull density of Example 4.1.

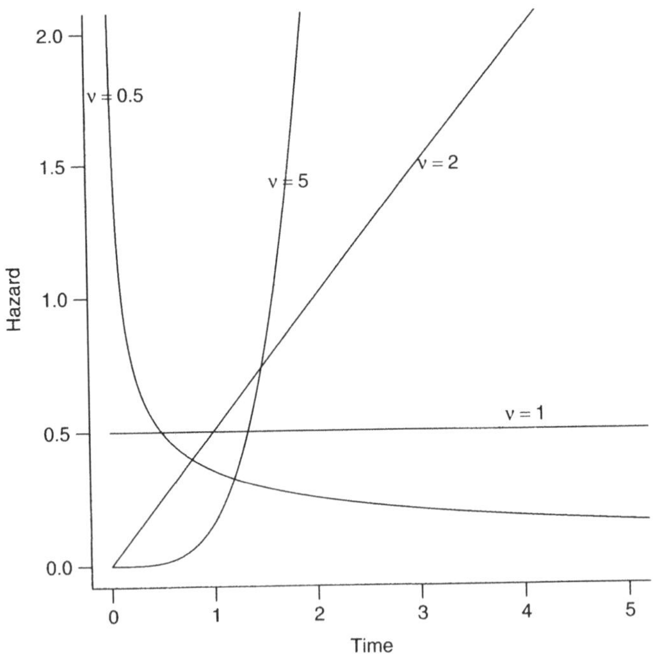

FIGURE 4.4 Weibull hazard functions for $\omega = 0.5$ with varying shape parameter ν for Example 4.1.

The hazard function for the Weibull is

$$h(t) = \nu\omega^\nu t^{\nu-1}. \tag{4.7}$$

The last section of the R code in Figure 4.3 computes and plots the hazard for Example 4.1.

The cumulative hazard or the integrated hazard is the quantity

$$H(t) = \int_0^t h(u)\,du = -\log S(t). \tag{4.8}$$

This quantity is used for model checking and identification for time to event data. For example, data that have a roughly linear estimated cumulative hazard support an exponential model.

Interpreting hazard function shapes is important in both reliability and survival analysis. In general, hazard shapes are divided into three main types:

- *Decreasing failure rate* (DFR). This is a hazard function that is decreasing. In survival analysis, this phase is called *infant mortality* because the risks associated with dying or falling ill are the greatest for a newborn immediately after birth and then drop off. In reliability, this is called the *burn-in period.* In the context of manufacturing a new product, this sort of hazard is associated with low-quality units that fail immediately after being put on line, while higher-quality units continue to work, failing at a later time. Good product testing is designed to eliminate substandard product so that this will not occur on the product that is sold.
- *Constant hazard.* This hazard corresponds to no aging and arises from an exponential model. Failures in this phase are called the *chance or random failure*, in that failures are caused by sudden stresses on the system, say breakage due to a sudden increase in temperature. In the medical context, this corresponds to an accidental death or injury.
- *Increasing failure rate* (IFR). This is a hazard function that is increasing. This is appropriate when the person or unit is subject to aging. This is due to equipment deterioration, fatigue, and general wear-out. This can be applied with the same terms in the medical context.

Putting these three basic shapes together gives the most famous shape in reliability, called the *bathtub hazard.* This is a hazard that is decreasing, constant, and then increasing. It can be used to describe the life of a person as follows: At birth, the hazard is the highest and the child is in the infant mortality phase. The hazard decreases slowly as the child grows up and finally, hits a constant phase. For a successful parent, this means that he or she raised the child to age 18 years and kept it alive. For the next 50 or so years, the adult is in the constant hazard phase. This is also called the *useful life* period in the reliability literature. In the last phase, the adult is aging and subject to general fatigue and illness related to age, and the hazard is increasing.

Another hazard shape that is important is a hazard function that is first increasing and then decreasing. In the medical context, such a hazard arises in patients recuperating after surgery. Immediately after surgery, their risk for complications is increasing and, after a few days, it starts to decrease and tapers off. In the engineering context, this works for new systems that are just put on line. Suppose that we have just put a new aircraft into service. The hazard is initially increasing on early sorties while the crew are learning the aircraft and working out any kinks. After this *learning curve*, the hazard decreases and is constant for its useful-life period until aging effects set in. Aalen and Gjessing (2001) provide a framework for understanding hazard function shapes that is applicable to time to event data.

Most reliability and survival analysis books contain descriptions of various probability models used for lifetime data, and their survivor or reliability functions, and hazard functions. Some introductory books on survival analysis are Collett (1994), Cox and Oakes (1984), Klein and Moeschberger (1997), or Therneau and Grambsch (2000). In the reliability context, see Ansell and Phillips (1994), Ascher and Feingold (1984), Barlow and Proschan (1975), Crowder et al. (1991), Hoyland and Rausand (1994), Martz and Waller (1982), Meeker and Escobar (1998), Nelson (1990), or Wolstenholme (1999). In addition, most of the common lifetime distributions are available in statistics programs such as Minitab, R, S-plus, MATLAB, SAS, MAPLE, and with some skill, even in EXCEL. Readers might find it useful to plot densities, survivor or reliability functions, and hazard functions for various parameter values for different distributions.

4.4 KAPLAN–MEIER ESTIMATOR

Time-to-event data for which we observe every event are called *complete data* or *uncensored data*. When data are uncensored, survivor functions can be estimated using the empirical CDF. The empirical CDF is

$$\hat{F}(t) = \frac{\text{no. lifetimes} \leq t}{n}, \tag{4.9}$$

so that

$$\hat{S}(t) = \hat{R}(t) = 1 - \hat{F}(t) = \frac{\text{no. lifetimes} > t}{n}. \tag{4.10}$$

However, in practice, data are generally censored. The most commonly used survivor function estimate for right-censored survival data is the Kaplan–Meier (KM) estimator due to Kaplan and Meier (1958). The texts in survival analysis and reliability mentioned previously all present a description of the KM estimator.

We can construct the survivor function estimator as follows. Let T be the survival time of interest. To compute the Kaplan–Meier estimate, put all of the observations, censored and uncensored, in increasing order. If a censored time and an observed survival time occur simultaneously, convention is to treat the uncensored observation as preceding the corresponding censored time. Let t_i, $i = 0, \ldots, r$, be r ordered observed *death times* out of a total N observations. Let n_i be the number of people known to be alive just before time t_i and let d_i be the number of people who die at time t_i. For convenience, let $t_0 = 0$, $n_0 = N$, and $d_0 = 0$. For each interval, the estimated proportion of people alive just before t_i and that survive beyond t_i is $(n_i - d_i)/n_i$. This leads to the Kaplan–Meier estimate of the survival curve:

$$\hat{S}(t) = \hat{P}(T > t) = \prod_{i=1}^{k} \frac{(n_i - d_i)}{n_i}, \quad t_k \le t < t_{k+1} \text{ and for } k = 0, 1, \ldots, r. \tag{4.11}$$

In the absence of censoring, (4.11) reduces to the empirical survivor function (4.10). This survivor function estimator is commonly used and is easily computed with standard statistical software. In reliability analysis it is denoted by $\hat{R}(t)$, the random time T is called the *event time*, n_i is the number of components that have not experienced the event at time t_i, and d_i is the number of components that have experienced the event at time t_i.

Example 4.2: SURVIVOR FUNCTION ESTIMATION. We return to the HIV blood transfusion model of Example 2.10. The transfusion data, from the San Francisco Men's Health Study (SFMHS), consist of people who were given HIV-infected blood products. Detailed descriptions of the data are in Huzurbazar and Huzurbazar (1999) and Longini et al. (1989). The data were collected from 1978 to 1980 and have been used to model the early stages of progression for people with HIV infection. Figure 2.13 presented the flowgraph model for these data. State 0 is HIV infection with negative antibody status, state 1 is HIV infection with positive antibody status but without AIDS symptoms, and state 2 is pre-AIDS symptoms. The transfusion data are informative about the early stages of HIV infection because the exact time of transfusion is known. Huzurbazar and Huzurbazar (1999) use the transfusion data to illustrate saddlepoint methods in a frequentist setting and present a censored data histogram as part of their analysis. The data consist of 90 people who received HIV-infected blood or blood products through transfusion. Seventy people made the transition from state 0 to state 1. For 35 of these 70 people, there is no information beyond state 1. For the remaining 35 people, 22 made the transition from state 1 to state 2, and 13 were right-censored in state 1. In other words, these 13 people were last seen in state 1 and they were not observed to make the transition to the end state, state 2.

For this example, we will only use the time spent in state 1, which is HIV infection, antibody positive status. The data consist of 35 people. Of these, 22 made the transition from state 1 to a progressively worse state 2 which is characterized by pre-AIDS symptoms. Thirteen people were observed for a period of time in state 1 and then right-censored while in state 1. Figure 4.5 shows the Kaplan–Meier survivor function estimate for these data. The death times and survivor function are also given following the R program commands of Figure 4.6. R outputs a survivor function value for every data point, without regard to whether an event occurred at that time point. Users of S-plus or SAS should note that these packages

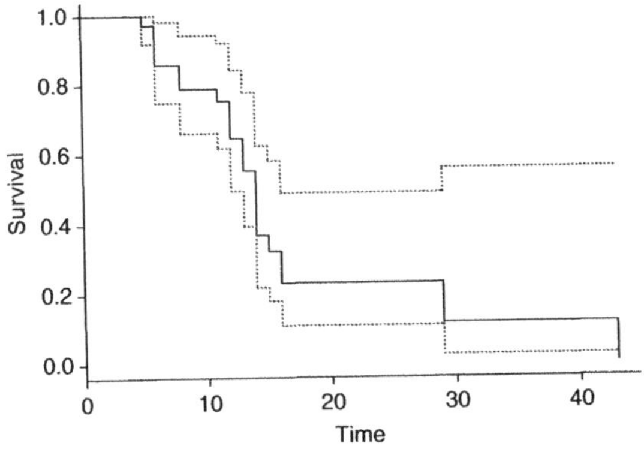

FIGURE 4.5 Kaplan–Meier survivor function for the waiting time in state 1 for the transfusion data. Dotted lines are 95% confidence bands for the survivor function.

```
R program command:
>survfit(formula = Surv(V1, V2),
 + data = transcensor, na.action = na.exclude,
 + conf.int = 0.95, se.fit = T,
 + type = "kaplan-meier", error = "greenwood",
 + conf.type = "log", conf.lower = "usual")

death time    survivor
     5        0.97143
     6        0.85714
     7        0.85714
     8        0.78857
    11        0.75273
    12        0.64519
    13        0.55302
    14        0.36868
    15        0.32260
    16        0.23043
    19        0.23043
    20        0.23043
    28        0.23043
    29        0.11521
    43        0
```

FIGURE 4.6 R program code for a Kaplan–Meier survivor function.

will only output the survivor function at the death times observed. In Figure 4.5, the two dotted lines appearing on either side of the survivor function are the 95% confidence bands for the survivor function that are given by Greenwood's formula (cf. Collett, 1994) for the standard error of the Kaplan–Meier survivor function. They are generated by the command line `error="greenwood"` within the `survfit` function call.

A predecessor of the Kaplan–Meier estimator is the lifetable estimator, which is also called the *actuarial estimate*. Like the Kaplan–Meier, it allows for right censoring. However, it breaks up the time period into intervals, usually equally spaced, and then computes the survivor function. □

4.4.1 Estimating the Hazard

A natural way to approximate the hazard is to extend the Kaplan–Meier survivor function estimate to the hazard. Using the same notation as in the Kaplan–Meier estimate, this is

$$\hat{h}(t) = \frac{d_i}{n_i(t_{i+1} - t_i)} \qquad \text{for } t_i \le t < t_{i+1},\ i = 0, 2, \ldots, r-1, \tag{4.12}$$

where t_i are the r ordered death times or event times and d_i are the number of people who die at time t_i. An alternative estimator of the hazard function that is commonly used in reliability is Nelson's (1969) estimate, which begins with the cumulative hazard. The Nelson estimator of the cumulative hazard is

$$\hat{H}(t) = \sum_{i=1}^{k} \frac{1}{n_i} \qquad \text{for } t_k \le t < t_{k+1},\ k = 0, 1, \ldots, r. \tag{4.13}$$

Viewed this way, this is a first-order approximation to the Kaplan–Meier estimator. We can convert this to a reliability function approximation

$$\hat{R}(t) = \exp[-\hat{H}(t)].$$

A generalization of this estimator is the Nelson–Aalen estimator (cf. Aalen, 1978) for the cumulative intensity function for counting processes.

4.5 HISTOGRAM FOR CENSORED DATA

One approach to constructing a histogram for censored survival data involves basing it on the Kaplan–Meier estimate of the survivor function (cf. Huzurbazar and Huzurbazar (1999) and Huzurbazar (2005)). The censored data histogram (CDH) converts the Kaplan–Meier estimate of the survivor function to a histogram of the data accounting for both censored and uncensored observations.

To construct a censored data histogram, create a set of equal-width intervals. Suppose that we create w such intervals and index the interval limits as $\tilde{t}_j$, $j = 0, \ldots, w$. Then for each interval, the corresponding estimated probability is

$$P(\tilde{t}_j < T \le \tilde{t}_{j+1}) = P(T > \tilde{t}_j) - P(T > \tilde{t}_{j+1}) \doteq \hat{S}(\tilde{t}_j) - \hat{S}(\tilde{t}_{j+1}) \tag{4.14}$$

for $j = 0, \ldots, w-1$, where $\hat{S}(\cdot)$ is the Kaplan–Meier estimate of the survivor function. This estimate includes the right endpoint of each histogram bin. Alternatively, we can also define the censored data histogram estimated probability for each interval by including the left endpoint of the interval,

$$P(\tilde{t}_j \leq T < \tilde{t}_{j+1}) = P(T \geq \tilde{t}_j) - P(T \geq \tilde{t}_{j+1}) \doteq \hat{S}(\tilde{t}_j^-) - \hat{S}(\tilde{t}_{j+1}^-) \tag{4.15}$$

for $j = 0, \ldots, w-1$, where $\hat{S}(\cdot)$ is the Kaplan–Meier estimate of the survivor function and $\hat{S}(t^-)$ is the left-hand limit of $\hat{S}(t)$ at time t. If the interval endpoints do not fall on a death time, the results from (4.14) and (4.15) are identical; otherwise, the censored data histograms will differ slightly. If the method chosen is (4.14) and the left endpoint coincides with the smallest death time, an extra bin is added to the left end of the histogram, accommodating the fact that the endpoint of the additional bin cannot be less than zero. Similarly, if the method chosen is (4.15), and the largest death time coincides with the upper limit of the histogram, then an additional bin is added to the rightmost end of the histogram. The CDH is intended as a rough guide to assist in choosing parametric models for waiting times. Many refinements of such a histogram are possible (cf. Barnett and Cohen, 2000); the CDH as computed is both practical and easy to implement.

Example 4.3: Computation of the Censored Data Histogram. The computation of (4.14) and (4.15) using the survivor function of Example 4.2 is given in Tables 4.1 and 4.2. Note that there is little difference between the histograms. □

Before presenting the details of the program, we show the censored data histogram results from the program for the transfusion data.

Example 4.4: Censored Data Histogram with Overlay for HIV Blood Transfusion Data. This example shows the censored data histogram for both

TABLE 4.1 Censored Data Histogram Heights Using (4.14)

Bin $(a, b]$	Total Probability (TP)	CDH Height (TP/Bin), Eq. (4.14)
0–5	$1 - 0.97143 = 0.02857$	0.005714
5–10	$0.97143 - 0.78857 = 0.18286$	0.036572
10–15	$0.78857 - 0.3226 = 0.46597$	0.093194
15–20	$0.3226 - 0.23043 = 0.09217$	0.018434
20–25	$0.23043 - 0.23043 = 0$	0
25–30	$0.23043 - 0.11521 = 0.11522$	0.023044
30–35	$0.11521 - 0.11521 = 0$	0
35–40	0	0
40–45	$0.11521 - 0 = 0.11521$	0.023042
45–50	0	0
50–55	0	0
55–60	0	0

TABLE 4.2 Censored Data Histogram Heights Using (4.15)

Bin $[a, b)$	Total Probability (TP)	CDH Height (TP/Bin), Eq. (4.15)
0–5	$1 - 1 = 0$	0
5–10	$1 - 0.78857 = 0.21143$	0.042286
10–15	$0.78857 - 0.36868 = 0.41989$	0.083978
15–20	$0.36868 - 0.23043 = 0.13825$	0.02765
20–25	$0.23043 - 0.23043 = 0$	0
25–30	$0.23043 - 0.11521 = 0.11522$	0.023044
30–35	$0.11521 - 0.11521 = 0$	0
35–40	0	0
40–45	$0.11521 - 0 = 0.11521$	0.023042
45–50	0	0
50–55	0	0
55–60	0	0

of the waiting times involved in the transfusion data of Example 4.2. Let T_0 be the waiting time in state 0 until transition to state 1 and let T_1 be the waiting time in state 1 until transition to state 2. The total waiting time for passage from $0 \to 2$ is $T = T_0 + T_1$. Previous analyses have taken T_0 and T_1 to be exponentially distributed. Figures 4.7 and 4.8 present censored data histograms of observations

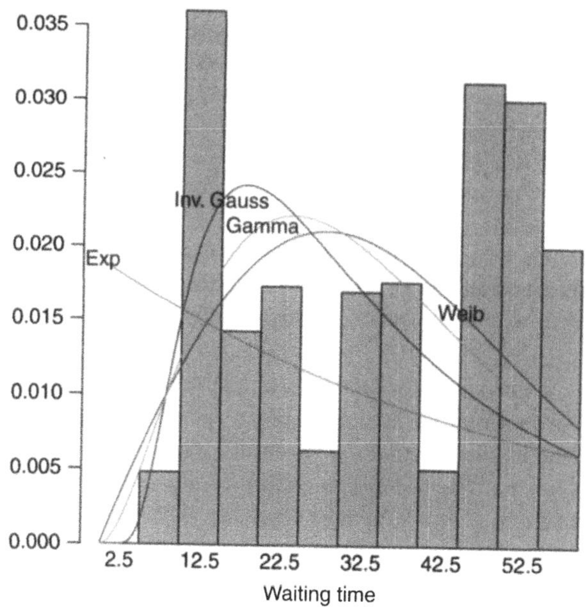

FIGURE 4.7 Censored data histogram of T_0, the waiting time from state 0 to state 1, for Example 4.4.

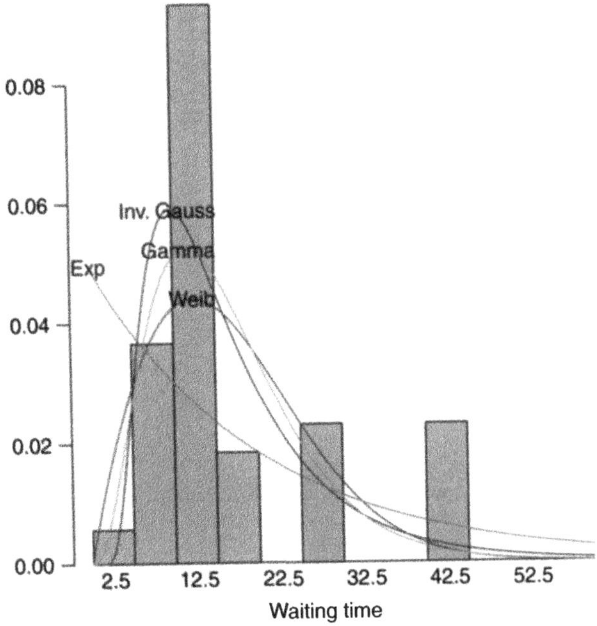

FIGURE 4.8 Censored data histogram of T_1, the waiting time from state 1 to state 2, for Example 4.4.

on T_0 and T_1. Some candidate parametric models are overlaid on the censored data histograms. The end result from the program is a histogram for each state-to-state waiting time. The user has the option of overlaying estimates of four parametric models, individually or as a group, on each censored data histogram. The choice of densities is the exponential, gamma, Weibull, and inverse Gaussian, and they are estimated using maximum likelihood on the censored and uncensored observations. They are intended as a guide toward choosing a parametric model for each branch of the flowgraph. The user can modify the program to omit either some or all of the estimated densities or add more densities as desired. The histogram of observations on T_0 suggests that an exponential model may be adequate. However, the histogram of observations on T_1 suggests that a distribution with a positive mode, such as a gamma, would be a more appropriate distribution. □

Example 4.5: Censored Data Histogram for Hydraulic Pump System with Feedback. We return to the hydraulic pump system with feedback presented in Example 2.8. The flowgraph model is given in Figure 4.9. We simulate 100 observations on this system. The waiting time from $0 \to 1$ is Exp(0.01) with mean 100, the waiting time from $1 \to 0$ is inverse Gaussian with parameters $\lambda_1 = 22.36$ and $\lambda_2 = 2.236$, and the waiting time from $1 \to 2$ is Gamma(1000, 5). The probability that a pump is repaired and the system is returned to full operating condition is the transition probability from $1 \to 0$, $p_{10} = 0.7$. In the data, 27 realizations were censored: 16 observations were censored in state 0 and 11 observations were censored

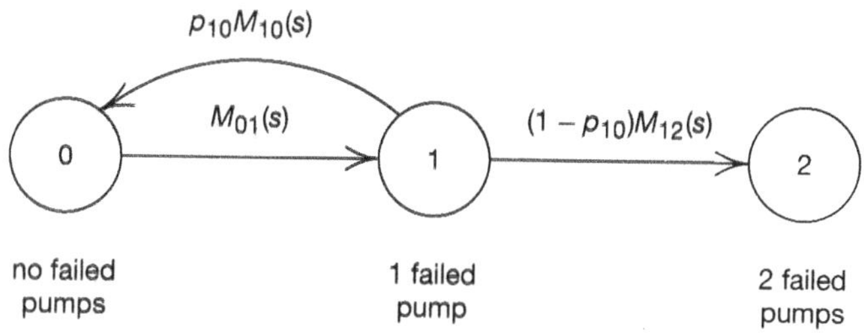

FIGURE 4.9 Flowgraph model for the hydraulic pump system.

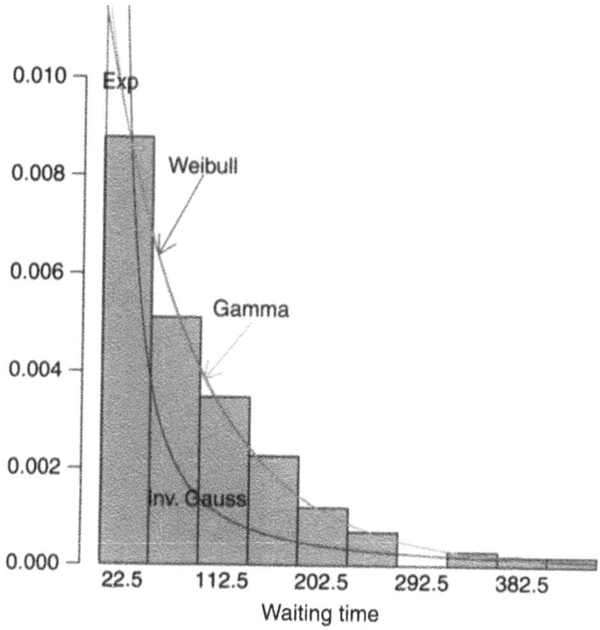

FIGURE 4.10 Censored data histogram for state $0 \to 1$ for the hydraulic pump model of Figure 4.9.

in state 1. There were 274 complete observations on the $0 \to 1$ transition, 190 complete observations on the $1 \to 0$ transition, and 73 complete observations on the $1 \to 2$ transition. The censored data histograms for states $0 \to 1$, $1 \to 0$, and $1 \to 2$ are given in Figures 4.10, 4.11, and 4.12, respectively. Since Figure 4.10 represents data that are truly exponential, the Weibull, gamma, and exponential models are all close, as expected. The exponential is a special case of both the gamma and the Weibull. The inverse Gaussian, generally a rich family, is not able to mimic the exponential shape. In Figure 4.11, the data are truly inverse Gaussian and we see that the gamma is very close to the inverse Gaussian and that the Weibull also provides an adequate fit. It should be noted that for Figure 4.11, the

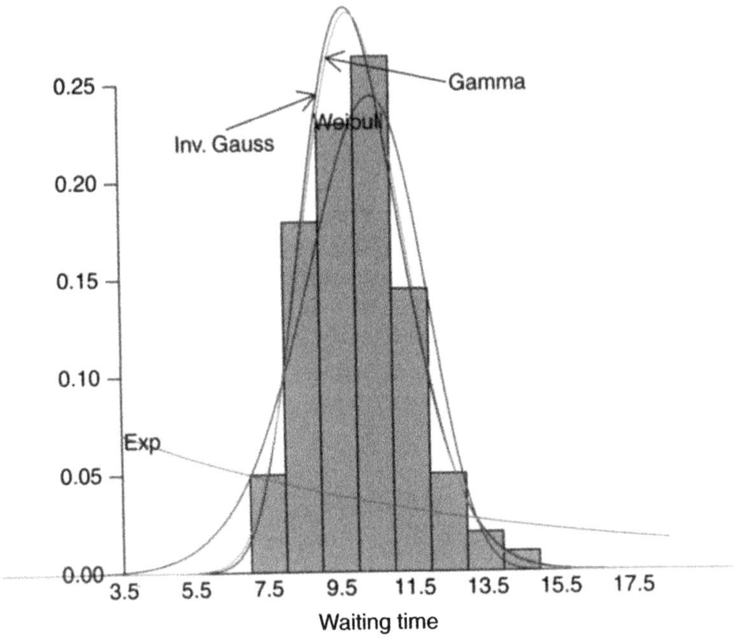

FIGURE 4.11 Censored data histogram for state 1 → 0 for the hydraulic pump model of Figure 4.9.

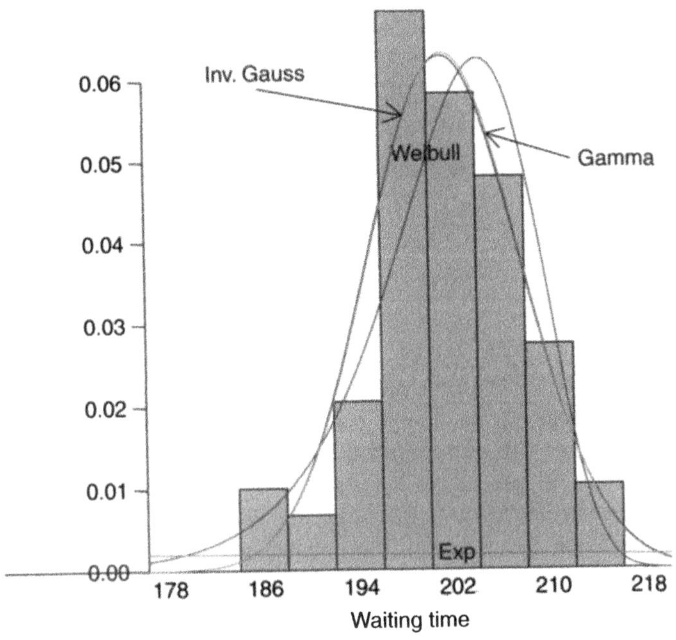

FIGURE 4.12 Censored data histogram for state 1 → 2 for the hydraulic pump model of Figure 4.9.

censored observations in state 1 could not be used for estimating the overlay distributions for the $1 \to 0$ transition. This is because they are all substantially larger than the largest event time for this transition, leading to numerical instabilities in the likelihood optimization. In Figure 4.12, where the data are truly gamma, the gamma and inverse Gaussian are very close. The Weibull is again a close second. Note that observations censored in a given state are used in estimating the overlay distribution parameters marginally for each of the possible transitions from that state. It would be most appropriate to jointly estimate the overlay distributions for all of the possible transitions from a given state when censored observations are present, as these contribute a mixture over the survivor functions corresponding to the possible transitions. However, the overlaid distributions are intended only as a rough guide, and therefore the marginal estimation method is used. □

EXERCISES

4.1 Verify the relationships between the density, CDF, survivor or reliability function, and hazard function stated in (1.1).

4.2 Prove that the requirement that $S(\infty) = 0$ implies that

$$\lim_{t \to \infty} \int_0^t h(u)\,du = \infty.$$

4.3 Crowder et al. (1991) present data on a type of braided cord. The goal of the experiment is to determine the strength of braided cord after weathering. There were 48 pieces of cord, of which seven were damaged during the course of the experiment, giving right-censored strength values. The data consist of 41 uncensored observations and seven censored, indicated by an asterisk.

36.3	52.4	54.8	57.1	60.7	41.7	52.6	54.8	57.3	43.9
55.1	57.7	49.9	53.1	55.4	57.8	50.1	53.6	55.9	58.1
53.6	56.0	58.9	51.9	53.9	56.1	59.0	52.1	53.9	56.5
52.3	54.1	56.9	59.6	52.3	54.6	57.1	60.4	52.7	50.8
59.1	26.8*	29.6*	33.4*	35.0*	40.0*	41.9*	42.5*		

(a) Consider just the uncensored observations. Using statistical software of your choice, plot a histogram of the data.

(b) Construct a survivor or reliability function estimator for the entire data set using statistical software of your choice.

(c) Construct an estimated hazard for these data using statistical software of your choice.

(d) Using the results of part (b) and (4.14), construct a censored data histogram for these data.

4.4 Collett (1994) gives data on the survival times of women with breast cancer. A portion of the data on the survival times of women who underwent surgery for tumor removal is given below. Right-censored observations are indicated by an asterisk.

5	8	10	13	18	24	26	26	31	35
40	41	48	50	59	61	68	71	76*	105*
107*	109*	113	116*	118	143	154*	162*	188*	212*
217*	225*								

(a) Consider just the uncensored observations. Using statistical software of your choice, plot a histogram of the data.

(b) Construct a survivor or reliability function estimator for the entire data set using statistical software of your choice.

(c) Construct an estimated hazard for these data using statistical software of your choice.

(d) Using the results of part (b) and (4.14), construct a censored data histogram for these data.

CHAPTER 5

Bayesian Prediction for Flowgraph Models

In this chapter we are concerned with computing Bayesian predictive densities, cumulative distribution functions, survivor or reliability functions, and hazard functions for flowgraph models. Information about the process allows us to draw the flowgraph as seen in Chapter 2. Observations on the various transitions allow us to consider parametric models for the branch waiting times of the flowgraph. In practice, there are many candidate models that are plausible models for the data. Generally, we pick one of these candidate models that adequately explains the data. Our interest lies in making inferences about the unknown parameters in the model, updating our knowledge of the mechanism that generated the data, evaluating branch models, and ultimately using our data and the flowgraph model for prediction. Bayesian analysis formally incorporates subjective information about a problem into the analysis. There are many excellent texts on Bayesian methods. Savage (1954), Jeffreys (1961), and DeFinetti (1972) are foundational. Other well-known texts include DeGroot (1970), Zellner (1971), Box and Tiao (1973), Aitchison and Dunsmore (1975), Berger (1993), and Bernardo and Smith (2000). Berry and Lindgren (1996) and DeGroot and Schervish (2001) are more recent introductory-level texts. Recent books with a computational emphasis are Gelman et al. (1995) and Carlin and Louis (1996). This chapter begins with a brief background of Bayesian inference.

5.1 BAYESIAN PREDICTIVE DENSITY

We can think about our model as the theory that drives what we can observe. For a given parameter vector $\boldsymbol{\theta}$, we expect things to behave in a certain way. We use the data to give us additional information about $\boldsymbol{\theta}$, but we can never know $\boldsymbol{\theta}$. We know that our theory (model) is incorrect, so the question then becomes:

Flowgraph Models for Multistate Time-to-Event Data, By Aparna V. Huzurbazar
ISBN 0-471-26514-4

How useful is it? What is often important is whether we can use the formulated model to make useful predictions about new observations, $Z = (Z_1, \ldots, Z_k)$, that are conditionally independent of $\mathcal{D}$, the observed data, given $\boldsymbol{\theta}$. In a flowgraph model, these future observations will generally be total or partial waiting times. In survival data, this can be the time to total failure of kidneys, the time until partial progression of a disease, such as full-blown AIDS in HIV-positive patients, or the time until severe deterioration of eyesight for a patient with diabetic retinopathy. In engineering systems data, this can be the time until total failure of hydraulic pumps, the time until a certain number of nodes (cells) are repaired in a cellular telephone network, or the time until there are a significant number of delays in a construction project.

The Bayes predictive density of a future observable Z given data $\mathcal{D}$ is

$$\begin{aligned} f_Z(z|\mathcal{D}) &= \int f_Z(z, \boldsymbol{\theta}|\mathcal{D})\, d\boldsymbol{\theta} \\ &= \int f_Z(z|\boldsymbol{\theta}, \mathcal{D})\pi(\boldsymbol{\theta}|\mathcal{D})\, d\boldsymbol{\theta} \\ &= \int f_Z(z|\boldsymbol{\theta})\pi(\boldsymbol{\theta}|\mathcal{D})\, d\boldsymbol{\theta} \\ &\equiv E_{\boldsymbol{\theta}|\mathcal{D}}[f_Z(z|\boldsymbol{\theta})], \end{aligned} \tag{5.1}$$

where Z has density $f_Z(z|\boldsymbol{\theta})$ and $\pi(\boldsymbol{\theta}|\mathcal{D})$ is the posterior distribution of $\boldsymbol{\theta}$.

Definition 5.1.1: BAYES THEOREM. The posterior distribution of $\boldsymbol{\theta}$ given the data, $\mathcal{D}$, is defined by

$$\pi(\boldsymbol{\theta}|\mathcal{D}) \propto \mathcal{L}(\boldsymbol{\theta}|\mathcal{D})\pi(\boldsymbol{\theta}), \tag{5.2}$$

where $\mathcal{L}(\boldsymbol{\theta}|\mathcal{D})$ is the likelihood function and $\pi(\boldsymbol{\theta})$ is the prior.

Although not in its original form, we refer to formula (5.2) as Bayes theorem (cf. Bayes, 1763). The likelihood function is defined as follows.

Definition 5.1.2. Suppose that we observe $(T_1, \ldots, T_n) = (t_1, \ldots, t_n)$. Then the *likelihood function* $\mathcal{L}(\boldsymbol{\theta}|t_1, \ldots, t_n)$ is

$$\mathcal{L}(\boldsymbol{\theta}|t_1, \ldots, t_n) = f_{T_1,\ldots,T_n}(t_1, \ldots, t_n|\boldsymbol{\theta}), \tag{5.3}$$

where $f_{T_1,\ldots,T_n}(t_1, \ldots, t_n|\boldsymbol{\theta})$ is the joint density function of $(T_1, \ldots, T_n)$ evaluated at $(t_1, \ldots, t_n)$.

Although (5.3) may lead us to think that the likelihood is a density, it is not a probability density function. A probability density function, $f(t|\boldsymbol{\theta})$, is a function of t for fixed $\boldsymbol{\theta}$. The likelihood, $\mathcal{L}(\boldsymbol{\theta}|t_1, \ldots, t_n)$, is a function of $\boldsymbol{\theta}$ for an observed (fixed) value of the data $(t_1, \ldots, t_n)$. Note that the likelihood function depends on the model that we assume. For independent observations on $T_1, \ldots, T_n$, the likelihood (5.3) is

$$\mathcal{L}(\boldsymbol{\theta}|t_1, \ldots, t_n) = \prod_{i=1}^{n} f_{T_i}(t_i|\boldsymbol{\theta}). \tag{5.4}$$

The prior distribution $\pi(\boldsymbol{\theta})$ describes a priori beliefs regarding model plausibility. This is our prior belief about $\boldsymbol{\theta}$, which is updated via the data through the likelihood function leading to a posterior probability distribution on $\boldsymbol{\theta}$. There are many types of priors that one can use. Priors can be elicited from experts or they can be conjugate (cf. Raiffa and Schlaifer, 1961), meaning that the posterior distribution is in the same distributional family as the prior, which aids computation substantially. Reference priors are defined as conventional priors that are used as a starting point for the analysis (cf. Kass and Wasserman, 1996). When there is a lack of prior information, one can use noninformative priors. These priors are also called *vague* or *flat priors*. These priors are attractive because they let the likelihood dominate the prior and result in a posterior that can be considered to arise mainly from the data. Noninformative priors can also be improper, meaning that they do not integrate to 1. However, noninformative priors are not necessarily invariant under transformation (i.e., a prior that is vague in one setting may be very informative under another parameterization). Jeffreys (1961) discusses criteria for categorizing choice of noninformative prior distributions and states that invariance is the most important criterion. He suggests a prior of the form

$$\pi(\boldsymbol{\theta}) \propto \det|I(\boldsymbol{\theta})|^{1/2}, \tag{5.5}$$

where $I(\boldsymbol{\theta})$ is the expected Fisher information in the model,

$$I_{ij}(\boldsymbol{\theta}) = -E_{T|\boldsymbol{\theta}}\left(\frac{\partial^2 \log[\mathcal{L}(\boldsymbol{\theta}|T_1, \ldots, T_n)]}{\partial\boldsymbol{\theta}_i \partial\boldsymbol{\theta}_j}\right). \tag{5.6}$$

The last equality holds under certain regularity conditions. The prior (5.5) is called *Jeffreys' prior* and is applicable to probability distributions that are differentiable with respect to the parameters in them. It is invariant under one-to-one transformation. Such invariant distributions can be constructed for many common problems. For example, Huzurbazar (1955) constructs invariants for distributions that have sufficient statistics for a continuous parameter. This includes distributions from the exponential family. In addition to the books on Bayesian methods already mentioned, further details on noninformative priors and invariance can be found in Hartigan (1964), Huzurbazar (1976), Jaynes (1980), Perks (1947), and Welch and Peers (1963). With flowgraphs, we will generally assume independent noninformative prior distributions for the components of $\boldsymbol{\theta}$. Some problems will use very wide uniform priors.

The flowgraph model gives us the MGF of the density $f_Z(z|\boldsymbol{\theta})$, so that for computational purposes, (5.1) can be written as

$$f_Z(z|\mathcal{D}) = \frac{\int f_Z(z|\boldsymbol{\theta})\mathcal{L}(\boldsymbol{\theta}|\mathcal{D})\pi(\boldsymbol{\theta})\,d\boldsymbol{\theta}}{\int \mathcal{L}(\boldsymbol{\theta}|\mathcal{D})\pi(\boldsymbol{\theta})\,d\boldsymbol{\theta}}. \tag{5.7}$$

Parametric assumptions about the flowgraph allow us to specify the likelihood function for complete data.

Definition 5.1.3. Complete data on the flowgraph model consists of having every intermediate transition for each observation. The associated waiting time for the transitions may be censored; however, the transition information is complete.

For example, in the flowgraph for cancer progression of Figure 2.1, complete data on a patient consist of a history of all transitions made by that patient, say, $I \to A \to R \to A \to D_1$. If the patient is censored in state A but we know that the patient eventually died from cancer, the waiting time for the transition $A \to D_1$ is censored but the transition information is still considered to be complete. In the context of flowgraphs, *incomplete data* refers to data with incomplete transition information in addition to censoring. We defer the discussion of incomplete data to Chapter 8. In this chapter we assume that data are complete and allow for the possibility of right censoring.

Example 5.1: BAYES PREDICTIVE DENSITY FOR A HYDRAULIC PUMP SYSTEM. Consider the hydraulic pump series system of Example 2.2. In Example 3.2 we derived the density of time to total failure of both pumps using exact inversion of the flowgraph MGF. Recall that each pump fails according to an Exp(λ) distribution and that the waiting time for the first pump to fail is the minimum of two exponentials, Exp(2λ), and the waiting time for the second pump to fail is Exp(λ). Let Y_0 be the waiting time distribution for the $0 \to 1$ transition and Y_1 be the waiting time distribution for the $1 \to 2$ transition. Suppose that we have data on 100 failure times of this system so that our data consist of 100 observations each on Y_0 and Y_1. Let $\mathbf{y}_0 = (y_{01}, y_{02}, \ldots, y_{0n})$ and $\mathbf{y}_1 = (y_{11}, y_{12}, \ldots, y_{1n})$ denote observations on the waiting times $0 \to 1$ and $1 \to 2$. The likelihood function is

$$\begin{aligned}\mathcal{L}(\lambda|\mathbf{y}_0, \mathbf{y}_1) &= \prod_{i=1}^{n} 2\lambda \exp(-2\lambda y_{0i})\,\lambda \exp(-\lambda y_{1i}) \\ &\propto \lambda^{2n} \exp\left[-\lambda\left(2\sum_{i=1}^{n} y_{0i} + \sum_{i=1}^{n} y_{1i}\right)\right].\end{aligned} \tag{5.8}$$

With a uniform prior on λ, $\pi(\lambda) = 1$ for $0 < \lambda \leq 1$, the posterior distribution on λ is

$$\begin{aligned}\pi(\lambda|\mathcal{D}) &\propto \mathcal{L}(\lambda|\mathcal{D})\pi(\lambda) \\ &\propto \lambda^{2n} \exp\left[-\lambda\left(2\sum_{i=1}^{n} y_{0i} + \sum_{i=1}^{n} y_{1i}\right)\right], \qquad 0 < \lambda \leq 1.\end{aligned} \tag{5.9}$$

The posterior is a truncated gamma distribution, Gamma$(2n+1, 2\sum_{i=1}^n y_{0i} + \sum_{i=1}^n y_{1i})$,

$$\pi(\lambda|\mathcal{D}) = \frac{\lambda^{2n}\exp[-\lambda(2\sum_{i=1}^n y_{0i} + \sum_{i=1}^n y_{1i})]}{\int_0^1 \lambda^{2n}\exp[-\lambda(2\sum_{i=1}^n y_{0i} + \sum_{i=1}^n y_{1i})]\,d\lambda}. \tag{5.10}$$

For this example, the sufficient statistics were calculated using the parameter value $\lambda = 1/1000$. We can write down the Bayes predictive density (5.7), up to the normalizing constant, for the future observable total failure time of the pumps, $T = Y_0 + Y_1$:

$$\begin{aligned} f_T(t|\mathbf{y}_0, \mathbf{y}_1) &= \int_0^1 f_T(t|\lambda)\,\pi(\lambda|\mathbf{y}_0, \mathbf{y}_1)\,d\lambda \\ &\propto \int_0^1 2\lambda[e^{-\lambda t} - e^{-2\lambda t}] \\ &\quad \times \lambda^{2n}\exp\left[-\lambda\left(2\sum_{i=1}^n y_{0i} + \sum_{i=1}^n y_{1i}\right)\right] d\lambda \end{aligned} \tag{5.11}$$

for $t > 0$, where $f_T(t|\lambda)$ was derived in (3.12). We can numerically integrate (5.11) for a given t over λ or we can estimate the integral using Monte Carlo sampling. We take a sample of $m = 400$ values from the posterior and approximate (5.11) with

$$\hat{f}_T(t|\mathbf{y}_0, \mathbf{y}_1) = \frac{\sum_{j=1}^m f_T(t|\lambda_j)}{m}. \tag{5.12}$$

For large m, this sum approaches the desired integral. Figure 5.1 gives the Bayes predictive density computed as in (5.12). The solid line in the figure is the predictive distribution, the posterior mean of $f_T(t|\lambda)$. The two dashed lines are the 5% and 95% pointwise prediction bounds. Figure 5.2 gives a histogram of posterior samples with the MLE indicated by a vertical line. □

In general, we do not know $f_Z(z|\boldsymbol{\theta})$ in (5.7). In this case, we use the saddlepoint approximation $\tilde{f}_Z(z|\boldsymbol{\theta})$ of (3.19) in place of $f_Z(z|\boldsymbol{\theta})$. Practically, this involves generating a random sample of $\boldsymbol{\theta}$ values from the posterior and performing the following computations for each $\boldsymbol{\theta}$ value:

1. Convert the flowgraph MGF, $M_{Z|\boldsymbol{\theta}}(s)$, to a density using the saddlepoint approximation for each value z of Z at which the density is desired.
2. Treat each density as a vector of values (evaluated at each value z in computation 1) indexed by $\boldsymbol{\theta}_j$.
3. For each z value, average the densities over the $\boldsymbol{\theta}$ values to give

$$\hat{f}_Z(z|\mathcal{D}) = \frac{\sum_{j=1}^m \tilde{f}_Z(z|\boldsymbol{\theta}_j)}{m}, \tag{5.13}$$

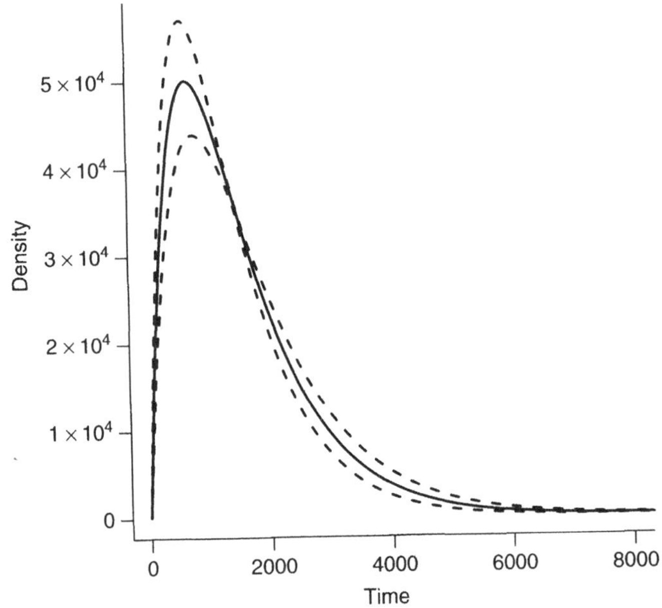

FIGURE 5.1 Bayes predictive density (via Monte Carlo sampling) for the time to total failure of hydraulic pumps for Example 5.1. The solid line is the predictive distribution. The dashed lines are the 5% and 95% pointwise prediction bounds.

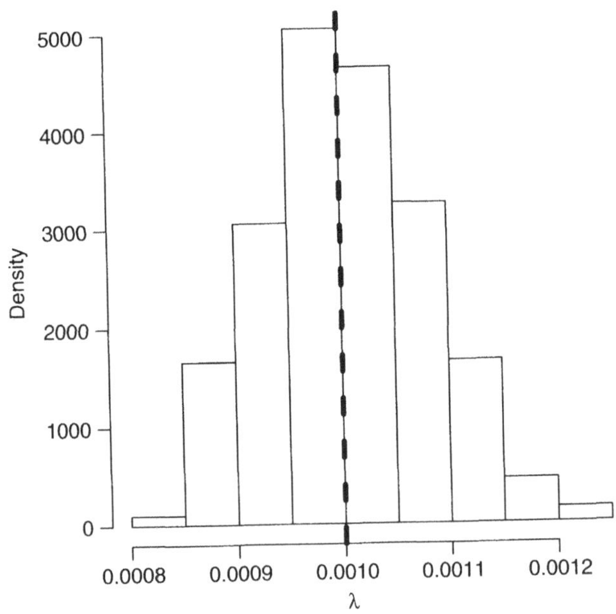

FIGURE 5.2 Histogram of posterior samples with MLE (vertical line) for the λ of Example 5.1.

where $\tilde{f}_Z(z|\boldsymbol{\theta}_j)$ is the density derived from the saddlepoint approximation. For large m, this sum approximates the integral (5.7).

We can also directly compute estimates of the mean and variance of the predictive distribution. The predictive distribution mean is estimated by $\overline{K'(0)}$,

$$E_{Z|\mathcal{D}}(Z) = \overline{K'(0)} = \frac{1}{m}\sum_{j=1}^{m} K'_{\boldsymbol{\theta}_j}(0), \tag{5.14}$$

where $K'_{\boldsymbol{\theta}_j}(0)$ is the derivative of the CGF of the waiting time Z computed as a function of each $\boldsymbol{\theta}_j$ sampled from the posterior and evaluated at 0 to give the mean. The variance of the predictive distribution is

$$\begin{aligned} V_{Z|\mathcal{D}}(Z) &= \overline{K''(0)} \\ &= \frac{1}{m}\sum_{j=1}^{m} K''_{\boldsymbol{\theta}_j}(0) + V\{[K'_{\boldsymbol{\theta}_j}(0)]_{j=1}^{m}\} \\ &= \frac{1}{m}\sum_{j=1}^{m} K''_{\boldsymbol{\theta}_j}(0) + \frac{1}{m}\sum_{j=1}^{m}[K'_{\boldsymbol{\theta}_j}(0) - \overline{K'(0)}]^2, \end{aligned} \tag{5.15}$$

where $K'_{\boldsymbol{\theta}_j}(0)$ and $K''_{\boldsymbol{\theta}_j}(0)$ are the first and second derivatives of the CGF of the flowgraph computed as a function of each $\boldsymbol{\theta}_j$ sampled from the posterior and $\overline{K'(0)}$ is given by (5.14).

Example 5.2: PROGRESSION OF CANCER. This example illustrates computation of the Bayes predictive density based on both the flowgraph model and the exact distribution in the case of a simple covariate. We continue with the cancer progression model due to Lagakos (1976) discussed in Examples 2.5 and 2.12. The flowgraph model for this example with the parametric assumptions discussed in Example 2.12 is given in Figure 5.3. Lagakos (1976) describes a randomized clinical trial of 62

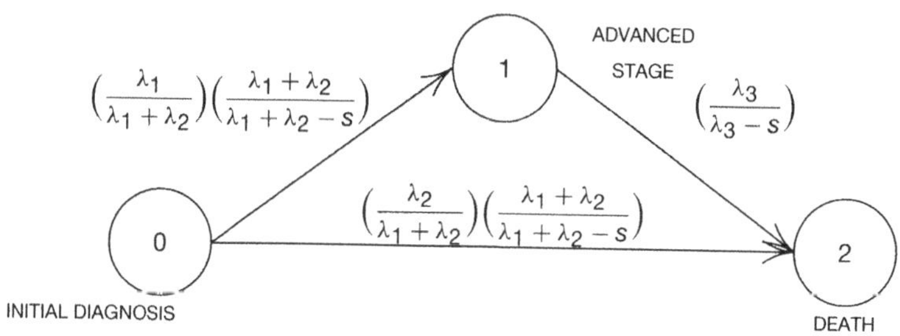

FIGURE 5.3 Flowgraph model for cancer progression for treatment 1.

male lung cancer patients involving two treatments. In this example, we consider treatment to be a simple covariate. The flowgraph of Figure 5.3 corresponds to treatment 1. Treatment 2 follows a similar flowgraph with exponential waiting times parameterized by λ_4, λ_5, and λ_6, in place of λ_1, λ_2, and λ_3.

The MGF of the distribution of the overall survival time of a cancer patient was given in (2.16). This is the MGF for a patient randomized to treatment 1. For a patient randomized to receive treatment 2, the analogous MGF is

$$M(s) = \left(\frac{\lambda_5}{\lambda_4+\lambda_5}\right)\left(\frac{\lambda_4+\lambda_5}{\lambda_4+\lambda_5-s}\right) + \left(\frac{\lambda_4}{\lambda_4+\lambda_5}\right)\left(\frac{\lambda_4+\lambda_5}{\lambda_4+\lambda_5-s}\right)\left(\frac{\lambda_6}{\lambda_6-s}\right)$$
$$\text{for } s < \min(\lambda_4+\lambda_5, \lambda_6). \qquad (5.16)$$

Due to the exponential waiting times used in this problem, we can compute the distribution of the exact survival time from state 0 to state 2 (cf. Lagakos, 1976). The exact density of T, the waiting time from $0 \to 2$ under treatment 1, is

$$f_T(t) = \frac{(\lambda_2-\lambda_3)(\lambda_1+\lambda_2)\exp[-(\lambda_1+\lambda_2)t] + \lambda_1\lambda_3\exp(-\lambda_3 t)}{\lambda_1+\lambda_2-\lambda_3} \qquad (5.17)$$

for $t \geq 0$ and $\lambda_1+\lambda_2 \neq \lambda_3$, and for $\lambda_1+\lambda_2=\lambda_3$ it is

$$f_T(t) = (\lambda_2 + \lambda_1\lambda_3 t)\exp(-\lambda_3 t). \qquad (5.18)$$

For treatment 2, the density is analogous with λ_4, λ_5, and λ_6 in place of λ_1, λ_2, and λ_3. The equivalent flowgraph analysis of this Markov model with exponential waiting times due to Lagakos (1976) is presented in Butler and Huzurbazar (1997). We begin with this model. One reason for the popularity of the Markov model with exponential waiting times is the ease with which it handles censoring.

Let p_I and p_{II} be the transition probabilities for direct transition from $0 \to 2$ for treatment 1 (I) and treatment 2 (II). Let T_{I0}, T_{I1}, and T_{I2} be the random waiting times for direct passage under treatment 1 from $0 \to 2$, $0 \to 1$, and $1 \to 2$, and let T_{II0}, T_{II1}, and T_{II2} be the random waiting times for direct passage under treatment 2 from $0 \to 2$, $0 \to 1$, and $1 \to 2$. We let n_{I0} and n_{II0} be the number of transitions observed on the $0 \to 2$ branch for treatments 1 (I) and 2 (II). Similarly, we let n_{I1} and n_{II1} be the number of transitions for branch $0 \to 1$, and n_{I2} and n_{II2} be the number of transitions for branch $1 \to 2$. Let n^*_{I0} denote the number of observations censored (indicated by $*$) in state 0 under treatment 1 (i.e., we do not know whether these patients progressed to state 1 or state 2). Let n^*_{I1} be the number of observations censored in state 1 under treatment 1. Let n^*_{II0} and n^*_{II1} be the analogous quantities under treatment 2.

Under treatment 1, the likelihood function for this model is

$$\begin{aligned}\mathcal{L}(\lambda_1, \lambda_2, \lambda_3 | \mathcal{D}_I) = &\prod_{i=1}^{n_{I0}} \left(\frac{\lambda_2}{\lambda_1 + \lambda_2}\right) (\lambda_1 + \lambda_2) \exp[-(\lambda_1 + \lambda_2) t_{I0i}] \\ &\times \prod_{i=1}^{n_{I1}} \left(\frac{\lambda_1}{\lambda_1 + \lambda_2}\right) (\lambda_1 + \lambda_2) \exp[-(\lambda_1 + \lambda_2) t_{I1i}] \\ &\times \prod_{i=1}^{n_{I2}} \lambda_3 \exp(-\lambda_3 t_{I2i}) \\ &\times \prod_{i=1}^{n^*_{I0}} \exp[-(\lambda_1 + \lambda_2) t^*_{I0i}] \\ &\times \prod_{i=1}^{n^*_{I1}} \exp(-\lambda_3 t^*_{I1i}),\end{aligned} \tag{5.19}$$

where the last two terms arise from censored observations. Censored observations contribute terms that are finite mixtures of the survivor functions. Recall that for an Exp(λ) distribution, the CDF is $1 - \exp(-\lambda t)$ so that the survivor function is $S(t) = \exp(-\lambda t)$. The last term represents patients that made the transition to the advanced stage 1 but were alive at the end of the study, so they contribute terms of the form $\exp(-\lambda_3 t^*_{I1i})$. The penultimate term in (5.19) arises from patients who are censored some time after initial diagnosis of cancer. Their censored waiting times are denoted by t^*_{I0i} for patient i. For these patients, we do not know whether their cancer would have progressed or if they would have died from other causes. In general, their contribution is the survivor function of a finite mixture, $S_M(t|\lambda_1, \lambda_2) = 1 - F_M(t|\lambda_1, \lambda_2)$, where $F_M(t|\lambda_1, \lambda_2)$ is the CDF of the mixture distribution. In this case, the CDF of the mixture distribution for one observation is

$$\begin{aligned}F_M(t|\lambda_1, \lambda_2) &= \left(\frac{\lambda_2}{\lambda_1 + \lambda_2}\right) \{1 - \exp[-(\lambda_1 + \lambda_2)t]\} \\ &\quad + \left(\frac{\lambda_1}{\lambda_1 + \lambda_2}\right) \{1 - \exp[-(\lambda_1 + \lambda_2)t]\} \\ &= 1 - \exp[-(\lambda_1 + \lambda_2)t]\end{aligned} \tag{5.20}$$

so that the corresponding survivor function contribution for one observation is

$$S_M(t|\lambda_1, \lambda_2) = \exp[-(\lambda_1 + \lambda_2)t].$$

The likelihood function of (5.19) simplifies to

$$\mathcal{L}(\lambda_1, \lambda_2, \lambda_3|\mathcal{D}_I) = \lambda_1^{n_{I1}} \exp\left[-\lambda_1\left(\sum_{i=1}^{n_{I0}} t_{I0i} + \sum_{i=1}^{n_{I1}} t_{I1i} + \sum_{i=1}^{n^*_{I0}} t^*_{I0i}\right)\right]$$
$$\times \lambda_2^{n_{I0}} \exp\left[-\lambda_2\left(\sum_{i=1}^{n_{I0}} t_{I0i} + \sum_{i=1}^{n_{I1}} t_{I1i} + \sum_{i=1}^{n^*_{I0}} t^*_{I0i}\right)\right]$$
$$\times \lambda_3^{n_{I2}} \exp\left[-\lambda_3\left(\sum_{i=1}^{n_{I2}} t_{I2i} + \sum_{i=1}^{n^*_{I1}} t^*_{I1i}\right)\right]. \tag{5.21}$$

With wide uniform priors on all the parameters, the posterior on $(\lambda_1, \lambda_2, \lambda_3)$ is explicit and independent in the parameters. The independent posteriors are

$$\lambda_1|\mathcal{D}_I \sim \text{Gamma}\left(n_{I1} + 1, \sum_{i=1}^{n_{I0}} t_{I0i} + \sum_{i=1}^{n_{I1}} t_{I1i} + \sum_{i=1}^{n^*_{I0}} t^*_{I0i}\right) \tag{5.22}$$

$$\lambda_2|\mathcal{D}_I \sim \text{Gamma}\left(n_{I0} + 1, \sum_{i=1}^{n_{I0}} t_{I0i} + \sum_{i=1}^{n_{I1}} t_{I1i} + \sum_{i=1}^{n^*_{I0}} t^*_{I0i}\right) \tag{5.23}$$

$$\lambda_3|\mathcal{D}_I \sim \text{Gamma}\left(n_{I2} + 1, \sum_{i=1}^{n_{I2}} t_{I2i} + \sum_{i=1}^{n^*_{I1}} t^*_{I1i}\right). \tag{5.24}$$

It is straightforward to compute the Bayes predictive density using the Monte Carlo technique of (5.13) and the data from Lagakos (1976). We can use the sufficient statistics from Lagakos (1976) to evaluate the terms of the posterior. The posterior distributions are $\lambda_1|\mathcal{D}_I \sim \Gamma(19, 300)$, $\lambda_2|\mathcal{D}_I \sim \Gamma(5, 300)$, $\lambda_3|\mathcal{D}_I \sim \Gamma(18, 254)$, $\lambda_4|\mathcal{D}_{II} \sim \Gamma(30, 680)$, $\lambda_5|\mathcal{D}_{II} \sim \Gamma(8, 680)$, and $\lambda_6|\mathcal{D}_{II} \sim \Gamma(26, 271)$. We consider treatment as a simple covariate and perform analyses for treatments 1 and 2. Figure 5.4 gives the Bayes predictive densities for both treatments. There does not appear to be any difference between the two treatments, a conclusion supported by Lagakos (1976). The solid lines are the predictive densities for treatment 1 and treatment 2 based on averaging the exact waiting time density given $\boldsymbol{\theta}$ of (5.17) over random samples from the posterior of $\boldsymbol{\theta}$. We can compute (5.17) only because we have exponential waiting times that allow this calculation. These predictive distributions are estimated with $m = 400$ samples of $\lambda_1, \ldots, \lambda_6$ from their posterior. In general, we cannot compute the exact conditional density $f_Z(z|\boldsymbol{\theta})$ and we use the saddlepoint approximation to compute the Bayes predictive density using (5.13). This computation, using $m = 400$ values from the posterior, gives the dashed lines in Figure 5.4. Note that we never actually get the exact predictive density; we estimate it via the random samples. The solid lines estimate the predictive density using

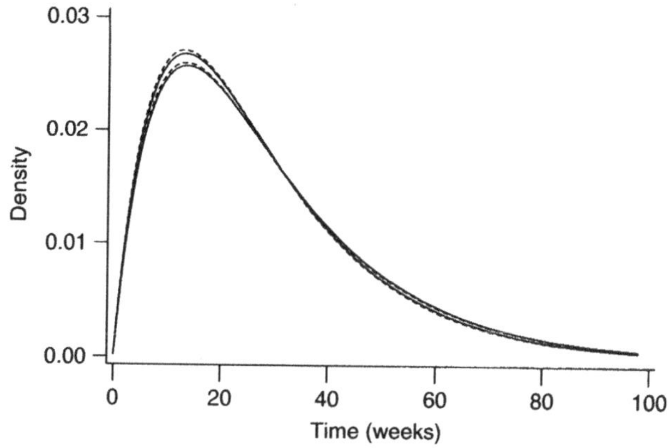

FIGURE 5.4 Comparison of predictive densities for treatment 1 and treatment 2 for the flowgraph model of Figure 5.3. The dashed lines are the saddlepoint approximations for the closely associated solid lines which are based on the exact densities. The predictive density for treatment 1 is the taller one.

the exact waiting time distribution given the parameters, and the dashed lines use the saddlepoint approximation of this conditional waiting time distribution. □

Example 5.3: CLOSED PARALLEL SYSTEM: PARTS TESTING STATION. This example illustrates computation of the Bayes predictive density for the overall waiting time until total failure of the parts testing process of Example 2.6. The parts being tested are bearing and seal assemblies for construction equipment. In this context, partial failure means that there are problems with the assemblies but that the assemblies are functioning in a manner that meets specifications. Partial failure can mean that the bearings are wearing unevenly or that the seals have minor cracks or deformation. Eventually, the seals will crack completely or the bearings will grind and the assembly will not be functioning as required. This is considered total failure. In Figure 5.5, state 0 represents the beginning of testing for the assemblies and states 1 and 2 represent partial and total failure of the assemblies. The probability of direct total assembly failure is represented by p so that $1 - p$ represents the probability of partial assembly failure. Let n_0, n_1, and n_2 be the number of assemblies that made the transitions $0 \to 2$, $0 \to 1$, and $1 \to 2$. Of the 30 assemblies, nine failed directly, so that $n_0 = 9$. The remaining 21 assemblies experienced partial failure followed by total failure, $0 \to 1 \to 2$, so that $n_1 = n_2 = 21$. Let T_0, T_1, and T_2 represent the random waiting times from $0 \to 2$, $0 \to 1$, and $1 \to 2$. We simulate data to mimic the real system with the waiting time from $0 \to 1$ as Gamma$(3, \beta_1)$, from $1 \to 2$ as Gamma$(7, \beta_2)$, and the waiting time from $0 \to 2$ as Exp(λ). We model the prior distribution on p with a Beta$(2, 2)$ distribution, β_1 with a Gamma(η_1, η_2) distribution, and β_2 with an Exp(η_3) distribution, with the hyperparameters set to be $\eta_1 = \eta_2 = 0.5$ and $\eta_3 = 2/7$. We use a Jeffreys' prior on λ,

$$\pi(\lambda) \propto \frac{1}{\lambda}. \tag{5.25}$$

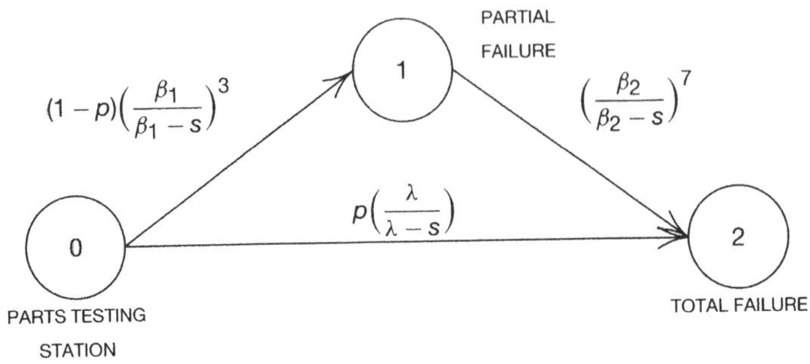

FIGURE 5.5 Flowgraph model for bearing and seal assembly system.

The likelihood function of $\boldsymbol{\theta} = (p, \lambda, \beta_1, \beta_2)$ given the data $\mathcal{D}$ is

$$\mathcal{L}(p, \lambda, \beta_1, \beta_2 | \mathcal{D}) = \left[\prod_{i=1}^{n_0} p\lambda \exp(-\lambda t_{0i})\right] \left[\prod_{i=1}^{n_1} (1-p) \frac{\beta_1^3 t_{1i}^2 \exp(-\beta_1 t_{1i})}{\Gamma(3)}\right] \times \left[\prod_{i=1}^{n_2} \frac{\beta_2^7 t_{2i}^6 \exp(-\beta_2 t_{2i})}{\Gamma(7)}\right] \tag{5.26}$$

for $0 \le p \le 1$, $\lambda > 0$, $\beta_1 > 0$, and $\beta_2 > 0$. The likelihood is separable in all of the parameters because we have complete data. We can compute the posteriors using (5.2) in three pieces, corresponding to each term of the likelihood and the terms involving p. The posterior on p given the data (n_0, n_1) is

$$\pi(p | n_0, n_1) \propto p^{n_0} (1-p)^{n_1} \, p(1-p), \tag{5.27}$$

which is the form of a Beta$(n_0 + 2, n_1 + 2)$ distribution. The parameter λ in the first term of (5.26), corresponding to the $0 \to 2$ transition, has the posterior given the data $\mathcal{D}_0 = \{t_{01}, \ldots, t_{0n_0}\}$,

$$\pi(\lambda | \mathcal{D}_0) \propto \lambda^{n_0} \exp\left(-\lambda \sum_{i=1}^{n_0} t_{0i}\right) \frac{1}{\lambda} = \lambda^{n_0 - 1} \exp\left(-\lambda \sum_{i=1}^{n_0} t_{0i}\right) \tag{5.28}$$

for $\lambda > 0$. Therefore, the posterior distribution of λ is Gamma$(n_0, \sum_{i=1}^{n_0} t_{0i})$. The posterior distribution of β_1, based on data $\mathcal{D}_1 = \{t_{10}, \ldots, t_{1n_1}\}$, is

$$\begin{aligned} \pi(\beta_1 | \mathcal{D}_1) &\propto \beta_1^{3n_1} \exp\left(-\beta_1 \sum_{i=1}^{n_1} t_{1i}\right) \beta_1^{\eta_1 - 1} \exp(-\eta_2 \beta_1) \\ &= \beta_1^{3n_1 + \eta_1 - 1} \exp\left[-\beta_1 \left(\eta_2 + \sum_{i=1}^{n_1} t_{1i}\right)\right], \end{aligned} \tag{5.29}$$

which is the form of a Gamma($3n_1 + \eta_1, \eta_2 + \sum_{i=1}^{n_1} t_{1i}$). The posterior distribution on β_2, based on data $\mathcal{D}_2 = \{t_{20}, \ldots, t_{2n_2}\}$, is

$$\pi(\beta_2|\mathcal{D}_2) \propto \beta_2^{7n_2} \exp\left(-\beta_2 \sum_{i=1}^{n_2} t_{2i}\right) \exp(-\eta_3\beta_2)$$

$$= \beta_2^{7n_2} \exp\left[-\beta_2 \left(\eta_3 + \sum_{i=1}^{n_2} t_{2i}\right)\right], \tag{5.30}$$

which is the form of a Gamma($7n_2 + 1, \eta_3 + \sum_{i=1}^{n_2} t_{2i}$).

In this example, it is much more difficult to write down the density of the overall waiting time from $0 \to 2$, $f_Z(z|\boldsymbol{\theta})$ in (5.7). The flowgraph gives us the MGF of this distribution, derived in (2.12). For this example it is

$$M(s) = p_{02}M_{02}(s) + p_{01}M_{01}(s)M_{12}(s)$$

$$= p\frac{\lambda}{\lambda - s} + (1-p)\left(\frac{\beta_1}{\beta_1 - s}\right)^3 \left(\frac{\beta_2}{\beta_2 - s}\right)^7. \tag{5.31}$$

The MAPLE program code to compute the MGF, the CGF, and its derivatives for the saddlepoint approximation inversion program is given in Figure 5.6.

```
with(codegen,makeproc,C):

ms := proc(s)
      local w01,w12,w02;
      w01 := (1-p)  *  ((b1/(b1-s))^3); # b1 is β1
      w12 := ((b2/(b2-s))^7); # b2 is β2
      w02 := p * (1/(1-s)); # l is λ

# The next line defines the MGF
      w01*w12+w02;
end proc:

# Define the CGF and its derivatives
k := makeproc(ln(ms(s)),[s]):
C(k,optimized,ansi,filename="c:/mplcode/assembly/k.c"):

dk := makeproc(diff(k(s),s),[s]):
C(dk,optimized,filename="c:/mplcode/assembly/dk.c"):

ddk := makeproc(diff(dk(s),s),[s]):
C(ddk,optimized,filename="c:/mplcode/assembly/ddk.c"):
```

FIGURE 5.6 MAPLE code for Example 5.3.

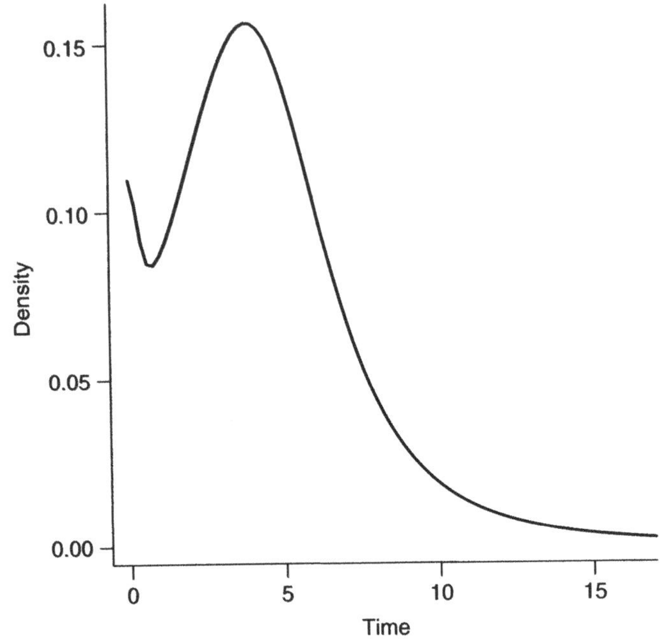

FIGURE 5.7 Bayes predictive density (via Monte Carlo sampling) for time to total failure (in months) of bearing and seal assembly of Example 5.3.

For each randomly sampled value of $\boldsymbol{\theta} = (p, \lambda, \beta_1, \beta_2)$ from the posterior, we can convert this MGF into a density $\tilde{f}_Z(z|\boldsymbol{\theta})$ using the saddlepoint approximation (3.19) for each value z of Z. Following the steps outlined previously, we can approximate the Bayes predictive density (5.13) using Monte Carlo integration for (5.7). □

Figure 5.7 presents the Bayesian predictive density of the time to total failure of the assembly system. The predictive density has a mean of 4.91 months and a standard deviation of 3.07 months. The 25th, 50th, 75th, 90th, 95th, and 99th percentiles of the predictive distribution are 2.47, 4.14, 6.04, 8.38, 10.41, and 16.37. The bump near $t = 0$ is reflective of the short exponential waiting times due to direct failure of some of the assemblies. This emphasizes the nature of the overall failure time distribution: that it is a finite mixture distribution with separate components.

In Example 5.3, the priors on p, β_1, and β_2 are conjugate, whereas the prior on λ is not conjugate. The prior chosen in Example 5.1 is not conjugate because the prior on λ is uniform and the posterior on λ is a truncated gamma distribution. Conjugate priors are a good starting point because they are computationally convenient and they provide intuition for how the data update the prior. They can also be viewed as providing a priori data or information. In higher dimensions with censored data, conjugacy is not generally possible and we will rely on computational rather than analytical methods for computing the posterior. Example 5.3 resulted in independent posteriors that were known distributions. It is relatively easy to sample from such posteriors. In general, the posterior distributions on the parameters will not be independent. This is especially true in the presence of censoring with competing risks. In the next section we describe general methods for sampling from such posteriors.

5.2 SAMPLING FROM THE POSTERIOR

To perform the integration in (5.7) with flowgraph models, we require a method that can be programmed as a routine application to a general posterior distribution. If the form of the posterior distribution is well known, we can sample directly from it using statistical software. Sometimes, the posterior can be factored into a known conditional posterior and a known marginal posterior. With complicated flowgraphs, the posterior is often not of a recognizable form. There are many methods for sampling from the posterior distribution, and some methods are better suited than others for specific problems. These methods include rejection sampling, importance sampling, Gibbs sampling, the Metropolis–Hastings algorithm, and other Markov chain Monte Carlo methods. Laplace's method can sometimes be used to estimate posterior distributions. We discuss these methods briefly as they pertain to flowgraphs. Gelman et al. (1995) provide an excellent introduction to these methods. For a rudimentary introduction to importance sampling and resampling, see Smith and Gelfand (1992); to Monte Carlo methods, see MacKay (1998); to Gibbs sampling, see Casella and George (1992); and to the Metropolis–Hastings algorithm, see Chib and Greenberg (1995).

5.2.1 Rejection Sampling

The simplest posterior sampling method is rejection sampling. We can apply this technique to sample from $\pi(\boldsymbol{\theta}|\mathcal{D})$ using its unnormalized form, $\mathcal{L}(\boldsymbol{\theta}|\mathcal{D})\pi(\boldsymbol{\theta})$. The density that we wish to sample from is called the *target density*. In rejection sampling, the target density is approximated by an "envelope" function. This probability density function, denoted by $g(\boldsymbol{\theta})$, is known and easily sampled. Generally, $g(\boldsymbol{\theta})$ is chosen to be roughly proportional to $\mathcal{L}(\boldsymbol{\theta}|\mathcal{D})\pi(\boldsymbol{\theta})$. Assume that the support of the prior is bounded. The simplest envelope function is a wide uniform distribution covering the prior support. Although this is generally a very inefficient way to sample, it is a method that works with complicated flowgraphs without burdening the user with choices of envelope functions as long as the prior is restricted to a bounded region. Much of the early work with flowgraph models used rejection sampling (cf. Butler and Huzurbazar, 1997). In addition to specifying $g(\boldsymbol{\theta})$, we must be able to determine the upper bound, M, for the importance ratio

$$\frac{\mathcal{L}(\boldsymbol{\theta}|\mathcal{D})\pi(\boldsymbol{\theta})}{g(\boldsymbol{\theta})}.$$

The rejection sampling algorithm works by sampling a value $\boldsymbol{\theta}$ from $g(\boldsymbol{\theta})$; then, with probability

$$\frac{\mathcal{L}(\boldsymbol{\theta}|\mathcal{D})\pi(\boldsymbol{\theta})}{Mg(\boldsymbol{\theta})}, \tag{5.32}$$

we accept $\boldsymbol{\theta}$ as a sample from $\pi(\boldsymbol{\theta}|\mathcal{D})$.

In general, the waiting times are not exponential and the analysis does not simplify neatly as it did in Example 5.2. For illustration, we modify Example 5.2 to include censored data in the presence of non-exponential waiting times.

Example 5.4: CANCER PROGRESSION WITH NONEXPONENTIAL WAITING TIMES. Consider the cancer model of Example 5.2 but suppose that the waiting times are not exponential. Suppose that the data indicate that a distribution such as a gamma or inverse Gaussian would be a better fit for the $0 \to 1$ waiting time. We retain the exponential assumptions on the $0 \to 2$ and $1 \to 2$ waiting times for this illustration; however, we change the parameters to avoid confusion with the previous analysis of Example 5.2. The waiting time distributions for transition from states $0 \to 2$ and $1 \to 2$ are Exp(β_1) and Exp(β_2), respectively. The waiting time distribution for transition from state $0 \to 1$ is assumed to be inverse Gaussian with parameters μ_1 and μ_2. The direct transition from $0 \to 2$ occurs with probability p and the transition from $0 \to 1$ occurs with probability $(1 - p)$. The flowgraph model, labeled with corresponding transmittances, is given in Figure 5.8. We assume that we have only one treatment, although analysis for a second treatment would be similar.

We simulate data for this illustration that is similar to the data of Example 5.2 except for the presence of an inverse Gaussian waiting time distribution. For simulation we used parameter values of $p = 0.17$, $\beta_1 = 0.013$, $\beta_2 = 0.067$, $\mu_1 = 7$, and $\mu_2 = 0.5$. We still maintain 62 patients, but now approximately 20% of the patients are censored. A summary of the data is as follows:

Uncensored			Censored	
$n_0 = 4$	$n_1 = 51$	$n_2 = 45$	$n_0^* = 7$	$n_1^* = 6$

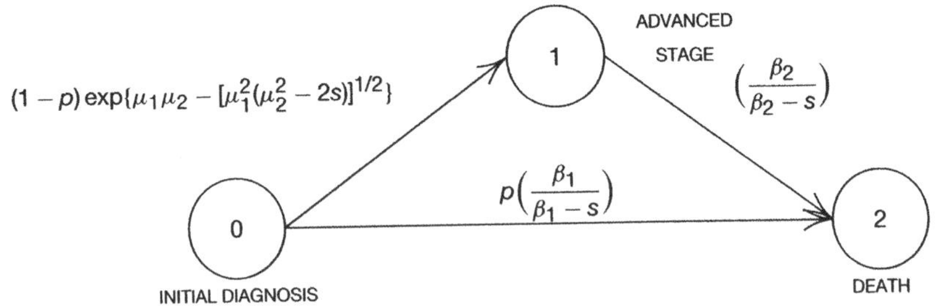

FIGURE 5.8 Flowgraph model for cancer progression (nonexponential waiting times).

Let T_0, T_1, and T_2 be the random waiting times for direct passage from $0 \to 2$, $0 \to 1$, and $1 \to 2$. We let n_0 be the number of transitions observed on the $0 \to 2$ branch, n_1 be the number of transitions on the $0 \to 1$ branch, and n_2 be the number of transitions on the $1 \to 2$ branch. Let n_0^* denote the number of observations censored (indicated by $*$) in state 0, and let n_1^* be the number of observations censored in state 1. Our parameter vector $\boldsymbol{\theta}$ is $\boldsymbol{\theta} = (p, \beta_1, \beta_2, \mu_1, \mu_2)$. The likelihood function is

$$\mathcal{L}(\boldsymbol{\theta}|\mathcal{D}) = \prod_{i=1}^{n_0} \beta_1 \exp(-\beta_1 t_{0i}) \prod_{i=1}^{n_1} f_{01}(t_{1i}|\mu_1, \mu_2) \prod_{i=1}^{n_2} \beta_2 \exp(-\beta_2 t_{2i})$$
$$\times \prod_{i=1}^{n_0^*} [1 - F_M(t_{0i}^*|p, \beta_1, \mu_1, \mu_2)] \prod_{i=1}^{n_1^*} \exp(-\beta_2 t_{1i}^*), \qquad (5.33)$$

where $f_{01}(\cdot|\mu_1, \mu_2)$ is the inverse Gaussian density and $F_M(\cdot|p, \beta_1, \mu_1, \mu_2)$ is the CDF of a mixture distribution. The inverse Gaussian density is

$$f_{01}(t|\mu_1, \mu_2) = \frac{\mu_1}{\sqrt{2\pi}\ t^{3/2}} \exp\left(-\frac{\mu_1^2/t + \mu_2^2 t}{2} + \mu_1\mu_2\right)$$

for $t > 0$, $\mu_1 > 0$, and $\mu_2 \geq 0$. The mixture distribution resulting from observations censored in state 0 is

$$F_M(t|p, \beta_1, \mu_1, \mu_2) = p[1 - \exp(-\beta_1 t)] + (1 - p)F_{01}(t|\mu_1, \mu_2), \qquad (5.34)$$

where $F_{01}(t|\mu_1, \mu_2)$ is the CDF of the inverse Gaussian distribution. The presence of the nonexponential waiting time coupled with censoring means that the likelihood in (5.33) is not separable. The MGF of the overall survival time is given by (2.12) using the MGFs specific to this problem,

$$M(s) = p\left(\frac{\beta_1}{\beta_1 - s}\right) + (1 - p)\exp\{\mu_1\mu_2 - [\mu_1^2(\mu_2^2 - 2s)]^{1/2}\}\left(\frac{\beta_2}{\beta_2 - s}\right). \qquad (5.35)$$

Figure 5.9 gives the MAPLE program code to compute the MGF, the CGF, and its derivatives for the saddlepoint approximation inversion program.

Uniform prior distributions are used.

$$\begin{aligned} \pi(p) &\sim \text{Unif}(0, 1/2) \\ \pi(\beta_1) &\sim \text{Unif}(0, 3/40) \\ \pi(\beta_2) &\sim \text{Unif}(0, 1/10) \\ \pi(\mu_1) &\sim \text{Unif}(4, 10) \\ \pi(\mu_2) &\sim \text{Unif}(0, 1). \end{aligned} \qquad (5.36)$$

```
with(codegen,makeproc,C):

ms := proc(s)
     local w01,w12,w02;

# mu1 is μ1, mu2 is μ2, b1 is β1, b2 is β2

     w01 := (1-p)*
            exp(mu1*mu2-((mu1^2)*((mu2^2)-2*s))^(.5));
     w12 := b2/(b2-s);
     w02 := p*(b1/(b1-s));

# The next line defines the MGF
     w01*w12+w02;
end proc:

# Define the CGF and its derivatives
k := makeproc(ln(ms(s)),[s]):
C(k,optimized,ansi,filename="c:/mplcode/generalcancer/k.c"):

dk := makeproc(diff(k(s),s),[s]):
C(dk,optimized,filename="c:/mplcode/generalcancer/dk.c"):

ddk := makeproc(diff(dk(s),s),[s]):
C(ddk,optimized,filename="c:/mplcode/generalcancer/ddk.c"):
```

FIGURE 5.9 MAPLE code for Example 5.4.

These distributions are also used for $g(\boldsymbol{\theta})$ in (5.32). The upper bound M was chosen so that

$$\log(M) = \log \mathcal{L}(\hat{\boldsymbol{\theta}}|\mathcal{D}) + \epsilon, \tag{5.37}$$

where $\epsilon = 10^{-6}$ and $\log \mathcal{L}(\hat{\boldsymbol{\theta}}|\mathcal{D}) = -383.21$.

The Bayes predictive density based on $m = 400$ samples is given in Figure 5.10. The density is very heavy-tailed, due primarily to the distributions chosen. The small kink near $t = 0$ is due to the mixture distribution: namely, that the $0 \to 2$ exponential will tend to β_1 while the $0 \to 1$ inverse Gaussian tends to 0 as $t \to 0$. Rejection sampling is very inefficient; in this case, only 3% of the samples are accepted. □

A generalization of rejection sampling is to use importance sampling to sample from a complicated high-dimensional posterior. Importance sampling is difficult to implement with complicated flowgraphs, especially so in the presence of censored data where choosing an envelope function can become difficult. Two other popular methods, Gibbs sampling and the Metropolis–Hastings algorithm, also

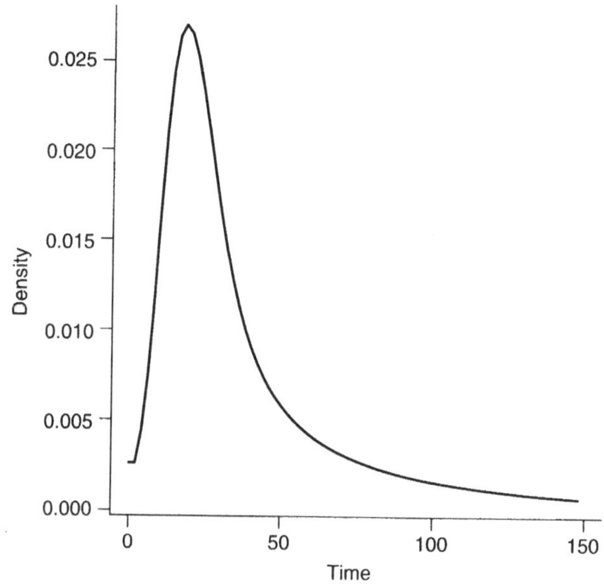

FIGURE 5.10 Bayes predictive density computed via rejection sampling for Example 5.4.

have shortcomings for application to flowgraphs. The problem with using Gibbs sampling with flowgraphs is that the required conditional full distributions cannot be determined easily when data are censored. One pitfall with the direct Metropolis–Hastings approach is that tuning the parameters can become tedious. The Metropolis–Hastings algorithm will work with flowgraphs if implemented carefully; however, a better alternative that requires far less tuning is available. We begin with an example for which Gibbs sampling is useful.

5.2.2 Gibbs Sampling

Gibbs sampling is a Markov chain Monte Carlo method that samples each full conditional distribution in sequence to generate samples from the joint posterior distribution. Suppose that the parameter vector $\boldsymbol{\theta} = (\theta_1, \theta_2, \ldots, \theta_k)$. At each iteration the Gibbs sampler draws samples from each θ_j conditional on the other $\boldsymbol{\theta}$'s and the data $\mathcal{D}$, that is, from the distribution

$$\pi(\theta_j | \theta_1, \theta_2, \ldots, \theta_{j-1}, \theta_{j+1}, \ldots, \theta_k, \mathcal{D}), \tag{5.38}$$

so that each θ_j is updated conditional on the latest value of $\boldsymbol{\theta}$ for the other components. This latest value will be at the current iteration for $\boldsymbol{\theta}$-values that have been updated previously, and at the previous iteration for $\boldsymbol{\theta}$-values that have not yet been updated. This process is repeated until the required number of samples from the posterior are obtained. An initial burn-in period of iterations is also required to achieve approximate convergence to the posterior. We illustrate the basics of using Gibbs sampling with flowgraphs for a simple series model.

Example 5.5: KIDNEY DISEASE PROGRESSION: BAYES PREDICTIVE DENSITY. We illustrate the computation of (5.7) using Gibbs sampling with the kidney series flowgraph of Figure 2.2 introduced in Example 2.1. Example 3.1 presented the exact inversion of the MGF for the waiting time distribution for the failure of both kidneys. Example 3.8 presented the maximum likelihood estimated density for this model. Y_0 is the random waiting time in state 0 until state 1 is reached, and Y_1 is the random waiting time in state 1 until state 2 is reached. Recall that $Y_0 \sim \text{Exp}(2\lambda_0)$ is independent of $Y_1 \sim \text{Exp}(\lambda_1)$. Suppose that we have complete observations on n kidney patients so that our data consist of transition time from $0 \to 1$ and $1 \to 2$ for each person. Let y_{0i} and y_{1i} denote the observations on the individual waiting times for patient i and summarize the observations on n patients as $\mathbf{y}_0 = (y_{01}, y_{02}, \ldots, y_{0n})$ and $\mathbf{y}_1 = (y_{11}, y_{12}, \ldots, y_{1n})$. We have the additional constraint on the problem from Gross et al. (1971) that $\lambda_0 \le \lambda_1$. This compensates for the additional burden on the second kidney after the first kidney fails. For purposes of illustration, we will assume a uniform prior on $\boldsymbol{\theta} = (\lambda_0, \lambda_1)$ given by

$$\pi(\lambda_0, \lambda_1) = \frac{1}{2.5} \qquad \text{for } 0 < \lambda_0 \le 1, \lambda_0 \le \lambda_1 \le 3, \tag{5.39}$$

where the limits are adjusted to accommodate the constraint on λ_0 and λ_1.

The likelihood function is

$$\begin{aligned} \mathcal{L}(\lambda_0, \lambda_1 | \mathbf{y}_0, \mathbf{y}_1) &= \prod_{i=1}^{n} 2\lambda_0 \exp(-2\lambda_0 y_{0i})\, \lambda_1 \exp(-\lambda_1 y_{1i}) \\ &\propto \lambda_0^n \exp\left(-2\lambda_0 \sum_{i=1}^{n} y_{0i}\right) \lambda_1^n \exp\left(-\lambda_1 \sum_{i=1}^{n} y_{1i}\right). \end{aligned} \tag{5.40}$$

The posterior distribution on (λ_0, λ_1) is proportional to the likelihood function (5.40) multiplied by the prior distribution,

$$\pi(\lambda_0, \lambda_1 | \mathbf{y}_0, \mathbf{y}_1) \propto \lambda_0^n \exp\left(-2\lambda_0 \sum_{i=1}^{n} y_{0i}\right) \lambda_1^n \exp\left(-\lambda_1 \sum_{i=1}^{n} y_{1i}\right),$$
$$\text{for } 0 < \lambda_0 \le 1 \text{ and } \lambda_0 \le \lambda_1 \le 3. \tag{5.41}$$

We can write down the Bayes predictive density (5.7), up to the normalizing constant, for the future observable total failure time of kidneys, $T = Y_0 + Y_1$:

$$\begin{aligned} f_T(t | \mathbf{y}_0, \mathbf{y}_1) &= \int_0^1 \int_{\lambda_0}^3 f_T(t | \lambda_0, \lambda_1) \pi(\lambda_0, \lambda_1 | \mathbf{y}_0, \mathbf{y}_1)\, d\lambda_1\, d\lambda_0 \\ &\propto \int_0^1 \int_{\lambda_0}^3 \left(\frac{2\lambda_0\lambda_1}{2\lambda_0 - \lambda_1}\right)(e^{-\lambda_1 t} - e^{-2\lambda_0 t}) \lambda_0^n \exp\left(-2\lambda_0 \sum_{i=1}^{n} y_{0i}\right) \\ &\quad \times \lambda_1^n \exp\left(-\lambda_1 \sum_{i=1}^{n} y_{1i}\right) d\lambda_1\, d\lambda_0 \end{aligned} \tag{5.42}$$

for $t > 0$, where $f_T(t|\lambda_0, \lambda_1)$ was given in (3.6). We can integrate (5.42) numerically for a given t over λ_0 and λ_1, or we can compute the integral using Monte Carlo sampling. For illustration, we sample from the posterior using Gibbs sampling. The full conditional distributions are derived as follows. The conditional distribution of λ_0 given λ_1 and the data $\mathcal{D} = (\mathbf{y}_0, \mathbf{y}_1)$ is

$$\pi(\lambda_0|\lambda_1, \mathcal{D}) = \frac{\pi(\lambda_0, \lambda_1|\mathcal{D})}{\pi(\lambda_1|\mathcal{D})}, \tag{5.43}$$

where the denominator is given by

$$\pi(\lambda_1|\mathcal{D}) \propto \lambda_1^n \exp\left(-\lambda_1 \sum_{i=1}^n y_{1i}\right) \int_0^{\min\{\lambda_1,1\}} \lambda_0^n \exp\left(-2\lambda_0 \sum_{i=1}^n y_{0i}\right) d\lambda_0. \tag{5.44}$$

Thus, the conditional distribution of λ_0 given $(\lambda_1, \mathcal{D})$ is proportional to a truncated gamma distribution, Gamma$(n+1, 2\sum_{i=1}^n y_{0i})$,

$$\pi(\lambda_0|\lambda_1, \mathcal{D}) = \frac{\lambda_0^n \exp(-2\lambda_0 \sum_{i=1}^n y_{0i})}{\int_0^{\min\{\lambda_1,1\}} \lambda_0^n \exp(-2\lambda_0 \sum_{i=1}^n y_{0i})\, d\lambda_0}, \qquad 0 < \lambda_0 \le \min\{\lambda_1, 1\}. \tag{5.45}$$

Similarly, the conditional distribution of λ_1 given λ_0 and the data $\mathcal{D}$ is proportional to a truncated gamma distribution, Gamma$(n+1, \sum_{i=1}^n y_{1i})$,

$$\pi(\lambda_1|\lambda_0, \mathcal{D}) = \frac{\lambda_1^n \exp(-\lambda_1 \sum_{i=1}^n y_{1i})}{\int_{\lambda_0}^3 \lambda_1^n \exp(-\lambda_1 \sum_{i=1}^n y_{1i})\, d\lambda_1}, \qquad \lambda_0 \le \lambda_1 \le 3. \tag{5.46}$$

We take a sample of m values of (λ_0, λ_1) from the posterior and approximate (5.42) with

$$\hat{f}_T(t|\mathbf{y}_0, \mathbf{y}_1) = \frac{\sum_{j=1}^m f_T(t|\lambda_{0j}, \lambda_{1j})}{m}. \tag{5.47}$$

For large m, this sum approaches the desired integral. The sufficient statistics, $\sum_{i=1}^n y_{0i} = 750$ and $\sum_{i=1}^n y_{1i} = 300$, are constructed from the MLEs for λ_0 and λ_1 reported in Gross et al. (1971) based on 100 samples of kidney failure times. The Gibbs sampler had 1000 burn-in samples of λ_0 and λ_1 and 2000 samples from the posterior. Of these, every fifth sample was taken to get $m = 400$ samples for the predictive density computation. Computing the autocorrelation function verifies that the samples are approximately uncorrelated. Figure 5.11 gives the Bayes predictive density of the time to total failure of the kidneys. The solid line in the figure is the predictive distribution, the posterior mean of $f_T(t|\lambda_0, \lambda_1)$. The two dashed lines are the 5% and 95% pointwise prediction bounds. Figure 5.12 gives histograms of posterior samples with the MLE indicated by a vertical line. □

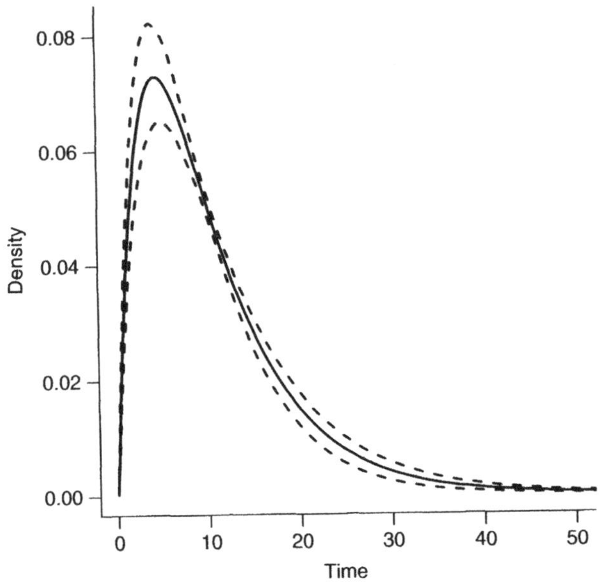

FIGURE 5.11 Bayes predictive density using Gibbs sampling for the time to total failure of kidneys for Example 5.5. Solid line is the predictive distribution. Dashed lines are the 5% and 95% pointwise prediction bounds.

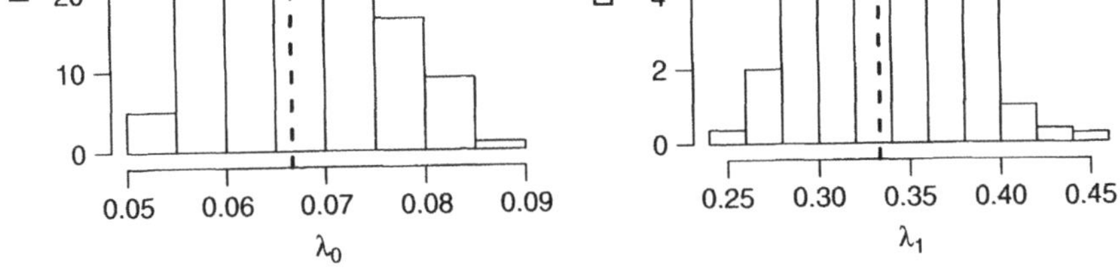

FIGURE 5.12 Histograms of posterior samples with MLE (vertical line) for λ_0 and λ_1 of Example 5.5, modeling the time to kidney failure.

5.2.3 Laplace's Method

Laplace's method can be used with flowgraphs that are relatively simple. The approximation works well if the integrand in (5.7) is a smooth unimodal function. Laplace's method was introduced by Leonard (1982) and is discussed in Ellis and

Rosen (1982). Tierney and Kadane (1986) introduce it in the context of Bayesian statistics for the computation of posterior modes and marginal densities. The main result from Ellis and Rosen (1982) is given below.

Result 5.2.1. Suppose that $g(\cdot)$ is a smooth, unimodal function on $\mathcal{R}^k$. Then

$$\int_{\mathcal{R}^k} \exp[ng(\boldsymbol{\theta})]\, d\boldsymbol{\theta} = \left(\frac{2\pi}{n}\right)^{k/2} |\det[D^2 g(\hat{\boldsymbol{\theta}})]|^{-1/2} \exp[ng(\hat{\boldsymbol{\theta}})][1 + O(n^{-1})], \quad (5.48)$$

where $\hat{\boldsymbol{\theta}}$ maximizes $g(\cdot)$ over $\mathcal{R}^k$ and $D^2 g(\hat{\boldsymbol{\theta}})$ is the matrix of second derivatives of $g(\boldsymbol{\theta})$ evaluated at $\hat{\boldsymbol{\theta}}$.

We refer the reader to Ellis and Rosen (1982) for a proof. In practice, unimodality can be relaxed to the case of one dominant mode. It is important that the function $g(\cdot)$ be stable and not change with n. The term *smooth* is used loosely to mean that at least two derivatives exist. Intuitively, Laplace's method takes advantage of the fact that as n gets larger, the integrand becomes more peaked about $\hat{\boldsymbol{\theta}}$. Since the integrand is always positive, we are integrating a positive function whose maximal value is growing with n. Therefore, only the local behavior of the function at its maximum determines the area at a high level of accuracy. Laplace's method is used for studying the asymptotic behavior of integrals of the form of (5.48) as $n \to \infty$. The leading order asymptotic behavior of the integral is determined by the largest value of the integrand. To apply Laplace's approximation with flowgraphs, we define $g(\cdot)$ to be

$$g(\boldsymbol{\theta}) = \log[\tilde{f}_Z(z|\boldsymbol{\theta})\pi(\boldsymbol{\theta}|\mathcal{D})]. \quad (5.49)$$

For each value z of Z for which the predictive density is desired, we find the $\hat{\boldsymbol{\theta}}$ that maximizes $g(\boldsymbol{\theta})$. Note that nested within Laplace's method is the saddlepoint approximation to the density, $\tilde{f}_Z(z|\boldsymbol{\theta})$, computed from the flowgraph MGF, $M_Z(s|\boldsymbol{\theta})$. We set $n = 1$ in (5.48). One of the derivations of the saddlepoint approximation is based on Laplace's method (cf. Barndorff-Nielsen, 1988; Reid, 1988), so we expect this approximation to work well. When used with simple flowgraphs, Laplace's method produces results that are identical to those based on posterior sampling. The primary drawback of this method is that it does not always work with complicated, high-dimensional models. In addition, the method is computationally slow relative to other methods. Butler and Huzurbazar (1993) present an example of Laplace's method using exponential waiting time distributions. The approximation may not work as well in the middle of the distribution with complex flowgraphs. As with rejection sampling, this method can have difficulty with high-dimensional posteriors and in the presence of censoring.

Example 5.6: LAPLACE'S APPROXIMATION WITH FLOWGRAPHS. Laplace's approximation will work with a complicated flowgraph as long as the branch waiting time distributions are exponentials. This is not a theoretical requirement but rather a practical one. Mainly, the overall flowgraph density must be well-behaved so that the function $g(\boldsymbol{\theta})$ is essentially unimodal. We illustrate Laplace's method using

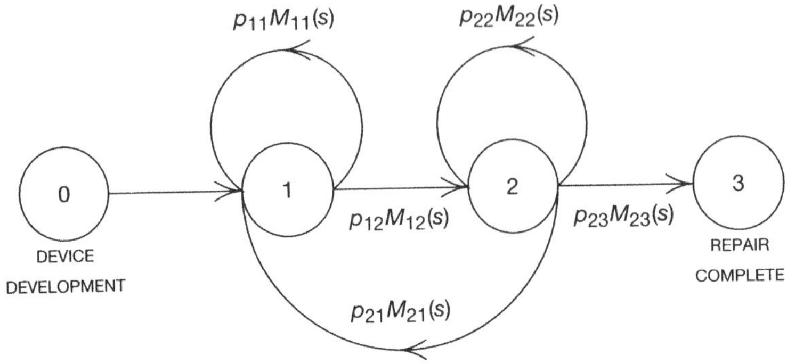

FIGURE 5.13 Flowgraph model for the reliability repair problem.

the reliability repair example of Section 2.6. The flowgraph is reproduced here as Figure 5.13. Recall that the device is being developed for a given application. The probability that the device fails on any given trial is p_{12}. If the device fails, it is worked on and there is a probability p_{23} that it is fixed. The states of the system are defined as follows: state 1 indicates that the trial is successful given that the device is faulty; state 2 indicates that the trial is a failure given that the device is faulty; and state 3 indicates that the trial is successful given that the device is fixed.

Recall that $M_{11}(s)$ and $M_{12}(s)$ represent MGFs of testing times and $M_{21}(s)$, $M_{22}(s)$, and $M_{23}(s)$ are MGFs of repair times. We make the following distributional assumptions for this illustration. State 0 is the start of the process and we assume that every device instantly reached state 1 (i.e., we assume a constant probability 1 and a constant waiting time of 0 to reach state 1). Let T_{ij} be the random waiting time from state i to state j for $i = 0, 1, 2$ and $j = 1, 2, 3$. We assume that $T_{11} \sim \text{Exp}(\theta_1)$, $T_{12} \sim \text{Exp}(\theta_4)$, $T_{22} \sim \text{Exp}(\theta_2)$, $T_{21} \sim \text{Exp}(\theta_3)$, and $T_{23} \sim \text{Exp}(\theta_5)$. The MGF of the distribution of the waiting time from $0 \to 3$, the time until the device is fixed, was found in (2.29).

The system was simulated for 10 realizations and complete data (without censoring) were recorded on each transition. With uniform priors on p_{11} and p_{23} and vague priors on all of the other parameters, the posterior distributions are

$$\begin{aligned}
\pi(\theta_1|\mathcal{D}) &\sim \text{Gamma}(3, 8.16)\\
\pi(\theta_2|\mathcal{D}) &\sim \text{Gamma}(4, 10.91)\\
\pi(\theta_3|\mathcal{D}) &\sim \text{Gamma}(3, 7.08)\\
\pi(\theta_4|\mathcal{D}) &\sim \text{Gamma}(9, 12.59)\\
\pi(\theta_5|\mathcal{D}) &\sim \text{Gamma}(10, 35.13)\\
\pi(p_{11}|\mathcal{D}) &\sim \text{Beta}(5, 15)\\
\pi(p_{23}|\mathcal{D}) &\sim \text{Beta}(10, 6).
\end{aligned}$$

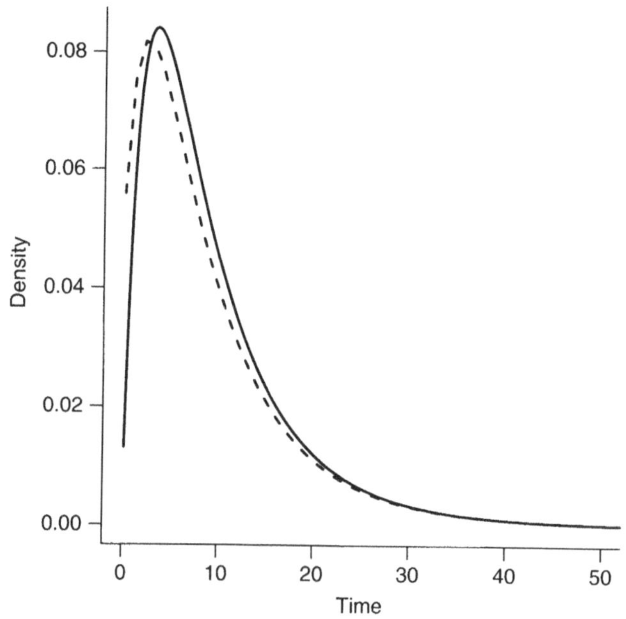

FIGURE 5.14 Bayes predictive density for Example 5.6 computed using Laplace's approximation for the flowgraph (dashed line) with a simulated "exact" density (solid line).

Using these, the resulting Bayes predictive density based on Laplace's method (5.48) is shown in Figure 5.14. Notice that the approximation is adequate but not as good as that of other methods. The "exact" density is computed via brute-force simulation of the system using the same posterior distributions. It is based on 10^7 realizations of the system. A kernel density estimate gives the resulting density. Note that such exact simulation is possible only because we have no censored or missing data. □

While the flowgraph of Figure 5.13 may appear complex, the waiting time distributions are simple and hence we can attempt to use Laplace's method. The method can be made to work on simpler flowgraphs with other distributions, but it requires specific fine tuning to the problem at hand. Hence we prefer alternative methods that are applicable more generally.

5.2.4 Slice Sampling

Our aim with flowgraphs is to have a method that is globally applicable, a method that will work well for most flowgraph models. Many of the methods discussed previously can be fine-tuned to work for specific flowgraphs and can work very well in those instances. However, our goal is to minimize the effort on the part of the user to sample from the posterior and to have a method that easily extends to each flowgraph model and data set. In general, with flowgraph models, we have high-dimensional posteriors whose shape is unknown to us. Censored data introduce more complexity. A recently developed technique that lends itself to sampling from the posterior with flowgraph models is a Markov chain Monte Carlo (MCMC) method called *slice sampling*. Slice sampling was introduced by Neal (1997) and is presented in detail in Neal (2003). Slice sampling works as a "black box" technique with flowgraphs and generally requires less tuning than does the Metropolis–Hastings algorithm.

The basic premise of slice sampling is that one can sample from any univariate distribution by sampling points uniformly under the graph of its density function. Multivariate slice sampling can be performed by applying a single variable slice sampling method to each variable in sequence, so we describe the method in the univariate case. We wish to sample from the posterior, $\pi(\theta|\mathcal{D})$, for $\theta \in \mathcal{R}$, which is proportional to $\mathcal{L}(\theta|\mathcal{D})\pi(\theta)$. We can accomplish this by sampling uniformly from the two-dimensional region under the graph of $\mathcal{L}(\theta|\mathcal{D})\pi(\theta)$. To carry out this computation, introduce an auxiliary variable Y such that the joint distribution of (θ, Y) is uniform on the graph $\{(\theta, y) : 0 < y < \mathcal{L}(\theta|\mathcal{D})\pi(\theta)\}$. The joint density of (θ, Y), $f(\theta, y)$, is

$$f(\theta, y) = \frac{1}{\int \mathcal{L}(\theta|\mathcal{D})\pi(\theta)\,d\theta} \qquad \text{for } \theta \in \mathcal{R} \text{ and } 0 < Y < \mathcal{L}(\theta|\mathcal{D})\pi(\theta) \tag{5.50}$$

and 0 otherwise. The marginal density for θ is then the posterior that we seek, $\pi(\theta|\mathcal{D})$:

$$\pi(\theta|\mathcal{D}) = \int_0^{\mathcal{L}(\theta|\mathcal{D})\pi(\theta)} \frac{1}{\int \mathcal{L}(\theta|\mathcal{D})\pi(\theta)\,d\theta}\,dy = \frac{\mathcal{L}(\theta|\mathcal{D})\pi(\theta)}{\int \mathcal{L}(\theta|\mathcal{D})\pi(\theta)\,d\theta}. \tag{5.51}$$

To sample θ, we can sample (θ, Y) jointly and discard Y.

To generate points that are drawn uniformly from $\{(\theta, y) : 0 < y < \mathcal{L}(\theta|\mathcal{D})\pi(\theta)\}$, we define a Markov chain that will converge to this uniform distribution. This can be done using Gibbs sampling. We can sample alternately from the conditional distribution $Y|\theta$ and the conditional distribution $\theta|Y$. The conditional distribution of $Y|\theta$ is uniform over the interval $(0, \mathcal{L}(\theta|\mathcal{D})\pi(\theta))$. The conditional distribution of $\theta|Y$ defines the *slice*. The slice is the set $\{\theta : Y < \mathcal{L}(\theta|\mathcal{D})\pi(\theta)\}$. Once the slice is identified, randomly sample a value of θ from the slice and use that to sample the next Y value. Iteratively sample θ and Y values to obtain the sequence $\theta_0 \rightarrow Y_1 \rightarrow \theta_1 \rightarrow Y_2 \rightarrow \theta_2 \rightarrow \cdots$. The main difficulty in implementing this technique is identifying the slice.

Neal (2003) presents two approaches for approximating the slice. The first approach is an additive method which can be summarized in three steps:

1. Starting with the initial value θ_0, draw a point y uniformly from $(0, \mathcal{L}(\theta_0|\mathcal{D})\,\pi(\theta_0))$. Let this point y set the vertical level. A horizontal line at this level, y, cuts the function $\mathcal{L}(\theta|\mathcal{D})\pi(\theta)$. This horizontal cut defines the slice.
2. Randomly position an interval of width w around θ_0. Then expand this interval in steps of size w until both ends of the interval are outside the slice.

3. A new point θ_1 is found by drawing uniformly from the expanded interval of step 2. If the point sampled is outside the slice, this point is used to shrink the interval of step 2. Sampling continues until a point θ_1 is determined to be inside the slice.

The initial interval of step 2 is expanded additively and then shrunk in step 3. Hence this procedure is called the *stepping-out and shrinkage procedure*, or just the *stepping-out procedure*. The stepping-out procedure is appropriate for any distribution as long as an estimate w for the typical width of the slice is available. In practice, the stepping-out procedure is limited to a finite number m of additive expansions, with the apportionment of expansions in either direction determined randomly.

The second approach for identifying the slice is multiplicative and is called the *doubling procedure*. The doubling procedure produces a sequence of intervals, each twice the size of the preceding one, until the resulting interval has both ends outside the slice. The interval is expanded on only one side, chosen at random, regardless of whether that side is currently outside the slice. If the initial interval's ends are already outside the slice, no doubling is done. This approach can expand the interval faster than the stepping-out procedure. As with the stepping-out procedure, a predetermined upper limit on the number of doublings allowed is often imposed. Unlike the stepping-out procedure, the doubling procedure requires additional checking to ensure that the choice of each new θ leaves the joint distribution of θ and Y invariant. The required acceptance test is easily automated and integrated into the sampling procedure.

Regardless of the approach used, efficiency is improved by choosing an interval at each step that contains as much of the slice as possible. At the same time, we want to avoid intervals that are much larger than the slice because this will make the sampling in step 3 inefficient.

Note that with these approaches, many updates will be needed to move from one part of the distribution to another. One way to improve sampling efficiency is by suppressing the random walk behavior of the sampling scheme by using an *overrelaxation method*. Overrelaxation methods update each variable in turn by drawing a new value for a variable to be on the opposite side of the mode from the current value. This differs from schemes such as Gibbs sampling, which draw a new value for a variable from its conditional distribution independent of the current value. There are many schemes for implementing overrelaxation in the Gaussian and non-Gaussian cases. Adler (1981) presents a scheme that is applicable when the conditional distributions are Gaussian. The method is discussed further in Barone and Frigessi (1990) and Green and Han (1992). Neal (1998) deals with the non-Gaussian case. Overrelaxation is most appropriate when the conditional distributions are unimodal. When using overrelaxation, every update must be made so as to produce few or no "rejections" that leave the state unchanged or the random walk behavior will not be suppressed.

One way to implement overrelaxation with slice sampling is to combine the stepping-out procedure with bisection to accurately locate the endpoints of the interval containing the slice. Assume that the slice is a single interval. The stepping-out procedure finds an interval containing the slice with the property that the two outermost steps locate the endpoints of the slice to within an interval of size w. Then a bisection procedure locates the endpoints of the interval more precisely. A new interval is formed from the improved endpoint locations. Let $\hat{L}$ and $\hat{R}$ be the left and right endpoints of this interval. To obtain an overrelaxed update, flip the current point θ_0 to a new point θ_1:

$$\theta_1 = \frac{\hat{L}+\hat{R}}{2} - \left(\theta_0 - \frac{\hat{L}+\hat{R}}{2}\right) = \hat{L} + \hat{R} - \theta_0. \tag{5.52}$$

The new point θ_1 is the same distance as the current point θ_0 from the middle of the interval. The new point θ_1 can be rejected if it is either outside the interval that was found prior to bisection or outside the slice. In this case, θ_1 is set equal to θ_0, the current point. The chance of rejection is greatly reduced by precisely locating the endpoints, $\hat{L}$ and $\hat{R}$. The initial interval prior to bisection can be determined using the doubling procedure. The new point θ_1 must pass the acceptance test alluded to previously.

The overrelaxation method remains valid for sampling from multimodal distributions. In addition, to ensure ergodicity of the Markov chain, an update drawn uniformly from the interval approximating the slice is conducted at each iteration with some small predefined probability α.

The main result of slice sampling can be summarized as follows:

Result 5.2.2. Suppose that $\theta \in \mathcal{R}$, and let Y be an auxiliary variable. If $(\theta, Y) \sim \text{Unif}\{(\theta, y) : 0 < y < \mathcal{L}(\theta|\mathcal{D})\pi(\theta)\}$, then

$$Y|\theta \sim \text{Unif}(0, \mathcal{L}(\theta|\mathcal{D})\pi(\theta)) \tag{5.53}$$

$$\theta|Y \sim \text{Unif}\{[\theta : Y < \mathcal{L}(\theta|\mathcal{D})\pi(\theta)]\} \tag{5.54}$$

$$\theta \sim \pi(\theta|\mathcal{D}). \tag{5.55}$$

Example 5.7: Slice Sampling. This example involves generating slice samples from a Gamma(2, 1/2) distribution. In all, there are 6000 samples. The first 1000 are discarded as burn-in. Figure 5.15 is a plot of this gamma distribution. The dashed vertical line is associated with the sample 2440, θ_{2440}. The point θ_{2441} is generated by:

1. Sampling a value for the horizontal slice from a uniform distribution covering the range of the vertical dashed line at θ_{2440}
2. Determining the slice associated with this sampled value
3. Drawing θ_{2441} uniformly from the slice

Figure 5.16 shows the results. A histogram of the 5000 slice samples is given. The dashed overlay represents a density estimate based on the slice samples, and the solid overlay is the gamma distribution being sampled. □

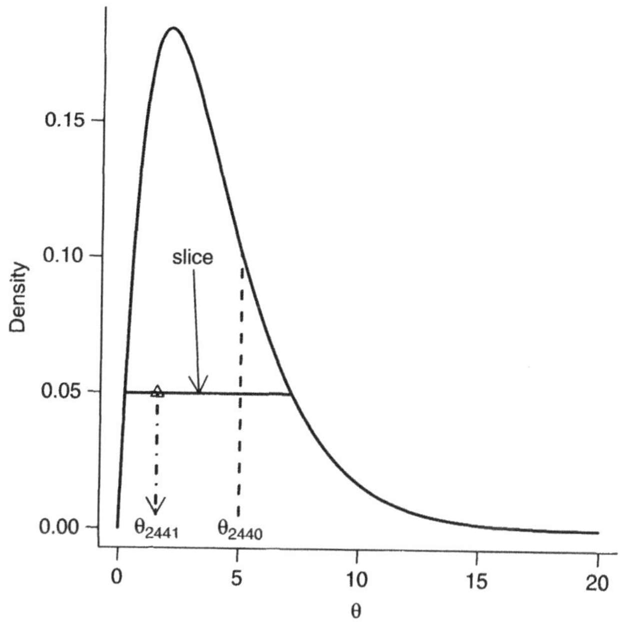

FIGURE 5.15 Gamma density used for slice sampling illustration in Example 5.7.

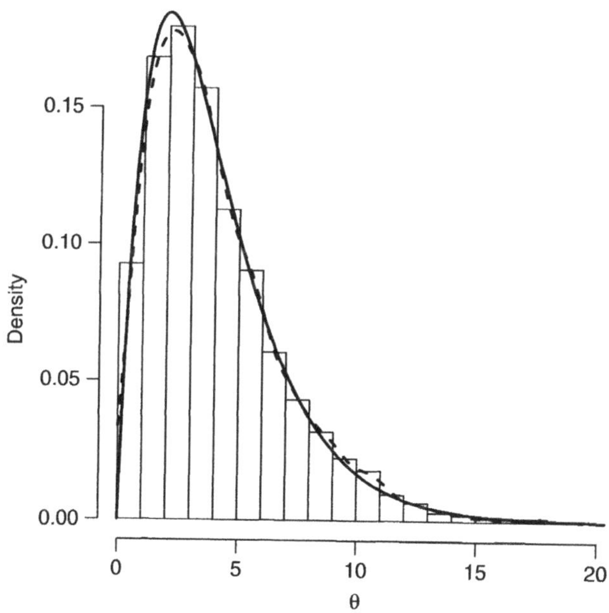

FIGURE 5.16 Slice sampling results for gamma density, Example 5.7.

5.3 USING SLICE SAMPLING WITH FLOWGRAPHS

In this section we demonstrate the use of slice sampling to compute the integral of (5.7) for flowgraph models. The first example concerns estimating the time to incubation for HIV/AIDS data and is a convolution of an exponential and a gamma distribution. Although the data are medical, the problem is also relevant for those interested in engineering applications. For example, the hydraulic pump system of Example 2.2 is a series model that could be analyzed in this manner with general waiting time distributions. The second example is a reanalysis of Example 5.4 using slice sampling. This is a more challenging example. The third example presents slice sampling results for the reversible illness death flowgraph model of Figure 2.22. Although presented in an engineering context, both of these examples are also relevant for medical applications.

Example 5.8: Transfusion Flowgraph Model. This example presents a full analysis of the Transfusion Data from the San Francisco Men's Health Study (SFMHS). These data consist of people who were given HIV-infected blood products. Example 2.10 presented a flowgraph model for this process, and Figure 5.17 reproduces this flowgraph. State 1 is HIV infection with negative antibody status, state 2 is HIV infection with positive antibody status but without AIDS symptoms, and state 3 is pre-AIDS symptoms. Example 4.4 constructed censored data histograms for these data. Based on these data, we make the modeling assumptions for the distributions on the branches of the flowgraph of Figure 5.17. Recall that T_1 is the waiting time in state 1 until transition to state 2 and T_2 is the waiting time in state 2 until transition to state 3. The total waiting time for passage from $1 \to 3$ is $T = T_1 + T_2$. The censored data histogram of observations on T_1 indicates that an exponential model is adequate. For T_2, the censored data histogram suggests that a distribution with positive mode, such as a gamma, would be more appropriate. We model $T_1 \sim \text{Exp}(\lambda)$ with mean $1/\lambda$ and density $f_{T_1}(\cdot|\lambda)$, and $T_2 \sim \text{Gamma}(\alpha, \beta)$ with mean α/β, density $f_{T_2}(\cdot|\alpha, \beta)$ and distribution function $F_{T_2}(\cdot|\alpha, \beta)$. The MGFs of these distributions appear on the branches of the flowgraph of Figure 5.17. For this illustration, we assume uniform priors on all of the parameters. The priors are independent and are $\lambda \sim \text{Unif}(0, 10)$, $\alpha \sim \text{Unif}(0, 10)$, and $\beta \sim \text{Unif}(0, 10)$.

The data contribute terms to the likelihood function as follows. Each person who made the transition from $1 \to 2$ contributes a term, $f_{T_1}(t_{1i}|\lambda)$, to the likelihood

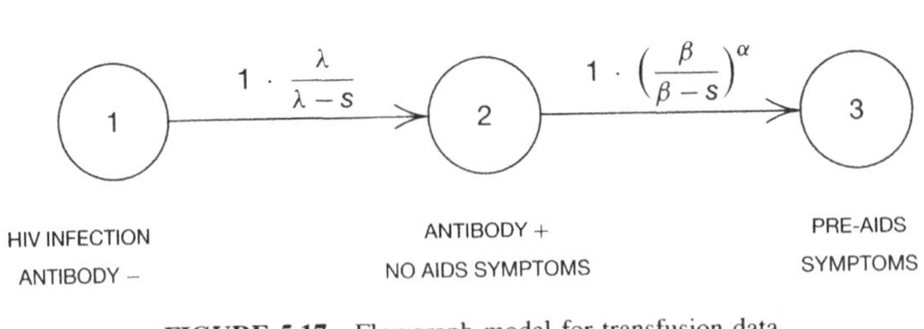

FIGURE 5.17 Flowgraph model for transfusion data.

function. The 70 people that made the transition from $1 \to 2$ contribute the first term in the likelihood (5.56). Each person who made the $2 \to 3$ transition contributes a term, $f_{T_2}(t_{2j}|\alpha, \beta)$, to the likelihood. Of the original 70 people, only 35 people were observed beyond state 2. The 22 people who made the $2 \to 3$ transition contribute the second term to the likelihood (5.56). Each person who is right-censored in state 2 contributes a term, $1 - F_{T_2}(t_{2j}|\alpha, \beta)$, to the likelihood as the person is last observed in state 2. The 13 people who are right censored in state 2 contribute the third term to the likelihood (5.56). The full likelihood function is

$$\mathcal{L}(\lambda, \alpha, \beta|\mathcal{D}) = \prod_{i=1}^{70} f_{T_1}(t_{1i}|\lambda) \prod_{j=1}^{22} f_{T_2}(t_{2j}|\alpha, \beta) \prod_{j=23}^{35} [1 - F_{T_2}(t_{2j}|\alpha, \beta)]. \quad (5.56)$$

With uniform priors on the parameters, the posterior is proportional to the likelihood. We use slice sampling to sample from this posterior to compute the Bayes predictive density, predictive survivor, and predictive hazard functions. A sample of 3000 $\boldsymbol{\theta} = (\lambda, \alpha, \beta)$ values was generated from this posterior, of which 1000 $\boldsymbol{\theta}$ values were discarded as burn-in. The doubling procedure was used to identify the slice with $w = 0.1$, $m = 100$, and $\alpha = 0.05$. Within the slice sampling procedure, w is the nominal interval size, m is the number of doublings, and α is a predefined probability of a uniform sampling update. For each $\boldsymbol{\theta}$ value the predictive density, survival, and hazard functions are computed using the flowgraph MGF. The MGF of the distribution of the waiting time from HIV infection to pre-AIDS symptoms is

$$M(s) = M_{12}(s)M_{23}(s) = \left(\frac{\lambda}{\lambda - s}\right)\left(\frac{\beta}{\beta - s}\right)^{\alpha} \qquad \text{for } s < \min(\lambda, \beta).$$

The necessary inputs to the saddlepoint approximation are the CGF, $K(s)$, and its first and second derivatives. The CGF is

$$K(s) = \log(\lambda) - \log(\lambda - s) + \alpha \log(\beta) - \alpha \log(\beta - s) \qquad \text{for } s < \min(\lambda, \beta). \quad (5.57)$$

The saddlepoint equation, $K'(\hat{s}) = t$, is given by

$$(\lambda - \hat{s})^{-1} + \alpha(\beta - \hat{s})^{-1} = t \qquad \text{for } \hat{s} < \min(\lambda, \beta). \quad (5.58)$$

The second derivative at the saddlepoint is

$$K''(\hat{s}) = (\lambda - \hat{s})^{-2} + \alpha(\beta - \hat{s})^{-2}.$$

Figures 5.18, 5.19, and 5.20 give the Bayes predictive density, predictive survivor, and predictive hazard functions, respectively. Table 5.1 gives the predictive CDF values. The "exact," presented for comparison, is computed via numerical

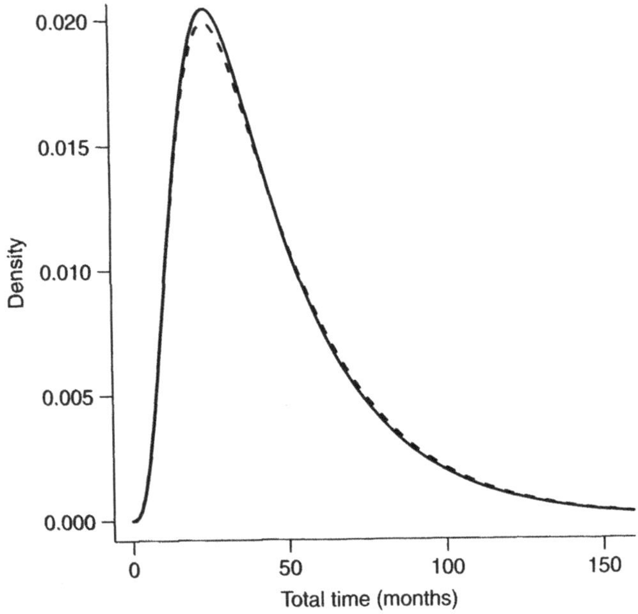

FIGURE 5.18 Bayes predictive density via slice sampling for the time to incubation. The solid line is the "exact"; the dashed line is the saddlepoint approximation.

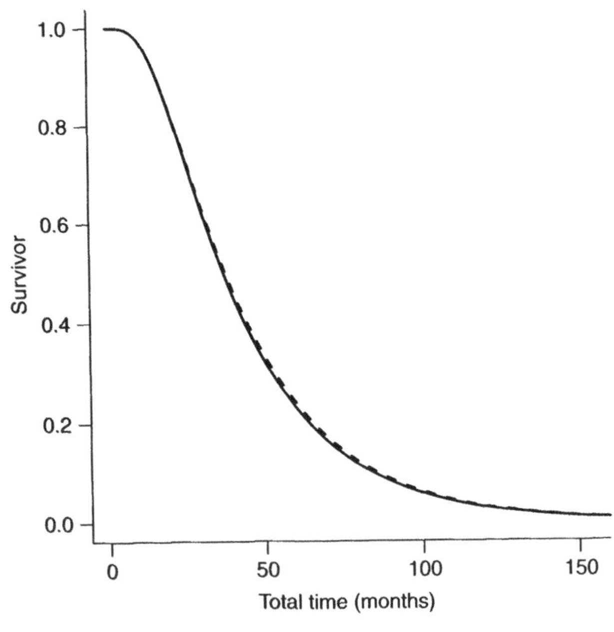

FIGURE 5.19 Bayes predictive survivor function via slice sampling for the time to incubation. The solid line is the "exact"; the dashed line is the saddlepoint approximation.

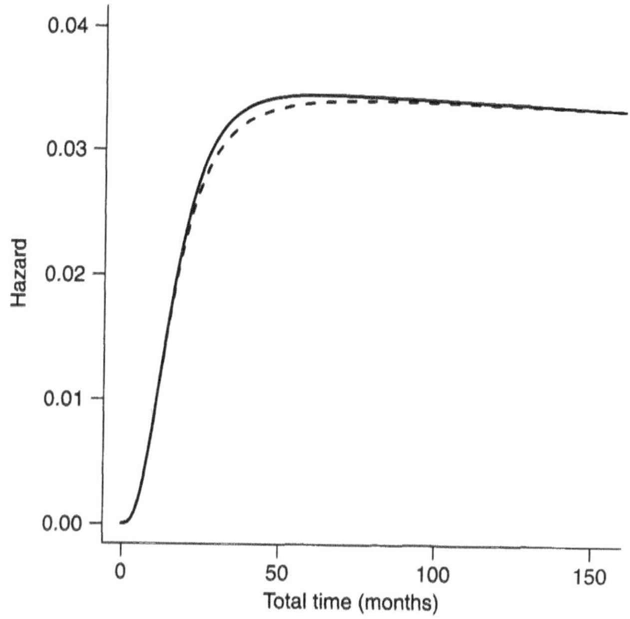

FIGURE 5.20 Bayes predictive hazard via slice sampling for the time to incubation. The solid line is the "exact"; the dashed line is the saddlepoint approximation.

TABLE 5.1 Predictive CDF Values for AIDS/HIV Transfusion Data (Time in Months)

Percentile	"Exact"	$F(t)^a$
0.25	23.86	24.06
0.50	37.01	37.64
0.75	57.36	58.53
0.90	83.95	85.46
0.95	104.19	105.84
0.99	151.8	153.6

[a] $F(t)$ refers to the numerically integrated saddlepoint density approximation used with the flowgraph.

integration of the convolution. The exact distribution of T, the total waiting time from $1 \to 3$, is

$$\begin{aligned} P(T \le t) &= P(T_1 + T_2 \le t) \\ &= \int_0^t P_\lambda(T_1 + T_2 \le t | T_2 = u) f_{T_2(u|\alpha,\beta)}\, du \\ &= \int_0^t P_\lambda(T_1 \le t - u) f_{T_2}(u|\alpha, \beta)\, du. \end{aligned}$$

The CDF, density, and hazard function are

$$F_T(t|\lambda,\alpha,\beta) = \int_0^t F_{T_1}(t-u|\lambda) f_{T_2}(u|\alpha,\beta)\, du \tag{5.59}$$

$$f_T(t|\lambda,\alpha,\beta) = \int_0^t f_{T_1}(t-u|\lambda) f_{T_2}(u|\alpha,\beta)\, du \tag{5.60}$$

$$h_T(t|\lambda,\alpha,\beta) = \frac{f_T(t|\lambda,\alpha,\beta)}{1 - F_T(t|\lambda,\alpha,\beta)}. \tag{5.61}$$

We can do this computation only because we have a simple convolution. The "exact" is in quotes because we still use slice sampling to estimate the integral of (5.7). The mean of the predictive distribution is 44.91 months with a standard deviation of 30.32 months. Note that MAPLE cannot perform the inversion here because α is not an integer, nor can we use partial fraction expansion as described in Section 3.1. The predictive CDF plot is not given; however, the exact and saddlepoint approximated CDFs are in agreement. In Figure 5.20, the shape of the hazard function starts out like a gamma hazard and then approaches the value 0.02468 in the limit. In general, when T is the first passage time in a semi-Markov process, the asymptotic limit of the hazard approaches the upper bound of the convergence strip of the MGF (i.e., the upper limit of the saddlepoint). This result is presented without proof in Butler and Bronson (2002). In this case, it is the smallest pole of the MGF that is the smallest λ or β in the posterior slice samples [i.e., $s < \min(\lambda_i, \beta_i)$ for all $i = 1, \ldots, 2000$]. □

Example 5.9: SLICE SAMPLING FOR CANCER PROGRESSION. We return to the general model for cancer progression of Example 5.4 and illustrate the use of slice sampling. We consider two settings: one with the same prior distribution as the rejection sampling of Example 5.4, referred to as SS I; the second, referred to as SS II, has priors that are more vague. The priors for SS II are

$$p \sim \text{Uniform}(0,1) \tag{5.62}$$

$$\beta_1 \sim \text{Gamma}(1/100, 1/100) \tag{5.63}$$

$$\beta_2 \sim \text{Gamma}(1/100, 1/100) \tag{5.64}$$

$$\mu_1 \sim \text{Uniform}(0,100) \tag{5.65}$$

$$\mu_2 \sim \text{Uniform}(0,100). \tag{5.66}$$

Figure 5.21 is a plot of the predictive density for the flowgraph of Figure 5.8 based on slice sampling with 400 samples (1000 burn-in samples were discarded) for each set of priors. We use 400 samples so that the results can be compared to the rejection sampling results of Example 5.4. Recall that rejection sampling was very inefficient and it would take considerable effort to generate as many rejection samples as a large number of slice samples. For slice sampling, once again the

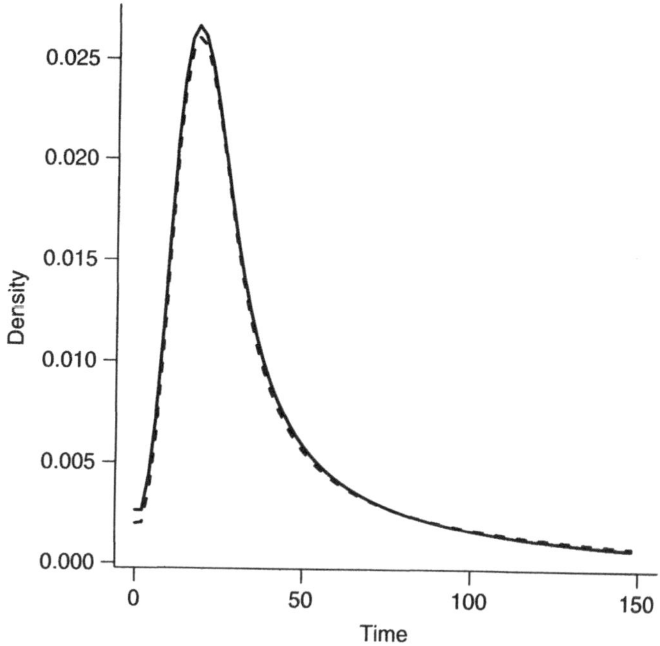

FIGURE 5.21 Bayes predictive density using slice sampling for Example 5.9. The solid line is SS I and the dashed line is SS II.

TABLE 5.2 Summary of Predictive Distribution Results for Examples 5.4 and 5.9, the Cancer Flowgraph with General Waiting Time Distributions

Percentile	Rejection Sampling	SS I	SS II
0.25	17.66	17.79	18.32
0.50	28.16	28.54	29.58
0.75	54.46	56.03	62.40
0.90	114.21	117.90	141.05
0.95	174.05	179.69	221.66
0.99	355.47	365.53	476.06
Mean	42.13	43.02	48.06
s.d.	59.92	60.29	81.44

doubling procedure was used with $w = 0.1$, $m = 100$, and $\alpha = 0.05$. Table 5.2 presents a comparison of summary statistics for the predictive distributions based on rejection sampling, used in Example 5.4, and on slice sampling. The distribution associated with the second prior (SS II) is heavier-tailed than that associated with the first prior (SS I). This slightly heavier tail for SS II has a noticeable effect on the mean, standard deviation, and upper quantiles of the predictive distribution

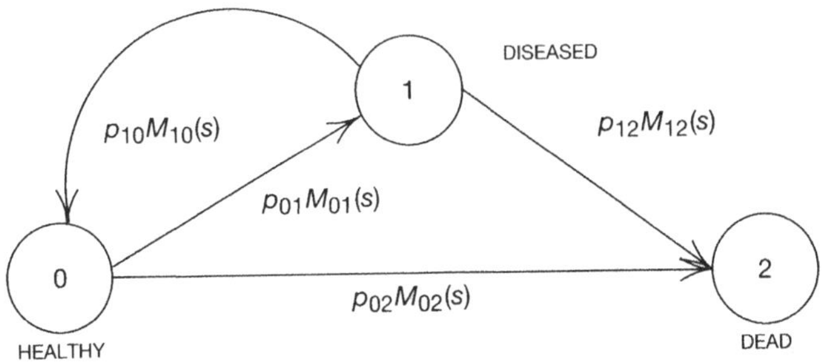

FIGURE 5.22 Flowgraph model for reversible illness–death multistate model.

and it is an interesting example of sensitivity to prior assumptions. This effect is not obvious on the density plot. □

Example 5.10: Reversible Illness–Death Model: Bayes Predictive Distribution. This model was presented in Figure 2.22 and discussed in Section 2.4.6. Figure 5.22 reproduces the flowgraph. For illustration, data on this flowgraph were simulated. The $0 \to 1$ waiting time is $IG(\lambda_1, \lambda_2)$ with mean λ_1/λ_2, the $0 \to 2$ and $1 \to 0$ waiting times are Gamma(α_1, β_1) (with mean α_1/β_1) and Gamma(α_2, β_2) (with mean α_2/β_2), and the $1 \to 2$ waiting time is $Exp(\mu)$ (with mean $1/\mu$). The corresponding MGFs for the parametric distributions are:

$$\begin{aligned}
M_{01}(s) &= \exp\{\lambda_1\lambda_2 - [\lambda_1^2(\lambda_2^2 - 2s)]^{1/2}\} && \text{for } s < \lambda_2^2/2 \\
M_{02}(s) &= \left(\frac{\beta_1}{\beta_1 - s}\right)^{\alpha_1} && \text{for } s < \beta_1 \\
M_{10}(s) &= \left(\frac{\beta_2}{\beta_2 - s}\right)^{\alpha_2} && \text{for } s < \beta_2 \\
M_{12}(s) &= \frac{\mu}{\mu - s} && \text{for } s < \mu.
\end{aligned} \tag{5.67}$$

Data were simulated for 50 people using this flowgraph with parameter values $p_{01} = 0.7$, $p_{10} = 0.4$, $\lambda_1 = 3$, $\lambda_2 = 5$, $\alpha_1 = 4$, $\beta_1 = 6$, $\alpha_2 = 2$, $\beta_2 = 3$, and $\mu = 0.2$. Data were randomly right-censored so that approximately 20% of the people were censored. For these data, all of the censored observations were in state 1. The data are summarized as follows:

Transition	$0 \to 1$	$0 \to 2$	$1 \to 2$	$1 \to 0$	Censored
No. Observations	54	24	15	28	11

Let $f_{gh}(\cdot)$ and $F_{gh}(\cdot)$ be the density and CDF, respectively, of the waiting time in state g until transition to state h. Let x_{igh} be the ith uncensored observation from state g to state h, and let x_{jg}^* be the jth censored observation where censoring occurred in state g. Let n_{gh} be the number of individual transitions from state g

to state h, and let n_g^* be the number of observations censored in state g. For this example, $n_0^* = 0$ and $n_1^* = 11$. Then the likelihood function is

$$\mathcal{L}(p_{01}, p_{10}, \lambda_1, \lambda_2, \alpha_1, \beta_1, \alpha_2, \beta_2, \mu|\mathcal{D})$$
$$= \prod_{i=1}^{n_{01}} p_{01} f_{01}(x_{i01}|\lambda_1, \lambda_2) \prod_{i=1}^{n_{12}} (1 - p_{10}) f_{12}(x_{i12}|\mu) \prod_{i=1}^{n_{10}} p_{10} f_{10}(x_{i10}|\alpha_2, \beta_2)$$
$$\times \prod_{i=1}^{n_{02}} (1 - p_{01}) f_{02}(x_{i02}|\alpha_1, \beta_1)$$

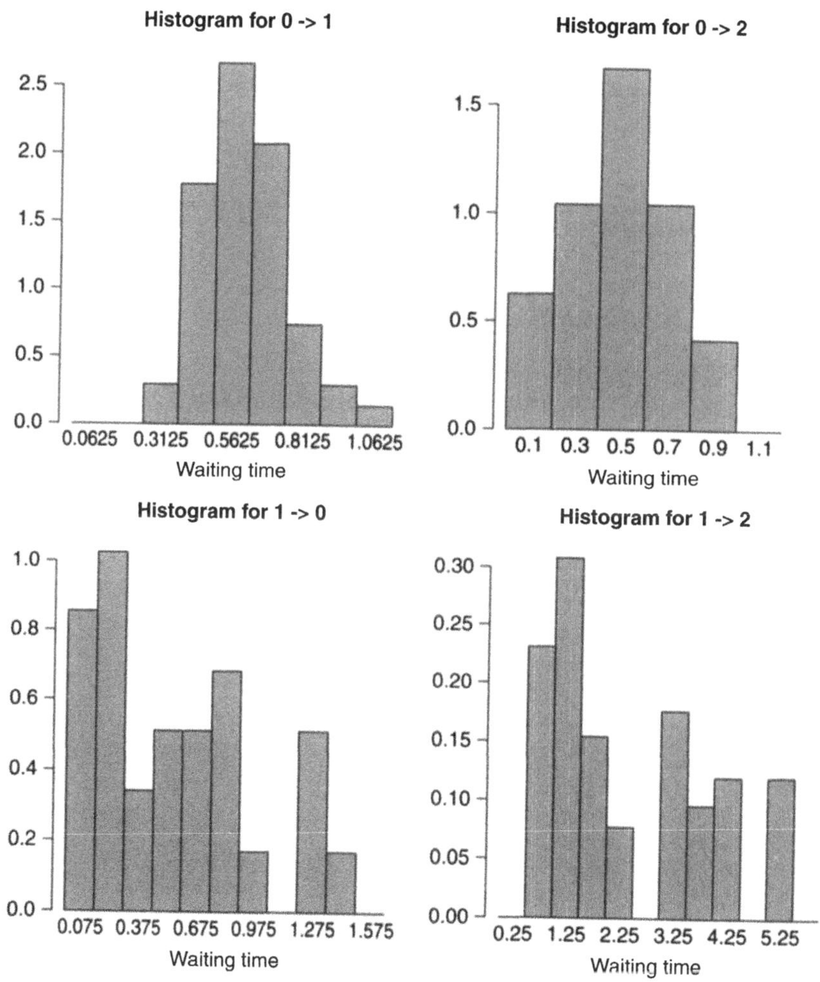

FIGURE 5.23 Censored data histograms for state-to-state waiting times in the reversible illness–death model of Figure 5.22.

$$\times \prod_{j=1}^{n_0^*} \{1 - [p_{01} F_{01}(x_{j0}^* | \lambda_1, \lambda_2) + (1 - p_{01}) F_{02}(x_{j0}^* | \alpha_1, \beta_1)]\}$$

$$\times \prod_{j=1}^{n_1^*} \{1 - [p_{10} F_{10}(x_{j1}^* | \alpha_2, \beta_2) + (1 - p_{10}) F_{12}(x_{j1}^* | \mu)]\}. \quad (5.68)$$

Figure 5.23 presents the censored data histograms for each state-to-state waiting time of the flowgraph in Figure 5.22. In this example, we excluded the overlay distributions to illustrate the histograms. If the data were not simulated, we would use the censored data histograms with the overlay distributions to suggest the parametric models, as in Example 5.8. The MGF of the overall survival time from $0 \rightarrow 2$ is found using the MAPLE code given in Figure 5.24.

We use noninformative priors on $\boldsymbol{\theta} = (p_{01}, p_{10}, \lambda_1, \lambda_2, \alpha_1, \beta_1, \alpha_2, \beta_2, \mu)$. A Unif(0, 1) prior is used for both p_{01} and p_{10}. Each of the remaining parameters has a Gamma(1/1000, 1/1000) prior. Slice sampling from the posterior was based

```
with(codegen,makeproc,C):

ms := proc(s)
       local w01,w10,w12,w02;

       # (l1,l2) are (λ1,λ2); mu is μ;
       # (a1, b1) are (α1,β1); (a2, b2) are (α2,β2)

       w01 := p01*exp(l1*l2-((l1^2)*((l2^2)-2*s))^(.5));
       w12 := (1-p10)*mu/(mu-s);
       w02 := (1-p01)*((b1/(b1-s))^a1);
       w10 := p10*((b2/(b2-s))^a2);

# The next line defines the MGF
       (w01*w12 + w02)/(1-w01*w10);
end proc:

# Define the CGF and its derivatives
k := makeproc(ln(ms(s)),[s]):
C(k,optimized,ansi,filename="c:/MSM/k.c"):

dk := makeproc(diff(k(s),s),[s]):
C(dk,optimized,ansi,filename="c:/MSM/dk.c"):

ddk := makeproc(diff(dk(s),s),[s]):
C(ddk,optimized,ansi,filename="c:/MSM/ddk.c"):
```

FIGURE 5.24 MAPLE code for Example 5.10.

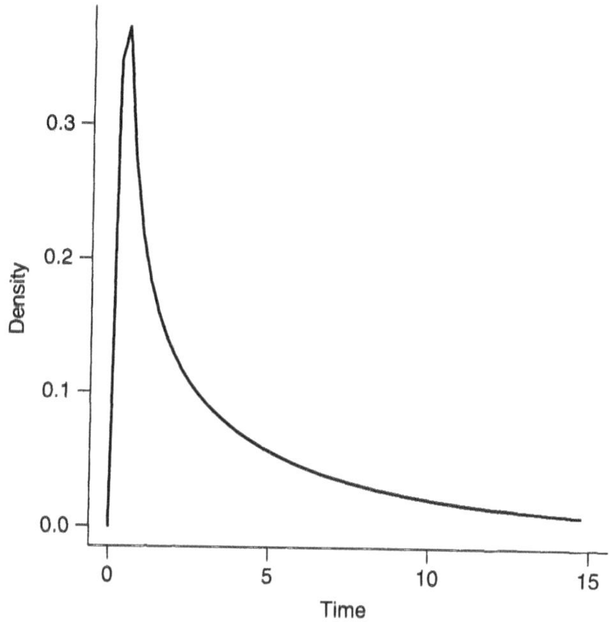

FIGURE 5.25 Bayes predictive density using slice sampling for the reversible illness–death model.

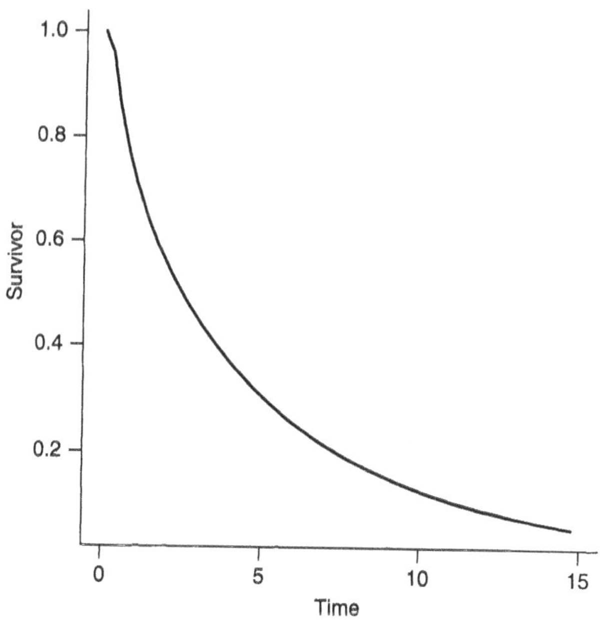

FIGURE 5.26 Bayes predictive survivor function using slice sampling for the reversible illness–death model.

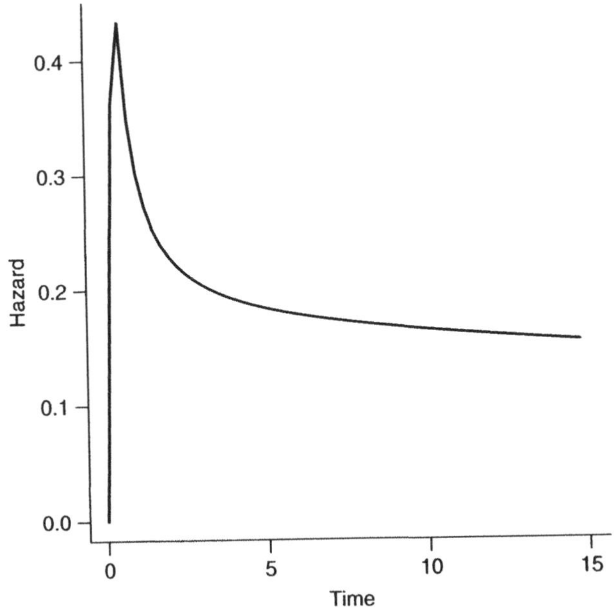

FIGURE 5.27 Bayes predictive hazard function using slice sampling for the reversible illness–death model.

on 2000 samples with an additional 1000 burn-in samples. The doubling procedure used an interval of size $w = 0.1$, $m = 100$ doublings, and uniform sampling rate $\alpha = 0.05$. The resulting Bayes predictive density, survivor, and hazard functions are presented in Figures 5.25, 5.26, and 5.27. Notice that the mixture of the inverse Gaussian and gamma distributions creates a very skewed overall waiting time distribution. The hazard function increases sharply and then drops off to approach its limit. In Chapter 6 we present the computational details for this example using the basic flowgraph programs. □

EXERCISES

5.1 Consider the cancer progression model presented in Example 5.2. Consider the parameters $(\lambda_1, \lambda_2, \lambda_3)$. Suggest conjugate priors for these parameters and rederive the posterior distributions.

5.2 Starting with the MGF, derive the formulas for the predictive mean and predictive variance given in (5.14) and (5.15).

CHAPTER 6

Computational Implementation of Flowgraph Models

In this chapter we describe an application of the computational programs for flowgraph analysis to the reversible illness–death model of Figure 5.22. The censored data histograms were given in Figure 5.23. In Section 6.1 we introduce the R program for computing censored data histograms in this example. Section 6.2 gives the modifications necessary to the MAPLE-generated code for the saddlepoint approximation program. In Section 6.3 we give the R program for Bayesian analysis using slice sampling, and in Section 6.4 we give the R code for a maximum likelihood analysis. In Section 6.5 we give the program for calculating the log-posterior for the example of Section 6.3, and Section 6.6 contains the code that is used to transform the parameter space for the optimization algorithm that estimates the posterior mode and MLEs. The programs presented here are the main programs that the user must modify for specific problems. They are presented in detail so that the user can make direct comparisons to the analytical work of Example 5.10. The remaining functions and routines are transparent to the user and do not require changes. These must be downloaded from the author's web site, as indicated in Section 1.4.

6.1 CODE FOR CENSORED DATA HISTOGRAMS

In this section we present the R program for constructing censored data histograms based on the Kaplan–Meier estimate of the survivor function. The program also computes rough estimates of candidate parametric models and overlays these estimates on a plot of the censored data histogram for each state-to-state waiting time. The main program is called `cdh.r` (`cdh` for "censored data histogram"). It calls R functions `kmhist.r` and `overlay.r`. The default histogram computed by the program is that of (4.14) and the default is set in `kmhist.r`. The function `kmhist.r` calls a C subroutine, `cdh.c`. These functions do not change with each

Flowgraph Models for Multistate Time-to-Event Data, By Aparna V. Huzurbazar
ISBN 0-471-26514-4

new analysis and the user should keep them in a root directory. For this illustration, the shared object directory is called `cdhist`. A subdirectory for each specific analysis is then created in this root directory. For illustration, we use data from the flowgraph of Figure 5.22. We create a subdirectory called `rid` (for "reversible illness death") within the `cdhist` directory. The root directory `cdhist` contains a `Makefile`. The user must modify the directory declaration line in this file to list the subdirectory of the current project. For our example, this line is "`DIR = rid`". The `rid` subdirectory contains the files `cdh.r` along with data files for each of the four state-to-state waiting times: `state01.txt`, `state02.txt`, `state10.txt`, and `state12.txt`. Each data file contains two columns of numbers. The first column consists of the actual waiting times, and the second column consists of a censoring indicator, which is 0 if the observation is censored and 1 if the event of interest occurred. The first row of each file has the column headings. In this case they are T for the waiting time and C for the censoring indicator.

In review, to use the programs, the user must create a directory in the `cdhist` directory for each new application, put the corresponding data in this directory, and make application-specific modifications to the `cdh.r` file. As mentioned previously, the end result from the program is a histogram for each state-to-state waiting time. The user has the option of overlaying rough estimates of four parametric models, individually or as a group, on each censored data histogram. The estimated densities are the exponential, gamma, Weibull, and inverse Gaussian. The user may modify the program to incorporate additional parametric families of overlay distributions. The estimation method is maximum likelihood and censored observations are included in the procedure. The overlaid distributions are intended as a guide toward choosing a parametric model for each branch of the flowgraph. The user may omit some or all of the estimated densities or add more overlay densities in `cdh.r` as desired. The main function `cdh.r` is discussed here with annotated R code. All of the functions required, including `kmhist.r` and `cdh.c`, can be downloaded from the author's web site.

```
#
# Source R functions
#

source("../kmhist.r"); source("../overlay.r");
source("../plots.r");


#
# Read the data
#

t01 <- read.table("state01.txt", header=TRUE)
t02 <- read.table("state02.txt", header=TRUE)
t10 <- read.table("state10.txt", header=TRUE)
t12 <- read.table("state12.txt", header=TRUE)
```

```
#
# Calculate Censored Data Histogram
#

# Create a vector of lower limits
ll <- numeric(4)

# Create a vector of upper limits
ul <- c(1.125, 1.4, 1.65, 6)

# Create a vector of bin widths.  These can be modified on
# subsequent runs.
bin <- c(0.125, 0.2, 0.15, 0.5)

# kmh is the list returned by kmhist.r
#
# ll = final lower limit
# old.ll = original lower limit
# mod.ll = indicator of modification to lower limit
# ul = final upper limit
# h = histogram heights

kmh <- kmhist(ll=ll[1], ul=ul[1], bin=bin[1], data=t01)
ll01 <- kmh$ll; old.ll01 <- kmh$old.ll; mod.ll01 <- kmh$mod.ll
ul01 <- kmh$ul; h01 <- kmh$h

kmh <- kmhist(ll=ll[2], ul=ul[2], bin=bin[2], data=t02)
ll02 <- kmh$ll; old.ll02 <- kmh$old.ll; mod.ll02 <- kmh$mod.ll
ul02 <- kmh$ul; h02 <- kmh$h

kmh <- kmhist(ll=ll[3], ul=ul[3], bin=bin[3], data=t10)
ll10 <- kmh$ll; old.ll10 <- kmh$old.ll; mod.ll10 <- kmh$mod.ll
ul10 <- kmh$ul; h10 <- kmh$h

kmh <- kmhist(ll=ll[4], ul=ul[4], bin=bin[4], data=t12)
ll12 <- kmh$ll; old.ll12 <- kmh$old.ll; mod.ll12 <- kmh$mod.ll
ul12 <- kmh$ul; h12 <- kmh$h

ll <- c(ll01, ll02, ll10, ll12)
ul <- c(ul01, ul02, ul10, ul12)
old.ll <- c(old.ll01, old.ll02, old.ll10, old.ll12)
mod.ll <- c(mod.ll01, mod.ll02, mod.ll10, mod.ll12)

#
# Prepare overlay distributions
#

eps <- 0.01
```

```
t <- list()
for (i in 1:4) {
   t[[i]] <- seq(from=ll[i]+eps, to=ul[i]+eps, by=0.1)
}

T <- t01$T[t01$C!=0]; C <- t01$T[!t01$C]
ip <- c(0, length(T), length(T)+length(C), 0); dp <- c(T, C)
dist01 <- overlay(ip, dp, t[[1]], dist="all")

T <- t02$T[t02$C!=0]; C <- t02$T[!t02$C]
ip <- c(0, length(T), length(T)+length(C), 0); dp <- c(T, C)
dist02 <- overlay(ip, dp, t[[2]], dist="all")

T <- t10$T[t10$C!=0]; C <- t10$T[!t10$C]
ip <- c(0, length(T), length(T), 0); dp <- c(T, C)
dist10 <- overlay(ip, dp, t[[3]], dist="all")

T <- t12$T[t12$C!=0]; C <- t12$T[!t12$C]
ip <- c(0, length(T), length(T)+length(C), 0); dp <- c(T, C)
dist12 <- overlay(ip, dp, t[[4]], dist="all")


#
# Plot Censored Data Histogram
#

h <- list(h01, h02, h10, h12)
trans <- c("0 -> 1", "0 -> 2", "1 -> 0", "1 -> 2")

postscript("rid_hist.ps", horizontal=F)
par(mfrow=c(2,2), pty="s")
for (j in 1:4) {
   if (mod.ll[j]) {
      breaks <- c(ll[j], seq(from=old.ll[j], to=ul[j],
                  by=bin[j]))
      w <- c(old.ll[j], rep(bin[j], length(h[[j]]) - 1))
   }
   else {
      breaks <- seq(from=ll[j], to=ul[j], by=bin[j])
      w <- rep(bin[j], length(h[[j]]))
   }
   mp <- sapply(1:length(h[[j]]),
                function (i) (breaks[i]+breaks[i+1])/2)
   barplot(h[[j]], width=w, space=0, names.arg=mp, col="grey",
           xlab="Waiting Time", main=paste("Histogram for",
           trans[j]), xlim=c(ll[j],ul[j]), ylim=c(0,max(h[[j]])),
           axisnames=TRUE)
}
graphics.off()
```

6.2 SADDLEPOINT APPROXIMATION CODE

Once parametric models for the branch transitions have been selected, the appropriate MGF can be computed using MAPLE. The C functions that are needed from MAPLE for the remaining analysis are the cumulant generating function, $K(s)$, and its first and second derivatives. These files must be called `k.c`, `dk.c`, and `ddk.c`. The `k.c` file for this example is given in Figure 6.1. The user modifications to the MAPLE-generated code are as follows:

1. Add "`double *th`" to the function call list "`double k(double s, double *th)`".
2. Assign each symbolic parameter in the model a corresponding $\boldsymbol{\theta}$ element. In this example, the lines typed in by the user are the next three lines, beginning with "`double p01=th[0], p10=th[1], l1=th[2]`".
3. Make the same modifications to `dk.c` and `ddk.c`.

```
#include <math.h>
double k(double s, double *th)
{
   double p01=th[0], p10=th[1], l1=th[2], l2=th[3], a1=th[4],
          b1=th[5], a2=th[6], b2=th[7], mu=th[8];
   double t10;
   double t2;
   double t21;
   double t27;
   double t3;
   double t7;
   double t9;
   {
      t2 = l1*l1;
      t3 = l2*l2;
      t7 = pow(t2*(t3-2.0*s),0.5);
      t9 = exp(l1*l2-t7);
      t10 = p01*t9;
      t21 = pow(b1/(b1-s),1.0*a1);
      t27 = pow(b2/(b2-s),1.0*a2);
      return(log((t10*(1.0-p10)*mu/(mu-s)+
            (1.0-p01)*t21)/(1.0-t10*p10*t27)));
   }
}
```

FIGURE 6.1 C function `k.c` for Figure 5.22.

6.3 CODE FOR BAYESIAN ANALYSIS

The R program, called `saddlept_ss.r`, to compute the Bayes predictive density, predictive CDF, predictive survivor or reliability function, and predictive hazard function using slice sampling is given below. By default, the end result from the program is a file called `summary_ss` that contains the following:

- Posterior mode and corresponding maximum log-posterior distribution value
- Number of burn-in and posterior samples used that can be specified by the user
- Predictive distribution mean and standard deviation
- Summary table of quantile values that can be specified by the user

Postscript plots of the predictive density, survivor, and hazard functions are written to files called `pd.ps`, `sf.ps`, and `hf.ps`. Histograms of posterior samples with a vertical line indicating the posterior mode are in `theta.ps` and line plots of the slice samples are in `conv.ps`. The user can modify the code for other output as desired. The setup for the program is as follows:

1. Create a directory for the main programs. Let us call it `msm` for "multistate models."
2. Put the required general-purpose R and C programs in `msm`.
3. Create a subdirectory for the current flowgraph project. Let us call it `fg` for flowgraph.
4. Modify the `Makefile` to set "`DIR = fg`".
5. Put programs specific to the current project in this directory. These are `saddlept_ss.r`, `post.c`, `transform.c`, `k.c`, `dk.c`, `ddk.c`, and the data files.

The two files that require effort on the part of the user are `post.c` and `transform.c`. The `post.c` file computes the user-specified log-posterior distribution. The form of the likelihood and the prior must be entered by the user. The `transform.c` file transforms the parameters for the optimization routine that finds the posterior mode. These files, as written for this example, are given in Sections 6.5 and 6.6.

The program `saddlept_ss.r` is the driver file for slice sampling. Modifications to portions of this program must be made for each new analysis.

```
#
# Source R functions
#

source("../transform.r")
source("../optimize.r"); source("../slice.r");
```

```
source("../saddle.r"); source("../sumstats.r");
source("../plots.r")

#
# Set-up
#

# Output summary file
outsum <- "summary_ss"

# Specify number of parameters
ntheta <- 9

# Specify number of slice samples
N <- 2000

# Number of burn-in samples
nburnin <- 1000

# Percentage of time slice samples are generated
# from a uniform dist'n over the slice vs. overrelaxation
al <- .05

# Initial lower and upper limits for waiting time
# dist'n of interest
m <- 1e-6; M <- 1e6;

# Upper limit of t for computing f(t), S(t), h(t) for plots
xlim.up <- 15

# Grid of t values for evaluation
t <- seq(m, xlim.up, 0.25)

#
# R input data. In this example, the censored and uncensored
# observations are in separate files. The uncensored
# observations are read first.
#

t01 <- read.table("state01.txt")
t02 <- read.table("state02.txt")
t10 <- read.table("state10.txt")
t12 <- read.table("state12.txt")

# Read the censored observations.
# Observations censored in state 0 are in "state0_cen.txt".
# Observations censored in state 1 are in "state1_cen.txt".

c0 <- read.table("state0_cen.txt")
```

```
c1 <- read.table("state1_cen.txt")

# Integer vector parameters containing the length of theta and
# the cumulative length of each vector of observed transition
# times and censored observations for passage to other routines

ip <- numeric(); ip[1] <- ntheta; ip[2] <- length(t01);
ip[3] <- ip[2]+length(t02); ip[4] <- ip[3]+length(t10);
ip[5] <- ip[4]+length(t12); ip[6] <- ip[5]+length(c0);
ip[7] <- ip[6]+length(c1)

#
# Double precision vector of all data passed to other routines
#

dp <- c(t01, t02, t10, t12, c0, c1)

#
# Specify priors.
#

# Set lower bound of each theta

lb <- numeric(ntheta)

# Set upper bound on each probability to be 1

ub <- c(1, 1)

# prior.p contains all the information necessary for computing
# the prior distribution. The first 2 vectors contain any
# finite lower bounds and upper bounds on the parameters. The
# remaining 7 vectors pass the values of the hyperparameters
# for the Gamma priors. These priors are Gamma(.001,.001) on
# all the waiting time distribution parameters.

prior.p <- c(lb, ub, rep(c(0.001,0.001), 7))

# From here on, there are no user changes required.
# The user may change any options to any of the routines called.

dp <- c(dp, prior.p)

#
# Find posterior mode (MAP)
#

MAP <- optimize(ntheta, ip, dp, maxiter=10000, outfile=outsum)
```

```
save.image()

#
# Slice sampling from posterior
#

Theta.ss <- slice(N, ntheta, ip, dp, start=MAP$theta, m=100,
                  w=0.1, alpha=al, nburn=nburnin, meth=2,
                  outfile=outsum)
Theta.ss <- Theta.ss[[1]]

write.table(round(Theta.ss, 5), file="Theta.ss", quote=FALSE,
            row.names=FALSE, col.names=FALSE)

if (nburnin > 0) { Theta.ss.burnin <- Theta.ss[1:nburnin,] }
Theta.ss <- Theta.ss[(nburnin+1):(nburnin+N),]

save.image()

#
# Set Theta for Saddlepoint Approximation and Plots
#

Theta <- Theta.ss

#
# Saddlepoint approximation
#

SP <- saddle(N, Theta, t, m, M)
pd <- SP$pd; nc <- SP$nc;
sp.lb <- SP$sp.lb; sp.ub <- SP$sp.ub;

save.image()

#
# Summary statistics
#

stats <- sumstats(N, Theta, t, nc, sp.lb, sp.ub, spd=pd,
                  outfile=outsum)
sf <- stats$sf; hf <- stats$hf; q <- stats$q;
xlim.up <- min(c(xlim.up, stats$xlim.up, q[length(q)]))

save.image()

#
# Plots
#
```

```
ps.options(width=6, height=6)
plots(t, pd, xlab="t", ylab="density", ulim=xlim.up,
      lwd=2, plot="pd")
plots(t, sf=sf, xlab="t", ylab="survivor", ulim=xlim.up,
      lwd=2, plot="sf")
plots(t, hf=hf, xlab="t", ylab="hazard", ulim=xlim.up,
      lwd=2, plot="hf")
ps.options(width=0, height=0)
plots(theta=Theta, plot.mode=MAP$theta, plot="theta")
plots(theta=Theta, theta.burnin=Theta.ss.burnin,
      plot.mode=MAP$theta,plot="conv")
```

6.4 CODE FOR MAXIMUM LIKELIHOOD ANALYSIS

The R program, called `saddlept_mle.r`, to compute the maximum likelihood estimated density, CDF, survivor or reliability function, and hazard function is given below. The end result from the program is a file called `summary_mle` that contains the following:

- MLEs
- Maximized log-likelihood value
- Maximum likelihood estimated mean and standard deviation of the waiting time distribution of interest
- Summary table of quantile values that can be specified by the user

Postscript plots of the estimated density, survivor, and hazard functions are written to files called `pd.ps`, `sf.ps`, and `hf.ps`. Again, the user can modify the code to obtain other output as desired. The setup for the program is very similar to the process described in Section 6.3. Required modifications are given in the code below. This program is a special case of the `saddlept_ss.r` program.

```
#
# Source R functions
#

source("../transform.r"); source("../optimize.r");
source("../saddle.r"); source("../sumstats.r");
source("../plots.r")

#
# Set-up
#

# Output summary file
outsum <- "summary_mle"
```

```
# Specify number of parameters
ntheta <- 9

# Set N=1 as one value of theta, the MLE, is obtained
N <- 1

# Initial lower and upper limits for waiting time
# distribution of interest
m <- 1e-6; M <- 1e6;

# Upper limit of t for computing f(t), S(t), h(t) for plots
xlim.up <- 15

# Grid of t values for evaluation
t <- seq(m, xlim.up, 0.25)

#
# R input data. In this example, the censored and uncensored
# observations are in separate files. The uncensored
# observations are read first.
#

t01 <- read.table("state01.txt")
t02 <- read.table("state02.txt")
t10 <- read.table("state10.txt")
t12 <- read.table("state12.txt")

# Read the censored observations.
# Observations censored in state 0 are in "state0_cen.txt".
# Observations censored in state 1 are in "state1_cen.txt".

c0 <- read.table("state0_cen.txt")
c1 <- read.table("state1_cen.txt")

# Integer vector containing the length of theta and
# the cumulative length of each vector of observed transition
# times and censored observations for passage to other routines

ip <- numeric(); ip[1] <- ntheta; ip[2] <- length(t01);
ip[3] <- ip[2]+length(t02); ip[4] <- ip[3]+length(t10);
ip[5] <- ip[4]+length(t12); ip[6] <- ip[5]+length(c0);
ip[7] <- ip[6]+length(c1)

#
# Double precision vector of all data passed to other routines
#

dp <- c(t01, t02, t10, t12, c0, c1)
```

```
# Create upper and lower bounds on probabilities
# for optimization. The program uses prior.p as a
# dummy to pass these bounds

lb <- c(0,0)
ub <- c(1,1)
prior.p <- c(lb, ub)

# From here on, there are no user changes required.
# The user may change any options to any of the routines called.

dp <- c(dp, prior.p)

#
# Find MLE
#

MLE <- optimize(ntheta, ip, dp, maxiter=10000, outfile=outsum)

save.image()

#
# Set Theta for Saddlepoint Approximation and Plots
#

Theta <- MLE$theta

#
# Saddlepoint approximation
#

SP <- saddle(N, Theta, t, m, M)
pd <- SP$pd; nc <- SP$nc;
sp.lb <- SP$sp.lb; sp.ub <- SP$sp.ub;

save.image()

#
# Summary statistics
#

stats <- sumstats(N, Theta, t, nc, sp.lb, sp.ub, spd=pd,
                  outfile=outsum)
sf <- stats$sf; hf <- stats$hf; q <- stats$q;
xlim.up <- min(c(xlim.up, stats$xlim.up, q[length(q)]))

save.image()

#
```

```
# Plots
#

ps.options(width=6, height=6)
plots(t, pd, xlab="t", ylab="density", ulim=xlim.up,
      lwd=2, plot="pd")
plots(t, sf=sf, xlab="t", ylab="survival", ulim=xlim.up,
      lwd=2, plot="sf")
plots(t, hf=hf, xlab="t", ylab="hazard", ulim=xlim.up,
      lwd=2, plot="hf")
```

6.5 CODE FOR post.c

The code below computes the log-posterior distribution for the example of Section 6.3. Compare the instructions to the likehood function given in (5.68) and the priors selected for this example. Note that this function is easily modified to calculate the log-likelihood in the example of Section 6.3 by deleting the components that correspond to the prior computations.

```
#include <math.h>
#include <stdlib.h>
#include <gsl/gsl_errno.h>
#include "../log_density.h"

void transform( double *, int *, double * );

double post( double *theta, int *ip, double *dp )
{

   int i, err, pp=ip[7]+ip[1]+2;
   double pdf, sf1, sf2, tmp, *ttheta, *p=theta;

   if( ip[0] ) {

/* Optimization */

      ttheta = (double *) malloc( ip[1] * sizeof( double ) );
      for( i=0; i<ip[1]; i++ ) {
         ttheta[i] = theta[i];
      }
      transform( ttheta, ip, dp );
      p = ttheta;
   }
   else {

/* Support for sampling */

      for( i=0; i<ip[1]; i++ ) {
```

```
        if( p[i] <= dp[ip[7]+i] ) {
           return( HUGE_VAL );
        }
      }
      if( p[0] >= dp[ip[7]+ip[1]] || p[1] >= dp[ip[7]+ip[1]+1] ){
         return( HUGE_VAL );
      }
   }

/* Calculate likelihood */

   tmp = 0.0;
   for( i=0; i<ip[2]; i++ ) {
      pdf = log_invgauss_pdf( dp[i], p[2], p[3] );
      tmp += pdf + log(p[0]);
   }
   for( i=ip[2]; i<ip[3]; i++ ) {
      pdf = log_gamma_pdf( dp[i], p[4], p[5] );
      tmp += pdf + log(1.0 - p[0]);
   }
   for( i=ip[3]; i<ip[4]; i++ ) {
      pdf = log_gamma_pdf( dp[i], p[6], p[7] );
      tmp += pdf + log(p[1]);
   }
   for( i=ip[4]; i<ip[5]; i++ ) {
      pdf = log_exponential_pdf( dp[i], p[8] );
      tmp += pdf + log(1.0 - p[1]);
   }
   for( i=ip[5]; i<ip[6]; i++ ) {
      invgauss_sf( dp[i], p[2], p[3], &sf1 );
      sf1 *= p[0];
      gamma_sf( dp[i], p[4], p[5], &sf2 );
      tmp += log( sf1 + (1.0-p[0])*sf2 );
   }
   for( i=ip[6]; i<ip[7]; i++ ) {
      gamma_sf( dp[i], p[6], p[7], &sf1 );
      sf1 *= p[1];
      sf2 = log_exponential_sf( dp[i], p[8] );
      sf2 = (1-p[1])*exp(sf2);
      tmp += log( sf1 + sf2 );
   }

/* Calculate prior */

   for( i=2; i<ip[1]; i++ ) {
      pdf = log_gamma_pdf( p[i], dp[pp+2*i-4], dp[pp+2*i-3] );
      tmp += pdf;
   }
```

```
    if( ip[0] ) {
       free( ttheta );
    }

    err = gsl_finite( tmp );
    if( !err ) {
       return( HUGE_VAL );
    }

    return( -tmp );
}
```

6.6 CODE FOR transform.c

The code below contains `transform.c` for the examples of Sections 6.3 and 6.4. Modifying this code for a new analysis is straightforward. The optimization program currently implemented requires each parameter being optimized to exist on the real line. However, most of the parameters encountered in practice have a restricted range, such as $(0, \infty)$ or (a, b). This code makes the necessary transformation to each parameter prior to evaluating the log-posterior distribution or log-likelihood function. For any parameter η such that $0 < \eta < +\infty$, we transform η to e^{η} in `transform.c`. If η is bounded so that $a < \eta < b$, we transform η as follows:

$$\eta \leftarrow \frac{a + be^{\eta}}{1 + e^{\eta}}.$$

The bounded parameters in this example are p_{01} and p_{10}. They are the first two parameters in the parameter list (see Figure 6.1), designated here as `x[0]` and `x[1]`. The remaining parameters are transformed in the for loop to the positive real axis.

```
#include <math.h>

void transform( double *x, int *ip, double *dp )
{
   int i, ld=ip[7];

/* Transformation to parameter support */

   x[0] = (dp[ld]+dp[ld+ip[1]]*exp(x[0])) / (1.0+exp(x[0]));
   x[1] = (dp[ld+1]+dp[ld+ip[1]+1]*exp(x[1])) / (1.0+exp(x[1]));
   for( i=2; i<ip[1]; i++ ) {
      x[i] = exp(x[i]);
   }
}
```

EXERCISES

6.1 Refer to the data on the strength of weathered braided cord given in Exercise 4.3. Construct a censored data histogram of these data using the R function presented in this chapter. Which parametric models appear to fit these data, if any?

6.2 Refer to the breast cancer data of Exercise 4.4. Construct a censored data histogram of these data using the R functions presented in this chapter. Which parametric models appear to fit these data, if any?

6.3 Download the program code for flowgraph analysis and practice. Create a simple flowgraph for application.

CHAPTER 7

Semi-Markov Processes

Semi-Markov models play an important role in the analysis of time-to-event data. However, in practice, data analysis for semi-Markov processes can be quite difficult and many simplifying assumptions are made. Markovian multistate models are popular for event history analysis and repeated events analysis for survival data. In a Markov process, the past and the future of the process are conditionally independent given the present. In a semi-Markov process, conditional on the sequence of states, the individual waiting times are independent (cf. Karlin and Taylor, 1975, pp. 207–208). This makes semi-Markov processes amenable to data analysis for multistate models as data are collected on the sequence of states and the individual waiting times. Flowgraph models are multistate models that inherently model semi-Markov processes and assist in the analysis of data from such processes. Although there is a great deal of literature on data analysis for Markov processes, there is little literature on data analysis for semi-Markov processes. Although detailed attention is paid to Markov processes in texts on stochastic processes, semi-Markov processes are treated as an advanced topic focusing mainly on definitions and limit theorems. This is primarily because data analysis for semi-Markov processes is quite difficult. Flowgraphs circumvent some of these difficulties by working in the MGF domain. There is a vast literature on stochastic processes in which Markov processes are discussed and some basics of semi-Markov processes considered. At the introductory level, we suggest Grimmett and Stirzaker (1992) or Taylor and Karlin (1994). Andersen et al. (1993), Bartlett (1955), Cox and Miller (1977), Doob (1953), Karlin and Taylor (1975, 1981), Prabhu (1965), and Resnick (1992) are more advanced. For application to population models and demography, see Biswas (1988). Limnios and Oprisan (2001) is a recent book dedicated to semi-Markov processes. The most common applications of Markov processes are the pure birth, pure death, and birth and death process. In Section 7.1 we review the basics of these and discuss the flowgraph analysis of such processes. In Section 7.2 we present an application of a Markov model for HIV. In Section 7.3 we discuss flowgraph models for semi-Markov processes. Applications to the progression of diabetic

Flowgraph Models for Multistate Time-to-Event Data, By Aparna V. Huzurbazar
ISBN 0-471-26514-4

retinopathy and a construction engineering project are presented. In Section 7.4 we present phase type distributions. Applications of flowgraphs with multiple exits and examples of arbitrary starting states are given. A closely related topic is that of Markovian and non-Markovian queuing systems, and we defer the reliability applications of flowgraph models to Chapter 9. In addition, the processes modeled in all of the examples discussed in Chapter 5 fall under the category of Markov or semi-Markov.

Before demonstrating the use of flowgraph models for Markov and semi-Markov processes, we review the definitions of these processes. Suppose that $\mathcal{S} = \{1, 2, 3, \ldots, m\}$ is a finite state space of a multistate process, and suppose that $h, j \in \mathcal{S}$. Denote the transition intensity (cf. Andersen et al., 1993) of the $h \to j$ transition by $\alpha_{hj}(\cdot)$. Statistical models are usually obtained by specifying a class of intensities $\alpha(t)$ for each individual or component.

Definition 7.0.1. Let X_n denote the state of the process after n transitions, and let T_n denote the time of transition to X_n. Then the *Markov property* states that

$$\begin{aligned} P[X_{n+1} &= j, T_{n+1} \in [t, t+\delta t) | \{(X_k, T_k)\}_{k=0}^{n-1}, (X_n = h, T_n), T_{n+1} \geq t] \\ &= P[X_{n+1} = j, T_{n+1} \in [t, t+\delta t) | X_n = h, T_{n+1} \geq t] \\ &= \alpha_{hj}(t)\,\delta t \qquad \text{for} \quad h, j \in \mathcal{S}. \end{aligned} \tag{7.1}$$

For a Markov process, the intensity function $\alpha_{hj}(t)$ is the hazard function.

The semi-Markov model incorporates the duration of time spent in the present state. The semi-Markov property can be written as follows.

Definition 7.0.2. Let X_n denote the state of the process after n transitions, and let T_n denote the time of transition to X_n. Then the *semi-Markov property* states that

$$\begin{aligned} P[X_{n+1} &= j, T_{n+1} \in [t, t+\delta t) | \{(X_k, T_k)\}_{k=0}^{n-1}, (X_n = h, T_n), T_{n+1} \geq t] \\ &= P[X_{n+1} = j, T_{n+1} \in [t, t+\delta t) | (X_n = h, T_n), T_{n+1} \geq t] \\ &= \alpha_{hj}(t - T_n)\,\delta t \qquad \text{for } h, j \in \mathcal{S}. \end{aligned} \tag{7.2}$$

The Markov property specifies that given that a process is in some state h, the transition time to a future state does not depend on the past history of the process. Put another way, the past and future of the process are conditionally independent given the present. In a semi-Markov process, the transition time to a future state depends on the duration of time spent in the present state. In practice, given that a Markov process is in some state h, the distribution of the transition time to state j depends only on j and on t, the time since origin. At the initial state, the transition time is a minimum of the waiting time distributions corresponding to all possible transitions, hence exponential waiting time distributions are often used for analytical convenience, due to their tractability in computing this minimum. Once beyond the initial state, the memoryless property of the exponential distribution enables waiting time distributions for all available transitions to be determined in closed form; therefore, again it is used for computational convenience. In contrast, flowgraph methodology for semi-Markov multistate models allows the use of a wide variety of waiting time distributions to model the data realistically. The data analysis proceeds by conditioning on the observed sequence of states for each individual. For a semi-Markov process, conditional on the sequence of states, the individual waiting times are independent. It is therefore possible to specify a tractable statistical model for the data that incorporates individual branch assessments of appropriate waiting time distributions and that justifies the application of flowgraph algebra for computing waiting times of interest.

An example of flowgraph analysis for a Markov process is the cancer progression model of Example 5.2 (Figure 5.3). In this model, the transition time from state 0 is the minimum of two independent exponential waiting times for the $0 \to 1$ and $0 \to 2$ transitions, and hence is also exponential. In Example 5.4, this model, shown in the flowgraph of Figure 5.8, is extended to include nonexponential waiting times, as it is assumed that transition times depend only on the duration in the current state. The same is true for the bearing and seal assembly model of Figure 5.5 discussed in Example 5.3.

7.1 BIRTH AND DEATH PROCESSES

Birth and death processes are one area of stochastic processes that have found rich applications in both medical and engineering settings. The traditional birth and death processes are by definition Markovian, as they necessarily assume Poisson arrivals, resulting in exponential interarrival times. In this section we review the basics of pure birth, pure death, and birth and death processes and present applications of flowgraph models for these processes primarily in the medical setting. Flowgraphs provide a general methodology for these problems and are quite easy to implement. The flowgraph analysis allows for different waiting time distributions and reduces to the Markovian analysis should the Markov assumptions hold.

Birth and death processes are examples of continuous-time Markov chains. Let $\{X(t)\}$ be a Markov process with stationary transition probabilities for transition from state i to state j given by

$$P[X(t+u) = j | X(u) = i] \qquad \text{for } i, j = 0, 1, 2, \ldots; t \geq 0.$$

Stationary transition probabilities do not depend on time u. This does not imply that the process itself is stationary.

Definition 7.1.1. A pure *birth process* is a Markov process that satisfies the following conditions for h small and $\lambda_k > 0$:

1. $X(0) = 0$
2. $P[X(t+h) - X(t) = 1 | X(t) = k] = \lambda_k h + o(h)$

3. $P[X(t+h) - X(t) = 0 | X(t) = k] = 1 - \lambda_k h + o(h)$
4. $P[X(t+h) - X(t) < 0 | X(t) = k] = 0$

$X(t)$ represents the number of "births" in the time interval $(0, t]$. The word *births* is used quite generally. Historically, such a process was used to model the population growth of certain organisms under steady-state conditions such as no mortality or immigration. In analysis of stochastic processes, the standard approach is to solve conditions 2 and 3 as differential equations for h small. The first condition is a boundary condition and the last condition merely ensures that there are no deaths (i.e., the process is continually growing). This is why the process is also sometimes referred to as a *right-shift process*. Letting $P_n(t) = P[X(t) = n]$, the differential equations that must be solved are

$$\begin{aligned} P_0'(t) &= -\lambda_0 P_0(t) \\ P_n'(t) &= -\lambda_n P_n(t) + \lambda_{n-1} P_{n-1}(t) \qquad \text{for } n \geq 1. \end{aligned} \tag{7.3}$$

The boundary conditions are $P_0(0) = 1$ and $P_n(0) = 0$ for $n > 0$. The solution to (7.3) is explicit:

$$\begin{aligned} P_0(t) &= \exp(-\lambda_0 t) \\ P_1(t) &= \lambda_0 \left[\frac{1}{\lambda_1 - \lambda_0} \exp(-\lambda_0 t) + \frac{1}{\lambda_0 - \lambda_1} \exp(-\lambda_1 t) \right], \ldots \\ P_n(t) &= \lambda_0 \lambda_1 \cdots \lambda_{n-1} [B_{0n} \exp(-\lambda_0 t) + \cdots + B_{nn} \exp(-\lambda_n t)], \end{aligned} \tag{7.4}$$

where

$$\begin{aligned} B_{0n} &= \frac{1}{(\lambda_1 - \lambda_0)(\lambda_2 - \lambda_0) \cdots (\lambda_n - \lambda_0)} \\ B_{kn} &= \frac{1}{(\lambda_0 - \lambda_k) \cdots (\lambda_{k-1} - \lambda_k)(\lambda_{k+1} - \lambda_k) \cdots (\lambda_n - \lambda_k)} \qquad \text{for } 0 < k < n \\ B_{nn} &= \frac{1}{(\lambda_0 - \lambda_n) \cdots (\lambda_{n-1} - \lambda_n)}. \end{aligned}$$

The parameters $\lambda_0, \lambda_1, \ldots, \lambda_n$ are called the *birth parameters* and this solution assumes that no two of the birth parameters are equal. This process accounts for censoring automatically. By definition, the transitions of the process must be sequential. For example, suppose that we observe the process in state 0 and then in state 2. This model assumes that the process must have transited through state 1 before reaching state 2, and the probability of this transition by time t is the two-step probability, $P_2(t)$, given in (7.4). *In terms of a flowgraph model, this pure birth process is a series model with exponential waiting times.* The flowgraph does not require that the $\{\lambda_i\}$ be distinct.

A pure death process is the complement of the pure birth process. In this case, the process begins in some state N, usually a population size, and then moves successively through states $N-1, N-2, \ldots, 2, 1, 0$ until extinction. The transition probabilities are derived analogously to those of the pure birth process. The death parameters $\mu_1, \mu_2, \ldots, \mu_N$ take the place of the birth parameters. The transition probabilities, $P_n(t) = P[X(t) = n|X(0) = N]$, are

$$\begin{aligned} P_N(t) &= \exp(-\mu_N t) \\ P_n(t) &= \mu_{n+1}\mu_{n+2}\cdots\mu_N[A_{nn}\exp(-\mu_n t) + \cdots + A_{Nn}\exp(-\mu_N t)] \end{aligned} \tag{7.5}$$

for $n = N-1, \ldots, 1, 0$, where

$$A_{Nn} = \frac{1}{(\mu_{N-1}-\mu_N)\cdots(\mu_n-\mu_N)}$$

$$A_{kn} = \frac{1}{(\mu_N-\mu_k)\cdots(\mu_{k+1}-\mu_k)(\mu_{k-1}-\mu_k)\cdots(\mu_n-\mu_k)} \qquad \text{for } n < k < N$$

$$A_{nn} = \frac{1}{(\mu_N-\mu_n)\cdots(\mu_{n+1}-\mu_n)}.$$

Flowgraph analysis for a pure death process is a series model with exponential waiting times and states $N-1, N-2, \ldots, 2, 1, 0$.

A birth and death process is a combination of these two processes. The process $X(t)$ is allowed to both increase and decrease in size. Let the stationary probabilities be

$$P_{ij}(h) = P[X(t+h) = j|X(t) = i] \qquad \text{for } t > 0. \tag{7.6}$$

The following assumptions apply:

1. $X(0) = 0$.
2. $P_{i,i+1}(h) = \lambda_i h + o(h)$ for $i \geq 0$.
3. $P_{i,i-1}(h) = \mu_i h + o(h)$ for $i \geq 1$.
4. $P_{ii}(h) = 1 - (\lambda_i + \mu_i)h + o(h)$ for $i \geq 0$.
5. $P_{ij}(0) = \delta_{ij}$.

Here $\mu_0 = 0$, $\lambda_0 > 0$, $\mu_i > 0$, and $\lambda_i > 0$ for $i = 1, 2, 3, \ldots$. The first and last conditions are boundary conditions. The differential equations for this process are derived using the Chapman–Kolmogorov equations,

$$P'_{0j}(t) = -\lambda_0 P_{0j}(t) + \lambda_0 P_{ij}(t) \tag{7.7}$$

$$P'_{ij}(t) = \mu_i P_{i-1,j}(t) - (\lambda_i + \mu_i)P_{ij}(t) + \lambda_i P_{i+1,j}(t). \tag{7.8}$$

While the transition probabilities for specific problems may be calculated from given birth and death parameters, they no longer have nice forms. At this point, the focus of traditional stochastic process analysis is on asymptotic behavior of the

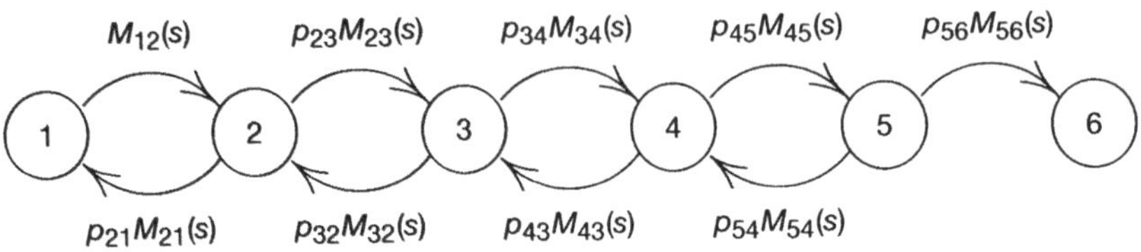

FIGURE 7.1 Flowgraph model for a birth and death process.

process as $t \to \infty$ and on finding quantities such as the stationary distribution of the process. There are many variants of these processes, such as linear birth (or death) processes, processes that allow some form of immigration, and so on. Markovian queuing models (cf. Chapter 9) are examples of birth and death processes. The flowgraph model for a six-state birth and death process is given in Figure 7.1. The probabilities and MGFs are in the general form.

The flowgraph is the same as the Markov model with exponential waiting times when the transition probabilities are set to be

$$\begin{aligned} P_{12} &= 1 \\ P_{i,i+1} &= \frac{\lambda_i}{\lambda_i + \mu_i} \qquad \text{for } i = 2, \ldots, 5 \\ P_{i,i-1} &= \frac{\mu_i}{\lambda_i + \mu_i} \qquad \text{for } i = 2, \ldots, 5 \end{aligned} \tag{7.9}$$

and the corresponding MGFs are

$$\begin{aligned} M_{12}(s) &= \frac{\lambda_1}{\lambda_1 - s} \\ M_{i,i+1}(s) = M_{i,i-1}(s) &= \frac{\lambda_i + \mu_i}{\lambda_i + \mu_i - s} \qquad \text{for } i = 2, \ldots, 5. \end{aligned} \tag{7.10}$$

The flowgraph analysis circumvents the difficulties of the traditional analysis for birth and death processes by modeling the observable waiting times directly and working in the MGF domain. In Section 7.2 we show an application of the pure birth process using a flowgraph model for a series system. In Section 7.3.1 we present a semi-Markov analysis of diabetic retinopathy data, the flowgraph model of which can be viewed as a birth and death process, and in Example 7.3 we analyze the flowgraph of Figure 7.1. We defer further discussion of birth and death processes to Chapter 9.

7.2 APPLICATION TO A MARKOV MODEL OF HIV INFECTION

The flowgraph in Figure 7.2 is a five-state Markov model representing the progressive states of AIDS for a patient. The data are described in Butler and Huzurbazar (1997) and Longini et al. (1989). State 1 represents HIV infection

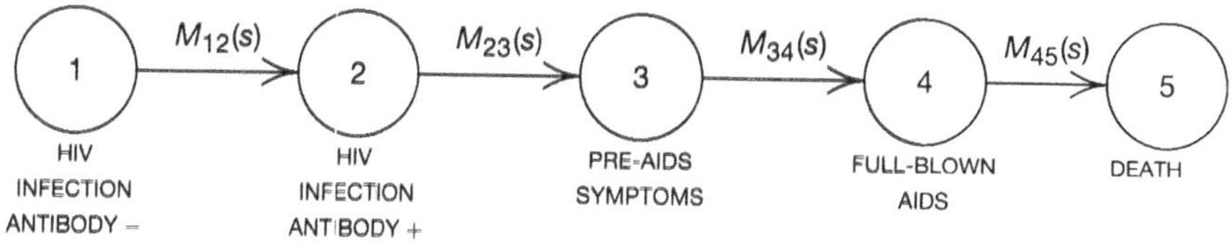

FIGURE 7.2 Flowgraph model for AIDS progression.

but with antibody-negative status, state 2 is antibody-positive status but without symptoms of AIDS, state 3 is pre-AIDS symptoms, state 4 is clinical AIDS, and state 5 is death.

The waiting time in each state i is Exp(λ_i) for $i = 1, 2, 3, 4$. The parameter vector is $\boldsymbol{\theta} = (\lambda_1, \lambda_2, \lambda_3, \lambda_4)$. This is a series model with MGF of the overall survival time, $1 \to 5$, given by

$$M(s) = M_{12}(s)M_{23}(s)M_{34}(s)M_{45}(s). \tag{7.11}$$

Since each waiting time is assumed to be exponential, the MGF (7.11) is

$$M(s) = \left(\frac{\lambda_1}{\lambda_1 - s}\right)\left(\frac{\lambda_2}{\lambda_2 - s}\right)\left(\frac{\lambda_3}{\lambda_3 - s}\right)\left(\frac{\lambda_4}{\lambda_4 - s}\right) \tag{7.12}$$

for $s < \min\{\lambda_1, \lambda_2, \lambda_3, \lambda_4\}$. Another quantity of interest is the time to incubation. This is the MGF of the distribution of the partial waiting time, $1 \to 4$,

$$M(s) = \left(\frac{\lambda_1}{\lambda_1 - s}\right)\left(\frac{\lambda_2}{\lambda_2 - s}\right)\left(\frac{\lambda_3}{\lambda_3 - s}\right) \tag{7.13}$$

for $s < \min\{\lambda_1, \lambda_2, \lambda_3\}$. The data consist of two cohorts. The first cohort, which we will refer to as the San Francisco cohort, consists of 513 homosexual and bisexual men from the San Francisco area who were HIV seropositive. Data for this cohort were collected from 1978 to 1980 with an average follow-up of 7 to 8 years. The second cohort, referred to as the transfusion cohort, received HIV-infected blood products via transfusion. This is the transfusion data that we discussed in Examples 2.10, 4.2, 4.3, 4.4, and 5.8. The follow-up for people in the transfusion cohort was only on average 3.3 years, so these data can only be used for the estimation of waiting times in early states of infection: namely, terms involving state 1. The combined data are what formulate the San Francisco AIDS study. Left, right, and interval censoring are all present in the data. Left censoring occurs for people who at the time of the first observation may have already been in their present state for an unspecified amount of time. Right censoring is present for people who at the time of the last observation may still be in one of the infected states. This includes persons lost to follow-up during the study or those who are censored due to the end of the study. Interval censoring occurs for persons who made a transition during some time interval between observations but the exact time of transition is not known.

The likelihood of $\boldsymbol{\theta}$ requires that we know the probability of transition from state i to j in time t for a pure birth Markov process. From (7.4), these are $P_{ij}(t)$, where

$$P_{ii}(t) = e^{-\lambda_i t}, \qquad i = 1, \ldots, 4$$

$$P_{ij}(t) = \lambda_i \cdots \lambda_{j-1} \sum_{k=i}^{j} B_{ikj} e^{-\lambda_k t}, \qquad i+1 \le j \le 4;\ i = 1,2,3,4 \quad (7.14)$$

$$P_{i5}(t) = 1 - \sum_{j=i}^{4} P_{ij}(t)$$

for

$$B_{ikj} = \prod_{l=i}^{j} \frac{1}{\xi(\lambda_l - \lambda_k)}, \qquad k = i, \ldots, j$$

$$\xi(x) = \begin{cases} x & x \neq 0 \\ 1 & x = 0. \end{cases}$$

The form of the likelihood presumes that all the λ-values are distinct; when a flat uniform prior is used, the set on which λ-values are not distinct has posterior probability zero and can be ignored. The contribution of the ith person to the likelihood is therefore

$$L_i(\boldsymbol{\theta}) = \prod_{j=0}^{m_i - 1} P_{y_{i,j}\, y_{i,j+1}}(\tau_{i,j+1} - \tau_{i,j}), \qquad (7.15)$$

where m_i is the number of transitions for person i, $\tau_{i,0} = 0, \tau_{i,1}, \ldots, \tau_{i,m_i}$ represent the times at which we observe person i to be in states $y_{i,0}, \ldots, y_{i,m_i}$, and $y_{i,0}$ is the initial state in which we observe person i. From the form of the likelihood, it is apparent that the Markov model handles censored data quite nicely through the transition probabilities. For example, the probability $P_{13}(t)$ is used for people that are observed in state 1 and again in state 3. The presumption of the pure birth process is that the person must have transited through state 2 although that transition was not observed. This likelihood is a product of transition probabilities, not waiting time distributions. This is an alternative method of computing the likelihood, and it is equivalent to writing down the likelihood as function of exponential waiting time distributions. Note that the exponential distributions come into play in (7.15) via the $\{B_{ikj}\}$. Since the state-to-state waiting times or interarrival times are independent exponential random variables, the waiting time distribution of transitions between two or more states without observing intervening states is a convolution of exponentials. Thus, while a likelihood could be written out in terms of waiting time distributions, it would be more complicated than (7.15). Therefore, this method is preferred for a Markov process with exponential waiting times.

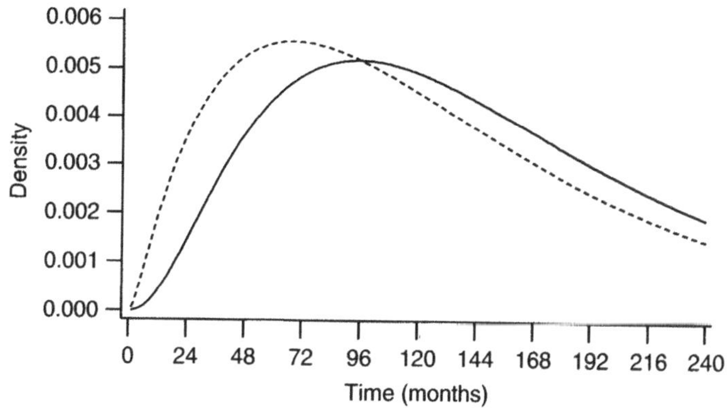

FIGURE 7.3 AIDS example. Predictive densities for overall survival time $1 \rightarrow 5$ (solid line) and time to incubation $1 \rightarrow 4$ (dashed line).

TABLE 7.1 Predictive Quantiles for the Overall Survival Time (in Months) Corresponding to Figure 7.3

Percentile	$\hat{F}^{-1}$	Percentile	$\hat{F}^{-1}$
0.10	55	0.70	188
0.20	76	0.80	226
0.30	96	0.90	287
0.40	115	0.95	345
0.50	136	0.99	475
0.60	159		

Combining this likelihood with a flat uniform prior on $\boldsymbol{\theta}$ produces a quite complicated four-dimensional posterior distribution. If we had observed the actual waiting times between states, the likelihood would have consisted of independent posteriors on the parameters; however, the presence of left and right censoring complicates the form of the posterior.

Figure 7.3 shows the Bayes predictive density, based on 400 $\boldsymbol{\theta}$-values from the posterior, of the time to incubation based on the flowgraph MGF (7.13). The density is depicted by the dashed line in Figure 7.3. The predictive distribution for the overall survival time of an AIDS patient based on the MGF of (7.12) is the solid line in Figure 7.3. Quantile values of this predictive CDF of survival time are given in Table 7.1. By comparison, Longini et al. (1989) use maximum likelihood to estimate the waiting time distribution for the time to incubation, $1 \rightarrow 4$, as the convolution of three independent exponentials. The mode of their approximation is 60 months. In contrast, the mode of the flowgraph predictive distribution is 66 months. The predictive density is more diffuse because it mixes over the posterior rather than being a point estimate at the MLE. Medley et al. (1987) use a

Weibull distribution rather than the convolutions of exponentials to model the incubation period for AIDS patients infected via blood transfusion. Brookmeyer and Gail (1988) also use a Weibull distribution to model the distribution of the time to incubation distribution used for predicting the course of the AIDS epidemic.

7.3 FLOWGRAPH ANALYSIS OF SEMI-MARKOV MODELS

Flowgraph models provide a link between traditional multistate models and semi-Markov processes. Flowgraphs model semi-Markov processes and provide a data analytic method for estimating densities, survivor and reliability functions, and hazard functions for waiting times of interest. There is very little work on non-Markov multistate models (cf. Hougaard, 1999). Semi-Markov analysis is restricted to cases where the waiting times between states all come from the same family of distributions, such as the Weibull, or each transition in the multistate model is modeled by a separate proportional hazards model. Although semi-Markov processes are a natural extension for multistate models, in practice, data analysis for semi-Markov processes can be quite difficult, especially when returns to a state are allowed (cf. Andersen and Keiding, 2002). Flowgraph models provide an extension of traditional multistate models to include semi-Markov processes. They allow vastly different distributions for waiting times between states, and they provide a closed form for the distribution of quantities of interest, such as total or partial waiting time in a stochastic network.

7.3.1 Flowgraph Model for Diabetic Retinopathy

Retinopathy is the name given to a disease of the retina due to diabetes. The retina is the film at the back of the eye, analogous to film in a camera. Light enters the eye and passes through to the retina. The messages about what is seen are then passed on to the brain. People with severe diabetes eventually develop retinopathy, hence the name *diabetic retinopathy*, which eventually causes blindness. Figure 7.4 is a flowgraph model for the degenerative stages of diabetic retinopathy due to Yau and Huzurbazar (2002). This model was first proposed as a Markov model by Marshall and Jones (1995). The stages of retinopathy in the model are classified according to a modified Airlie House classification scheme (cf. Diabetic Retinopathy Study Research Group, 1981). Grade I indicates no retinopathy. Grade II indicates microaneurysms only. In this stage there is damage to the macula, the central area of the retina. Grades III and IV indicate intermediate stages of background retinopathy. These involve early changes, such as an increase in the number of microaneurysms. Grades V and VI indicate preproliferative and proliferative retinopathy, respectively. In preproliferative retinopathy, the retina has been damaged by higher-than-normal sugar levels over several years. This is the stage before new blood vessels start growing. In proliferative retinopathy, very small blood vessels grow on the surface of the retina. In this condition, the sight is very badly affected and can lead to blindness. State 1 in Figure 7.4 corresponds to grade I in the Airlie House classification, state 2 is grades II and III, state 3

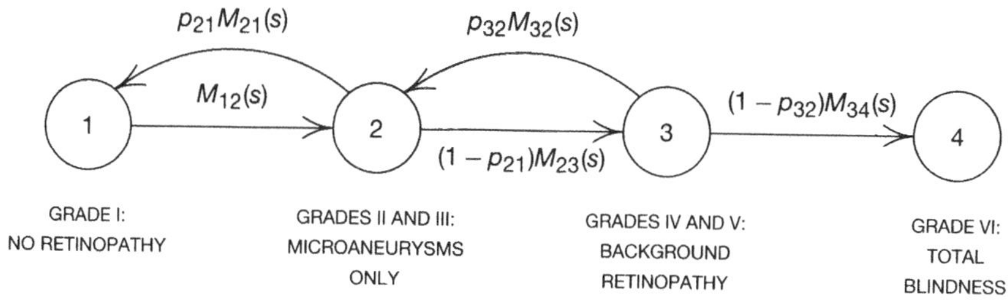

FIGURE 7.4 Flowgraph model for the progression of diabetic retinopathy.

combines grades IV and V, and state 4 is grade VI in the model. A patient with diabetes begins in state 1 having no retinopathy. The patient can progress to a degenerative state, 2. Once in state 2, the patient can improve and return to state 1, or progressively worsen to state 3. Once in state 3, the patient can improve and return to state 2, or worsen to state 4. A patient is considered blind when reaching state 4. Hence, state 4 is an absorbing state. Transitions are made to immediate neighboring states only. If transitions to nontouching states are made, the model assumes that unobserved transitions have occurred. For example, suppose that we observe a patient in state 1 at initial visit and state 3 at the second visit, we assume that the unobserved state 2 is reached between the two visits.

The MGF of the time to total blindness for the retinopathy flowgraph of Figure 7.4 is derived using (2.26). There is only one path from state 1 to state 4. The transmittance of the path is

$$P_1(s) = (1 - p_{21})(1 - p_{32})M_{12}(s)M_{23}(s)M_{34}(s).$$

There are two first-order loops, both of which touch the path. The transmittances of the loops are

$$l_1(s) = p_{21}M_{12}(s)M_{21}(s)$$

and

$$l_2(s) = p_{32}(1 - p_{21})M_{23}(s)M_{32}(s).$$

The sum of the transmittances of the loops is $L_1(s)$,

$$L_1(s) = l_1(s) + l_2(s).$$

The MGF of the time to total blindness is

$$\begin{aligned} M(s) &= \frac{P_1(s)}{1 - L_1(s)} \\ &= \frac{(1 - p_{21})(1 - p_{32})M_{12}(s)M_{23}(s)M_{34}(s)}{1 - [p_{21}M_{12}(s)M_{21}(s) + p_{32}(1 - p_{21})M_{23}(s)M_{32}(s)]}. \end{aligned} \quad (7.16)$$

The data consist of 277 patients having type I diabetes for five years or more. These patients visited the clinic at least twice during the study period. At each visit, a specialist graded retinal findings using a modified Airlie House classification. The worst eye grade for each visit was used to define the subject's state at the time of the visit. In a total of 891 patient visits and gradings, 368 transitions, as defined by the flowgraph model of Figure 7.4, between states of retinopathy occurred during the study period. It is possible for one patient to make the same transition more than once. Among the 368 observed transitions, 180 are censored (i.e., there was no change in state between the last two visits). In these cases, previous analyses assume that if a change were to be observed, it would be to the worse state (cf. Marshall and Jones, 1995; Yau and Huzurbazar, 2002). To illustrate the generality of the flowgraph, we allow the possibility that the patient may have improved and hence use a mixture distribution on the censored observations (i.e., if a patient is censored in state 2, we use a mixture distribution on the waiting times $2 \to 1$ and $2 \to 3$). The data, summarized in Table 7.2, are heavily censored. Eleven transitions are incomplete: Nine of the patients progressed to state 3 beginning at state 1 without being observed at state 2, and two of the patients arrived at state 4 from state 2 without being observed in state 3. We exclude these 11 observations in this analysis but consider them in Chapter 8, where we introduce methodology for incorporating incomplete data.

Censored data histograms of these data are used to suggest parametric models. The censored data histograms are given in Figures 7.5, 7.6, and 7.7. Based on the censored data histograms, the distributions of waiting times used in the analysis are:

- $1 \to 2$ Gamma(α_1, β_1) with mean α_1/β_1.
- $2 \to 1$ $IG(\lambda_1, \lambda_2)$.
- $2 \to 3$ Gamma(α_2, β_2) with mean α_2/β_2.
- $3 \to 2$ Exp(η) with mean $1/\eta$.
- $3 \to 4$ Gamma(α_3, β_3) with mean α_3/β_3.

TABLE 7.2 Summary of Diabetic Retinopathy Data

Transition	No Censoring	Right Censoring
$1 \to 2$	73	58
$2 \to 1$	39	
$2 \to 3$	48	
Censored in state 2		99
$3 \to 2$	14	
$3 \to 4$	3	
Censored in state 3		23
$1 \to 3$	9	
$2 \to 4$	2	
Total	188	180

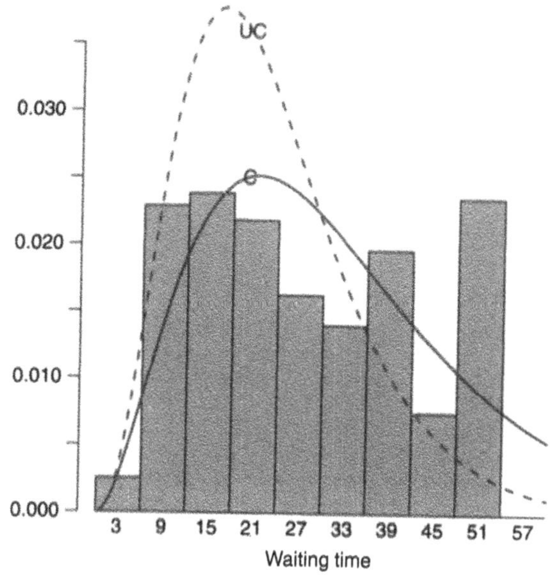

FIGURE 7.5 Censored data histogram for state 1 to state 2 of the retinopathy data for the model of Figure 7.4. Overlay plots of the complete and uncensored observations are included and are labeled C and UC.

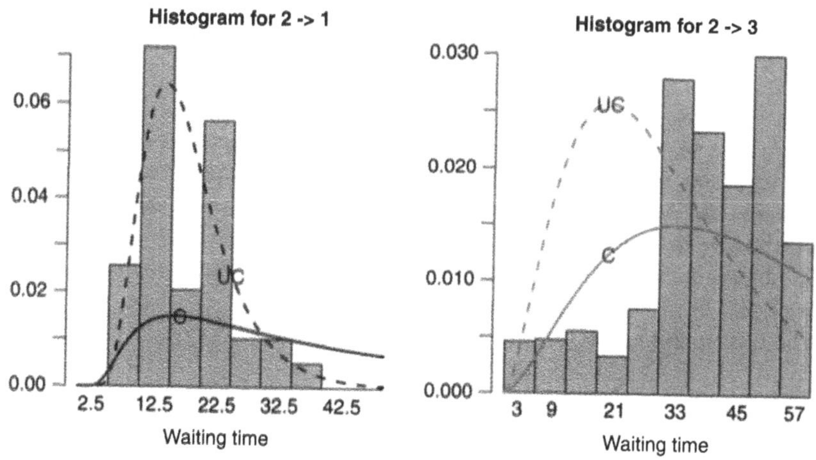

FIGURE 7.6 Censored data histograms for states 2 → 1 and 2 → 3 of the retinopathy data for the model of Figure 7.4. Overlay plots of the complete and uncensored observations are included and are labeled C and UC.

The choice of distributions as well as the parameterizations for the exponential and gamma distributions are slightly different that those used in Yau and Huzurbazar (2002). Each histogram in Figures 7.5, 7.6, and 7.7 is overlaid with two maximum likelihood estimated densities based on the distribution chosen for that

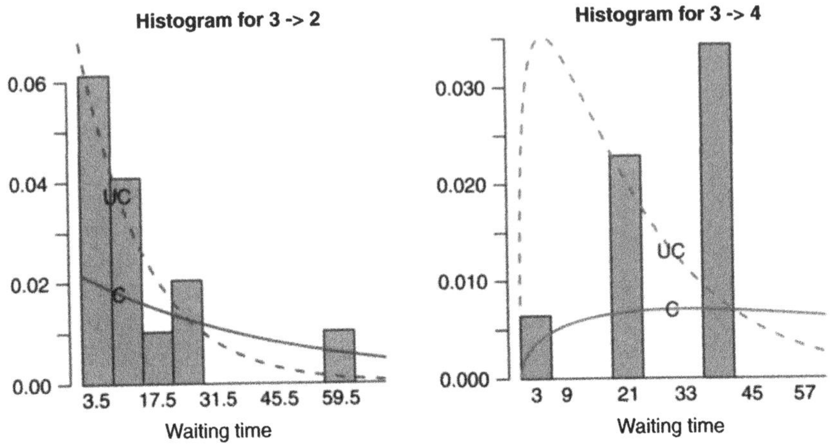

FIGURE 7.7 Censored data histograms for states 3 → 2 and 3 → 4 of the retinopathy data for the model of Figure 7.4. Overlay plots of the complete and uncensored observations are included and are labeled C and UC.

particular branch. The two overlaid densities are labeled C for complete and UC for uncensored. The density labeled C includes both the censored and uncensored observations. The density labeled UC is based only on the uncensored observations. For the purposes of these censored data histograms, observations that are censored in a state are counted in both the preceding and succeeding state (i.e., an observation censored in state 2 is counted in both the 2 → 1 and 2 → 3 transitions). Note that on the censored data histograms for states 2 → 1 and 3 → 2 in Figures 7.6 and 7.7, the uncensored density appears to provide a better fit, lending support to the assumptions of previous analyses that censored observations would only have transited to the worst state. The complete density fits better on the censored data histograms for the 1 → 2 and 2 → 3 transitions, which is also consistent with censored observations transiting to the worst state. The 3 → 4 transition does not have enough data to make any distinction.

The corresponding MGFs are

$$
\begin{aligned}
M_{12}(s) &= \left(\frac{\beta_1}{\beta_1 - s}\right)^{\alpha_1} && \text{for } s < \beta_1 \\
M_{21}(s) &= \exp\{\lambda_1\lambda_2 - [\lambda_1^2(\lambda_2^2 - 2s)]^{1/2}\} && \text{for } s < \lambda_2^2/2 \\
M_{23}(s) &= \left(\frac{\beta_2}{\beta_2 - s}\right)^{\alpha_2} && \text{for } s < \beta_2 \\
M_{32}(s) &= \left(\frac{\eta}{\eta - s}\right) && \text{for } s < \eta \\
M_{34}(s) &= \left(\frac{\beta_3}{\beta_3 - s}\right)^{\alpha_3} && \text{for } s < \beta_3.
\end{aligned}
\tag{7.17}
$$

We choose a variety of distributions for illustrative purposes and recognize that other distributions may work just as well. For example, we chose a gamma distribution for the $3 \rightarrow 4$ transition, whereas Yau and Huzurbazar (2002) choose an exponential. For some of these waiting times, such as states $3 \rightarrow 4$ or $3 \rightarrow 2$, there are not many observations. We perform a Bayesian analysis with diffuse priors on all of the parameters. The priors on p_{21} and p_{32} are Unif(0, 1) and the priors on all of the other parameters are Unif(0, 20).

To derive the likelihood function, let f_{gh}, F_{gh}, and S_{gh} denote the density, CDF, and survivor function, respectively, of the waiting time in state g until transition to stage h. Let x_{igh} be the ith uncensored observation from state g to state h, and let x^*_{jg} be the jth censored observation where censoring occurred in state g.

Data on waiting time in state 1 (no retinopathy) until transition to 2 (microaneurysms only) are available from 131 observations. Among these, 73 made the transition to state 2. Their contribution to the overall likelihood function is

$$\prod_{i=1}^{73} f_{12}(x_{i12}|\alpha_1, \beta_1). \tag{7.18}$$

The remaining 58 observations that were censored in state 1 contribute the following term to the likelihood function:

$$\prod_{j=1}^{58} S_{12}(x^*_{j1}|\alpha_1, \beta_1) = \prod_{j=1}^{58} [1 - F_{12}(x^*_{j1}|\alpha_1, \beta_1)]. \tag{7.19}$$

There are 186 observations providing information on the waiting time in state 2 until transition to either state 1 or state 3. In state 2, 39 observations were observed to return to state 1 (no retinopathy); 48 deteriorated to state 3 (background retinopathy); and 99 had no change until the end of the study. The likelihood given the 87 uncensored observations is

$$\prod_{i=1}^{39} p_{21} f_{21}(x_{i21}|\lambda_1, \lambda_2) \times \prod_{i=1}^{48} (1 - p_{21}) f_{23}(x_{i23}|\alpha_2, \beta_2). \tag{7.20}$$

For the 99 observations censored in state 2, we do not know what outcome they would have had, either no retinopathy (state 1) or background retinopathy (state 3). Thus the censoring distribution is a finite mixture distribution and the contribution to the likelihood of these observations is

$$\prod_{j=1}^{99} \{1 - [p_{21} F_{21}(x^*_{j2}|\lambda_1, \lambda_2) + p_{23} F_{23}(x^*_{j2}|\alpha_2, \beta_2)]\}. \tag{7.21}$$

For the 40 observations in state 3, 14 recovered to state 2 (microaneurysms only), 3 reached total blindness, the absorbing state 4, and 23 were censored. They are used to estimate the true underlying proportions p_{32} and p_{34}. The contribution of the uncensored observations to the total likelihood is

$$\prod_{i=1}^{14} p_{32} f_{32}(x_{i32}|\eta) \times \prod_{i=1}^{3}(1 - p_{32}) f_{34}(x_{i34}|\alpha_3, \beta_3), \tag{7.22}$$

and the contribution of the censored observations is

$$\prod_{j=1}^{23}\{1 - [p_{32}F_{32}(x_{j3}^*|\eta) + p_{34}F_{34}(x_{j3}^*|\alpha_3, \beta_3)]\}. \tag{7.23}$$

For example, suppose that we have observations on patient i with transitions $1 \rightarrow 2 \rightarrow 1 \rightarrow 2 \rightarrow 3$ at month 0, 5, 24, 40 and 56. Then patient i contributes data $x_{i_1 12} = 5 - 0 = 5$ months, $x_{i_1 21} = 24 - 5 = 19$ months, $x_{i_2 12} = 40 - 24 = 16$ months, and $x_{i_1 23} = 56 - 40 = 16$ months, where $x_{i_2 12}$ denotes the second transition from $1 \rightarrow 2$ for patient i. There is no censoring in this case. Suppose that we were to observe this patient one more time at month 62, and assume that he or she had made no change and remained in state 3; then the patient would contribute a piece of censored data $x_{j3}^* = 62 - 56 = 6$ months. The contribution to the likelihood for this patient is

$$\begin{aligned} L(\boldsymbol{\theta}|\text{patient } i) = {} & f_{12}(5|\alpha_1, \beta_1) \times p_{21} f_{21}(19|\lambda_1, \lambda_2) \times f_{12}(16|\alpha_1, \beta_1) \\ & \times (1 - p_{21}) f_{23}(16|\alpha_2, \beta_2) \\ & \times \{1 - [p_{32}F_{32}(6|\eta) + p_{34}F_{34}(6|\alpha_3, \beta_3)]\}. \end{aligned}$$

Putting equations (7.18)–(7.23) together gives the overall likelihood

$$\begin{aligned} & L(p_{21}, p_{32}, \alpha_1, \beta_1, \lambda_1, \lambda_2, \alpha_2, \beta_2, \eta, \alpha_3, \beta_3|\mathcal{D}) \\ & = \prod_{i=1}^{73} f_{12}(x_{i12}|\alpha_1, \beta_1) \prod_{j=1}^{58}[1 - F_{12}(x_{j1}^*|\alpha_1, \beta_1)] \prod_{i=1}^{39} p_{21} f_{21}(x_{i21}|\lambda_1, \lambda_2) \\ & \quad \times \prod_{i=1}^{48}(1 - p_{21}) f_{23}(x_{i23}|\alpha_2, \beta_2) \prod_{j=1}^{99}\{1 - [p_{21}F_{21}(x_{j2}^*|\lambda_1, \lambda_2) \\ & \qquad\qquad + p_{23}F_{23}(x_{j2}^*|\alpha_2, \beta_2)]\} \\ & \quad \times \prod_{i=1}^{14} p_{32} f_{32}(x_{i32}|\eta) \prod_{i=1}^{3}(1 - p_{32}) f_{34}(x_{i34}|\alpha_3, \beta_3) \\ & \quad \times \prod_{j=1}^{23}\{1 - [p_{32}F_{32}(x_{j3}^*|\eta) + p_{34}F_{34}(x_{j3}^*|\alpha_3, \beta_3)]\}. \end{aligned} \tag{7.24}$$

```
# Include package to generate C code
with(codegen,makeproc,C):

# Define the MGF as a procedure function of s
ms := proc(s)

# Define transmittances using w's for transmittances
local w12, w23, w34, w32, w21;

# Enter each transmittance

   w12 := (b1/(b1-s))^a1;
   w21 := p21*exp(l1*l2-((l1^2)*((l2^2)-2*s))^(.5));
   w23 := (1-p21)*((b2/(b2-s))^a2);
   w32 := p32*(eta/(eta-s));
   w34 := (1-p32)*((b3/(b3-s))^a3);

# Calculate M(s)
   (w12*w23*w34)/(1-(w12*w21)-(w23*w32));

end proc;

# Make the MGF
mx := makeproc(ms(s),[s]):
```

FIGURE 7.8 MAPLE code for the retinopathy flowgraph.

The MAPLE program code to compute the MGF of the time to total blindness is given in Figure 7.8. The MGF of the flowgraph from (7.16) using MAPLE is

$$M(s) = \left(\frac{\beta_1}{\beta_1 - s}\right)^{\alpha_1} (1 - p_{21}) \left(\frac{\beta_2}{\beta_2 - s}\right)^{\alpha_2} (1 - p_{32}) \left(\frac{\beta_3}{\beta_3 - s}\right)^{\alpha_3}$$
$$\times \left[1 - \left(\frac{\beta_1}{\beta_1 - s}\right)^{\alpha_1} p_{21}\, e^{\lambda_1 \lambda_2 - (\lambda_1^2(\lambda_2^2 - 2s))^{0.5}} - (1 - p_{21}) \left(\frac{\beta_2}{\beta_2 - s}\right)^{\alpha_2} p_{32}\, \eta(\eta - s)^{-1}\right]^{-1} \tag{7.25}$$

for $s \in (c_1, c_2)$, an open neighborhood of 0. We determine c_1 and c_2 using the numerical algorithm of Section 3.5. To compute the saddlepoint approximation for the distribution of T, the waiting time from state 1 (no retinopathy) to state 4 (blindness), we must evaluate the CGF $K(s) = \log[M(s)]$ and its derivatives, $K'(s)$ and $K''(s)$, at the saddlepoint. Figure 7.9 gives additional MAPLE program code for this computation.

```
# Define the CGF, K(s)
k := makeproc(ln(ms(s)),[s]):

# Write the CGF as a C function to a file called k.c
C(k,optimized,ansi,filename="c:/k.c"):

# Define K'(s) and write it as a C function
dk := makeproc(diff(k(s),s),[s]):
C(dk,optimized,ansi,filename="c:/dk.c"):

# Define K''(s) and write it as a C function
ddk := makeproc(diff(dk(s),s),[s]):
C(ddk,optimized,ansi,filename="c:/ddk.c"):
```

FIGURE 7.9 MAPLE code for $K(s)$, $K'(s)$, and $K''(s)$.

The saddlepoint equation (3.20), $K'(s) = t$, is solved for each value of t at which a density estimate is desired. The function $K'(s)$ is (from MAPLE)

$$
\begin{aligned}
&K'(s)\\
&= \left\{ \left(\frac{\beta_1}{\beta_1 - s}\right)^{\alpha_1} \alpha_1 (1 - p_{21}) \left(\frac{\beta_2}{\beta_2 - s}\right)^{\alpha_2} (1 - p_{32}) \left(\frac{\beta_3}{\beta_3 - s}\right)^{\alpha_3} (\beta_1 - s)^{-1} \right.\\
&\quad \times \left[1 - \left(\frac{\beta_1}{\beta_1 - s}\right)^{\alpha_1} p_{21} e^{\lambda_1 \lambda_2 - [\lambda_1^2(\lambda_2^2 - 2s)]^{0.5}} - (1 - p_{21}) \left(\frac{\beta_2}{\beta_2 - s}\right)^{\alpha_2} \right.\\
&\quad \times \left. p_{32} \eta (\eta - s)^{-1} \right]^{-1} + \left(\frac{\beta_1}{\beta_1 - s}\right)^{\alpha_1} (1 - p_{21}) \left(\frac{\beta_2}{\beta_2 - s}\right)^{\alpha_2} \alpha_2 (1 - p_{32})\\
&\quad \times \left(\frac{\beta_3}{\beta_3 - s}\right)^{\alpha_3} (\beta_2 - s)^{-1} \left[1 - \left(\frac{\beta_1}{\beta_1 - s}\right)^{\alpha_1} p_{21} e^{\lambda_1 \lambda_2 - [\lambda_1^2(\lambda_2^2 - 2s)]^{0.5}} \right.\\
&\quad - \left. (1 - p_{21}) \left(\frac{\beta_2}{\beta_2 - s}\right)^{\alpha_2} p_{32} \eta (\eta - s)^{-1} \right]^{-1} + \left(\frac{\beta_1}{\beta_1 - s}\right)^{\alpha_1} (1 - p_{21}) \left(\frac{\beta_2}{\beta_2 - s}\right)^{\alpha_2}\\
&\quad \times (1 - p_{32}) \left(\frac{\beta_3}{\beta_3 - s}\right)^{\alpha_3} \alpha_3 (\beta_3 - s)^{-1} \left[1 - \left(\frac{\beta_1}{\beta_1 - s}\right)^{\alpha_1} p_{21} e^{\lambda_1 \lambda_2 - [\lambda_1^2(\lambda_2^2 - 2s)]^{0.5}} \right.\\
&\quad - \left. (1 - p_{21}) \left(\frac{\beta_2}{\beta_2 - s}\right)^{\alpha_2} p_{32} \eta (\eta - s)^{-1} \right]^{-1}\\
&\quad - \left(\frac{\beta_1}{\beta_1 - s}\right)^{\alpha_1} (1 - p_{21}) \left(\frac{\beta_2}{\beta_2 - s}\right)^{\alpha_2} (1 - p_{32}) \left(\frac{\beta_3}{\beta_3 - s}\right)^{\alpha_3}\\
&\quad \times \left[- \left(\frac{\beta_1}{\beta_1 - s}\right)^{\alpha_1} \alpha_1 p_{21} e^{\lambda_1 \lambda_2 - [\lambda_1^2(\lambda_2^2 - 2s)]^{0.5}} (\beta_1 - s)^{-1} \right.\\
&\quad - \left(\frac{\beta_1}{\beta_1 - s}\right)^{\alpha_1} p_{21} \left[\lambda_1^2 \left(\lambda_2^2 - 2s\right)\right]^{-0.5} \lambda_1^2 e^{\lambda_1 \lambda_2 - [\lambda_1^2(\lambda_2^2 - 2s)]^{0.5}}
\end{aligned}
$$

$$
\begin{aligned}
&- (1-p_{21})\left(\frac{\beta_2}{\beta_2-s}\right)^{\alpha_2} \alpha_2\, p_{32}\, \eta\, (\beta_2-s)^{-1}\, (\eta-s)^{-1} \\
&- (1-p_{21})\left(\frac{\beta_2}{\beta_2-s}\right)^{\alpha_2} p_{32}\, \eta\, (\eta-s)^{-2}\Big] \\
&\times \left[1-\left(\frac{\beta_1}{\beta_1-s}\right)^{\alpha_1} p_{21}\, e^{\lambda_1\lambda_2-[\lambda_1^2(\lambda_2^2-2s)]^{0.5}} - (1-p_{21})\left(\frac{\beta_2}{\beta_2-s}\right)^{\alpha_2}\right. \\
&\times \left. p_{32}\, \eta\, (\eta-s)^{-1}\right]^{-2}\Bigg\}\left[1-\left(\frac{\beta_1}{\beta_1-s}\right)^{\alpha_1} p_{21}\, e^{\lambda_1\lambda_2-[\lambda_1^2(\lambda_2^2-2s)]^{0.5}}\right. \\
&\left.- (1-p_{21})\left(\frac{\beta_2}{\beta_2-s}\right)^{\alpha_2} p_{32}\, \eta\, (\eta-s)^{-1}\right]\left[\left(\frac{\beta_1}{\beta_1-s}\right)^{\alpha_1}\right]^{-1} \\
&\times (1-p_{21})^{-1}\left[\left(\frac{\beta_2}{\beta_2-s}\right)^{\alpha_2}\right]^{-1} (1-p_{32})^{-1}\left[\left(\frac{\beta_3}{\beta_3-s}\right)^{\alpha_3}\right]^{-1}.
\end{aligned}
\tag{7.26}
$$

While the expressions for $M(s)$ and $K'(s)$ of (7.25) and (7.26) may appear complicated, in the scheme of analysis for flowgraphs, the form of the functions is not important as long as the functions exist. Ultimately, the only requirement is to have C subroutines that calculate these quantities that are usable by a program that calculates saddlepoints. $K''(s)$ is calculated similarly. The flowgraph program introduced in Chapter 5 for Bayesian analysis also gives maximum likelihood estimates, if desired. The MLEs are $\hat{p}_{21} = 0.2765$, $\hat{p}_{32} = 0.7594$, $\hat{\alpha}_1 = 2.9706$, $\hat{\beta}_1 = 0.0923$, $\hat{\lambda}_1 = 9.8695$, $\hat{\lambda}_2 = 0.4951$, $\hat{\alpha}_2 = 2.6907$, $\hat{\beta}_2 = 0.0517$, $\hat{\eta} = 0.0271$ $\hat{\alpha}_3 = 1.3897$, and $\hat{\beta}_3 = 0.0261$. Using (7.26) and these MLEs, we can estimate the mean time to total blindness as $K'(0) = 500.92$ months and its standard deviation $\sqrt{K''(0)} = 415.74$ months. The value of c_2, the upper bound for the saddlepoints, as a function of the MLE $\hat{\theta}$ is $c_2 = 0.002\,41\,49$.

Figure 7.10 gives both the Bayes predictive density (solid line) and the maximum likelihood estimated density (dashed line). The Bayes predictive density is computed using slice sampling to acquire 2000 posterior samples, an initial 1000 burn-in samples are discarded. The priors used were Unif(0, 1) for p_{21} and p_{32} and diffuse Unif(0, 20) on all of the other parameters. The maximum likelihood estimated saddlepoint density approximation, $\hat{f}(t)$, is given in Figure 7.10. The survivor function estimate, $\hat{S}(t)$, is given in Figure 7.11 and the hazard function estimate, $\hat{h}(t)$, is given in Figure 7.12. The hazard rises fairly steeply until $t \approx 250$ months; then it flattens out to asymptote at approximately 0.002 41 49, the upper limit of the radius of convergence of the MGF. The predictive hazard behaves quite differently from the maximum likelihood-based hazard, which is expected since it is mixing over all of the posterior samples. The smallest upper limit of all of the saddlepoints as a function of the posterior samples is 9.66×10^{-5}. Flowgraphs model the actual waiting time distributions and then convert those to the corresponding hazards. No direct assumptions are made about the hazard. Hazard shapes that are increasing and then decreasing are common with flowgraphs.

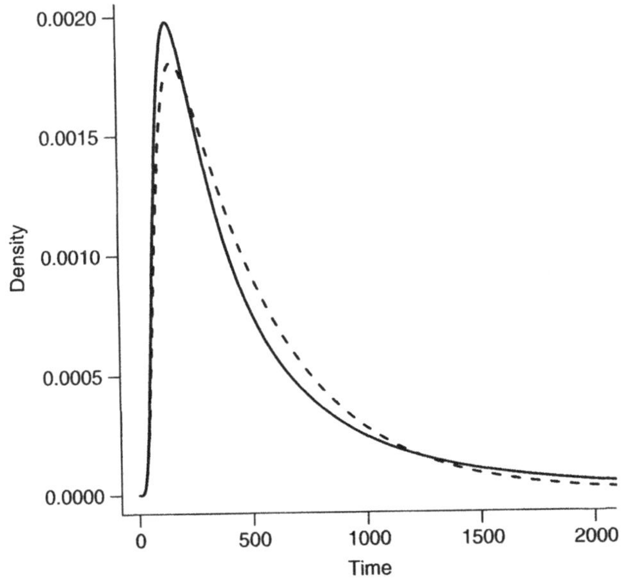

FIGURE 7.10 Probability density function of the time to total blindness for the diabetic retinopathy model of Figure 7.4. Bayes predictive density is the solid line, and the maximum likelihood estimated density is the dashed line.

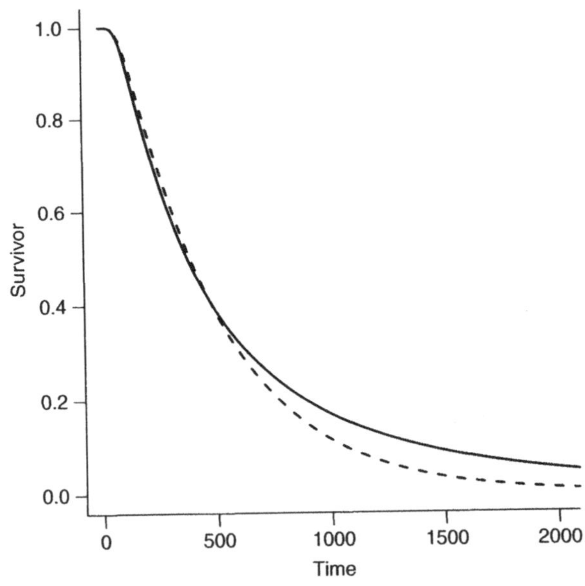

FIGURE 7.11 Survivor functions of the diabetic retinopathy data for the model of Figure 7.4. Bayes predictive survivor function is the solid line, and the survivor function based on maximum likelihood is the dashed line.

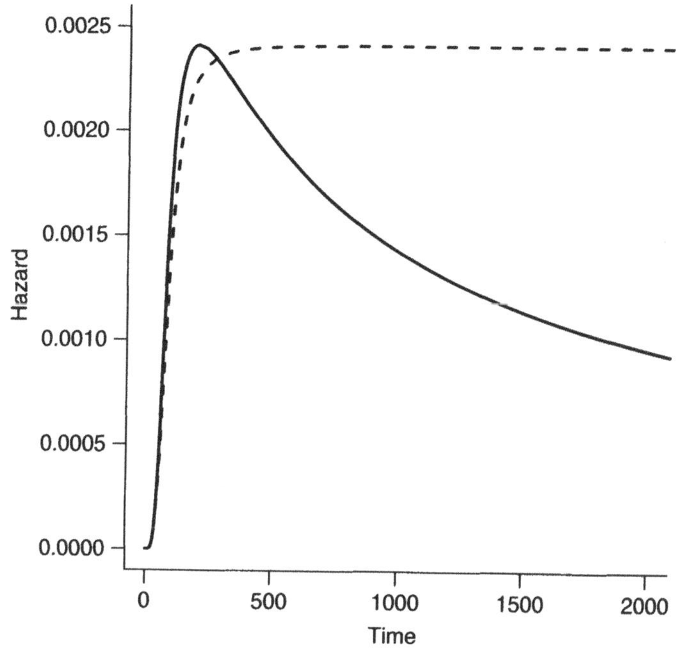

FIGURE 7.12 Bayes predictive hazard function (solid line) and the MLE estimated hazard (dashed line) of the time to total blindness for the diabetic retinopathy model of Figure 7.4.

TABLE 7.3 Selected Quantiles of the Density Functions of Figure 7.10 Based on Complete Data with Censoring

Percentile	Maximum Likelihood	Bayes Predictive Density
0.25	215.34	199.15
0.50	387.28	374.85
0.75	675.64	741.84
0.90	1055.42	1400.23
0.95	1342.48	2055.81
0.99	2008.83	4251.62
Mean	500.92	639.46
s.d.	415.74	925.69

The density function is heavy tailed, emphasizing the computational need to compute the saddlepoint approximation well into the tail of the distribution. Table 7.3 gives selected quantilc values.

Although Marshall and Jones (1995) and Yau and Huzurbazar (2002) focus on different aspects of the problem, we make some rough comparisons between our results. Marshall and Jones (1995) focus on model selection using models with covariates. Their resulting survival curve, adjusted for covariates, gives the following results: The probability of remaining free of total blindness for 5 years (60 months) beginning with no retinopathy is approximately 0.96, and for 10 years (120 months) it is 0.77. Yau and Huzurbazar (2002) use a flowgraph model with different assumptions. They retain the Marshall and Jones (1995) assumption that censored observations are assumed to have transited to the worst state. This affects their choice of branch waiting time distributions. In this analysis we have allowed censored observations to have a mixture distribution over the possible transitions to both the better and worse states. Yau and Huzurbazar (2002) present a flowgraph analysis based only on maximum likelihood.

Yau and Huzurbazar (2002) find that the probability of remaining free of total blindness for 5 years (60 months) beginning with no retinopathy is approximately 0.99, and for 10 years (120 months) it is just under 0.90. The maximum likelihood results here indicate that the probability of remaining free of blindness for 5 years is similar, 0.991 75, and for 10 years it is 0.918 74. Yau and Huzurbazar (2002) find that the corresponding survival times for 0.96 and 0.77 are 85 months and 183 months, respectively. The maximum likelihood analysis performed here has survival times of 93 months and 204 months for survivor function values of 0.96 and 0.77. The Bayesian analysis of the flowgraph model gives the probability of remaining free of blindness for 5 years beginning with no retinopathy as 0.990 52, and for 10 years as 0.902 46. The survival time corresponding to survivor function values of 0.96 and 0.77 are approximately 87 months and 188 months. Even if the comparison were to a Markov model without covariates, we would expect both the flowgraph survival times to be longer because our parametric models are based on much heavier-tailed distributions. In addition, the Bayes predictive distribution will have more spread. Although not directly comparable, the flowgraph model results are also in agreement with those presented in related studies of diabetic retinopathy such as Klein et al. (1984a, b).

Table 7.4 gives a comparison of the waiting times for each branch among the actual data (uncensored only) and the MLEs for the uncensored data and the

TABLE 7.4 Means of Branch Waiting Time Distributions for the Retinopathy Data

Transition	Actual Data (Uncensored)	MLE Uncensored Data Only	MLE Complete Data
$1 \to 2$	22.99	22.99	32.19
$2 \to 1$	18	17.99	19.93
$2 \to 3$	30.85	30.85	52.03
$3 \to 2$	14.79	14.82	36.86
$3 \to 4$	21.33	21.36	53.34
$\hat{p}_{21}$	0.4482	0.4479	0.2765
$\hat{p}_{32}$	0.8235	0.8237	0.7594

complete data. Although the individual transition means are generally much higher with censored data, there is compensation in smaller values of $\hat{p}_{21}$ and $\hat{p}_{32}$. This compensation results in overall means that are similar in the two cases. The $3 \to 2$ transition time for the complete data is more than twice that of the uncensored data, but there is a drop in the estimate of p_{32} for the complete data. Although the $2 \to 1$ transition means for the two cases are similar, the uncensored data MLE for $\hat{p}_{21}$ is about 1.5 times that of the complete data MLE. The mean of the overall maximum likelihood density based on uncensored data is 477.17, whereas the mean of the overall maximum likelihood density for the complete data is 500.92. Figure 7.13 gives comparative density plots for the complete data and the uncensored data for both the MLE and the Bayes predictive density. Table 7.5 gives corresponding quantile values. Both emphasize the effect of censoring on the estimates.

Additionally, if we wish to compute survival curves of remaining free of blindness, state 4, starting from state 2 (microaneurysms) or state 3 (intermediate background retinopathy), we would alter the flowgraph model to obtain the relevant MGFs, $M_{24}(s)$ for $2 \to 4$ and $M_{34}(s)$ for $3 \to 4$. Suppose that we are interested in the distribution of the waiting time from state 2 to state 4. Conceptually, we can redraw the original flowgraph of Figure 7.4 to obtain Figure 7.14.

The MGF of the time to total blindness for the retinopathy flowgraph of Figure 7.14 starting in state 2 is a modification of (7.16). The primary modification involves the starting and the ending states, changing the primary path through the flowgraph. The path from state 2 to state 4 has transmittance

$$P_1(s) = (1 - p_{21})(1 - p_{32})M_{23}(s)M_{34}(s).$$

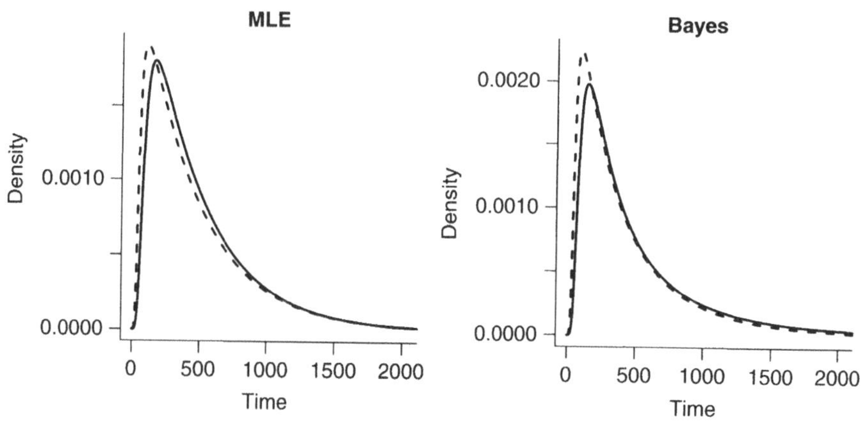

FIGURE 7.13 ML estimated densities and Bayes predictive densities for the time to total blindness for the diabetic retinopathy model of Figure 7.4. Each figure has the density based on complete data (solid) and the density based on the uncensored data only (dashed).

TABLE 7.5 Selected Quantiles of the Density Functions of Figure 7.13[a]

Percentile	Maximum Likelihood	Bayes Predictive Density
0.25	179.53	159.59
0.50	354.32	319.22
0.75	651.28	624.56
0.90	1043.42	1107.26
0.95	1339.99	1545.84
0.99	2028.56	2902.62
Mean	477.17	501.43
s.d.	428.32	628.31

[a] Based on uncensored data only.

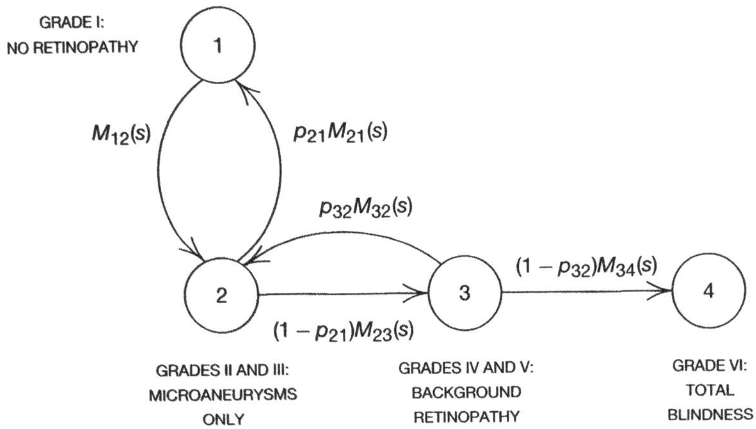

FIGURE 7.14 Flowgraph model for progression to state 4 starting in state 2 for the diabetic retinopathy data.

The two loops, both of which touch the path, are unchanged, so their transmittances remain

$$l_1(s) = p_{21}M_{12}(s)M_{21}(s)$$

and

$$l_2(s) = p_{32}(1 - p_{21})M_{23}(s)M_{32}(s).$$

The sum of the transmittances over the first-order loops is

$$L_1(s) = l_1(s) + l_2(s).$$

The MGF of the time to total blindness starting in state 2, microaneurysms, is

$$M_{24}(s) = \frac{P_1(s)}{1 - L_1(s)}$$
$$= \frac{(1 - p_{21})(1 - p_{32})M_{23}(s)M_{34}(s)}{1 - [p_{21}M_{12}(s)M_{21}(s) + p_{32}(1 - p_{21})M_{23}(s)M_{32}(s)]}. \quad (7.27)$$

The data and the analysis do not change. If our interest is in this new quantity, we simply replace the original flowgraph MGF of (7.16) with this one (7.27) and perform the analysis as before. Thus, we can take data on the original flowgraph and make inferences about any quantity within the flowgraph. This flowgraph allows for a return to state 1 starting in 1. If our interest were in direct transition from $2 \to 4$, returns to state 1 would not be allowed. In this case, we would not include the loop transmittance $l_1(s)$. The probability p_{23} is set to 1 and the MGF is

$$M(s) = \frac{(1 - p_{32})M_{23}(s)M_{34}(s)}{1 - p_{32}M_{23}(s)M_{32}(s)}. \quad (7.28)$$

7.3.2 Flowgraph Model for a Construction Engineering Project

A construction engineering project examined the impact of communication and planning on performance (cf. Garcia, 1997). Data on every aspect of operations were collected for 20 firms in New Mexico. Monthly performance is one of the main quantities of interest. Monthly performance is loosely defined as whether the project completed the month on schedule and whether it will complete the next month on schedule. Our aim is to evaluate the completion time for any given month using a flowgraph model. We use these data to illustrate a Bayesian predictive analysis for the flowgraph model, as in Huzurbazar (2004b).

Figure 7.15 gives a flowgraph model for the construction engineering project. State 0 is the planning state and state 1 is the design phase. From state 1, transition

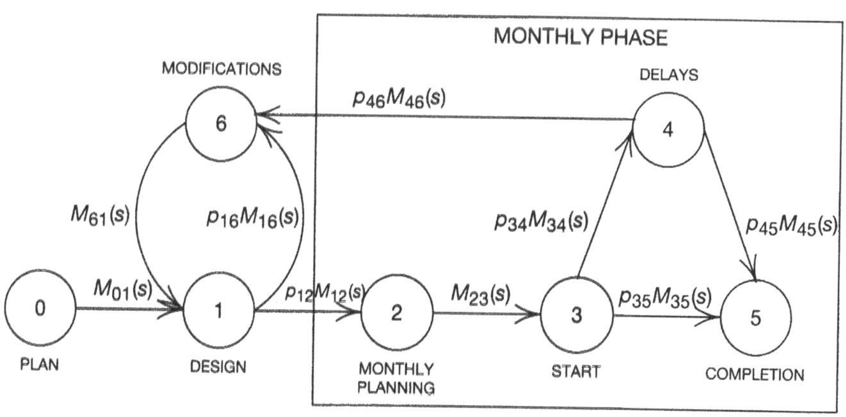

FIGURE 7.15 Flowgraph model for the construction engineering project.

is possible to state 6 (modifications) or into the monthly planning phase, state 2. The loop 1 → 6 → 1 represents the process of design with modifications. After the initial planning, the monthly phase begins, represented by the boxed subsystem in Figure 7.15. The time frame for the subsystem is one month (= 23 working days). State 2 is the monthly planning phase; state 3 is the start of the monthly work; state 4 represents delays in the monthly work which can result in modifications to the main project, indicated by a transition from 4 → 6, or lead eventually to completion, state 5.

Our data for the 20 firms is restricted to states 2, 3, 4, 5 and consist of observations only on the monthly phase, so our analysis assumes that the process begins in state 2 ($p_{12} = 1$) and from state 4, only transitions to state 5 are possible ($p_{45} = 1$). With appropriate data, it would be possible to analyze the entire overall schedule with the flowgraph model given in Figure 7.15. Six of the firms experienced no delays. For these firms, we observe transitions from MONTHLY PLANNING, to START, to COMPLETION, 2 → 3 → 5. Of the 14 firms that experienced delays, 10 were able to recover from these and complete the month on schedule, 4 → 5. The remaining four firms were irrecoverably delayed, indicated by the transition 4 → 6. We have no subsequent data for them on the current month and therefore, in the present analysis, we will not analyze the 4 → 6 transition further.

We use histograms of the data to suggest parametric models to be used with flowgraphs. For example, Figure 7.16 presents a histogram of data on the 3 → 4 transition. For illustration, a maximum likelihood estimated gamma density is overlaid in the figure. After examining histograms for each state-to-state transition of

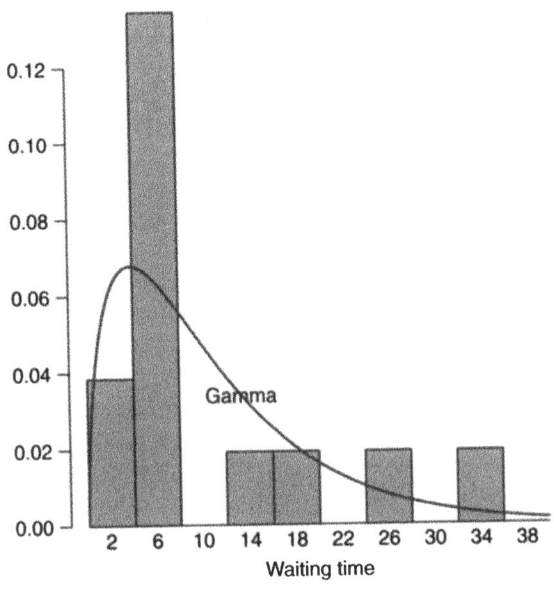

FIGURE 7.16 Histogram for data on the 3 → 4 transition.

the data, we make the following parametric assumptions about the branch transmittances:

- Branch $2 \to 3$, constant 4 days, MGF $M_{23}(s) = \exp(4s)$.
- Branch $3 \to 4$, Gamma(α, β) with mean α/β, MGF $M_{34}(s) = [\beta/(\beta - s)]^{\alpha}$ for $s < \beta$.
- Branch $4 \to 5$, constant 18 days, MGF $M_{45}(s) = \exp(18s)$.
- Branch $3 \to 5$, IG(λ_1, λ_2) with MGF

$$M_{35}(s) = \exp\left[\lambda_1 \lambda_2 - \sqrt{\lambda_1^2(\lambda_2^2 - 2s)}\right], \qquad s < \lambda_2^2/2.$$

Data on $2 \to 3$ are all within a small fraction of 4 days, and data on the $4 \to 5$ transition are clustered around 18 days. We treat these as roughly constant times to demonstrate how flowgraphs handle constant waiting times. The $3 \to 5$ transition has only six observations. For illustration, we use an inverse Gaussian distribution to allow for a heavy tail. This is motivated largely by the problem at hand. The $3 \to 5$ transition is made when work is completed without delays. In the construction industry, this is rarely the case. The heavy tail allows for longer waiting times when delays exist but are not recorded.

Using (2.26) we find the MGF of the overall waiting time distribution from $2 \to 5$. There are two paths through the system,

$$\begin{aligned} P_1(s) &= p_{34} M_{23}(s) M_{34}(s) M_{45}(s) \\ P_2(s) &= p_{35} M_{23}(s) M_{35}(s), \end{aligned} \tag{7.29}$$

where $p_{34} + p_{35} = 1$, and we take $p_{45} = 1$ for the monthly phase analysis. The MGF of the total waiting time distribution for monthly time to completion is

$$M_{25}(s) = e^{4s} p_{34} \left(\frac{\beta}{\beta - s}\right)^{\alpha} e^{18s} + e^{4s}(1 - p_{34}) \exp\left[\lambda_1 \lambda_2 - \sqrt{\lambda_1^2(\lambda_2^2 - 2s)}\right] \tag{7.30}$$

for $s < \min(\beta, \lambda_2^2/2)$, $\lambda_1 > 0, \lambda_2 > 0, \alpha > 0, \beta > 0, 0 \le p_{34} \le 1$. The model can accommodate any transition time distribution, discrete or continuous. In particular, if the $4 \to 5$ transition took on values 18 and 19 with probabilities 2/3 and 1/3, the term e^{18s} in (7.30) would be replaced by $(2/3)e^{18s} + (1/3)e^{19s}$. The MAPLE program code used to generate the MGF (7.30), as well as the CGF and its first two derivatives for saddlepoint computations, is given in Figure 7.17.

Our parameter vector $\boldsymbol{\theta}$ is $\boldsymbol{\theta} = (p_{34}, \alpha, \beta, \lambda_1, \lambda_2)$. To compute the Bayes predictive density of (5.7), we assign a Unif(0, 1) prior for p_{34} and we assume diffuse Unif(0, 20) priors on all of the other parameters. The likelihood is separable in p, (α, β), and (λ_1, λ_2). This results in a Beta(15, 7) posterior distribution on p_{34}, and

```
# Include package to generate C code
with(codegen,makeproc,C):

# Define the MGF as a procedure function of s
ms := proc(s)

# Define transmittances using w's for transmittances
local w23, w34, w35, w45;

# Enter each transmittance

   w23 := exp(4*s);
   w34 := p34*((b/(b-s))^a);
   w35 := (1-p34)*exp((l1*l2)-(sqrt(l1*l1*(l2*l2-2.0*s))));
   w45 := exp(18*s);

# The next line defines M(s)
   w23*(w34*w45+w35);

end proc;

# Make the MGF
mx := makeproc(ms(s),[s]):

# Define the CGF, K(s)
k := makeproc(ln(ms(s)),[s]):

# Write the CGF as a C function to a file called k.c
C(k,optimized,ansi,filename="c:/k.c"):

# Define K'(s) and write it as a C function
dk := makeproc(diff(k(s),s),[s]):
C(dk,optimized,ansi,filename="c:/dk.c"):

# Define K''(s) and write it as a C function
ddk := makeproc(diff(dk(s),s),[s]):
C(ddk,optimized,ansi,filename="c:/ddk.c"):
```

FIGURE 7.17 MAPLE code for the construction engineering flowgraph.

the posterior on the remaining parameters is given by

$$\pi(\lambda_1, \lambda_2, \alpha, \beta, |\text{data}) \propto \left(\frac{\beta^\alpha}{\Gamma(\alpha)}\right)^{14} \left(\prod_i^{14} t_i^{\alpha-1}\right) e^{-\beta \sum_i t_i}$$

$$\times \lambda_1^6 \exp\left(-\frac{\lambda_1^2}{2}\sum_j \frac{1}{t_j} - \frac{\lambda_2^2}{2}\sum_j t_j + 6\lambda_1\lambda_2\right), \quad (7.31)$$

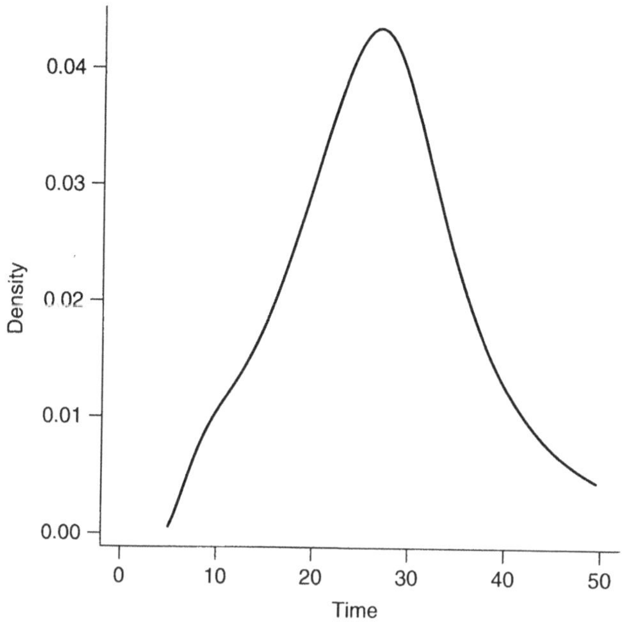

FIGURE 7.18 Bayes predictive density function for the completion time of the construction engineering project.

where $t_1, \ldots, t_{14}$ are the observed waiting times for the transition $3 \to 4$ and $t_{15}, \ldots t_{20}$ are the observed waiting times for the transition $3 \to 5$. Note that the waiting times for transitions $2 \to 3$ and $4 \to 5$ are constant. Figures 7.18, 7.19, and 7.20 show the Bayes predictive density, survivor, and hazard functions for this project. The predictive mean is 32.29 days and the standard deviation is 141.91 days, so we do not expect to remain on schedule. The large standard deviation is reflective of the very heavy tailed distribution for the $3 \to 5$ transition and the fact that once the project falls behind schedule it takes much longer to complete due to compounding problems. The hazard increases steadily until about 30 days and then drops off. This is reflective of the fact that project delays make it hard to complete in the 20 additional days shown in the figure. However, once the problems are resolved, the hazard for the completion time will start to increase again in the future. The 25th, 50th, 75th, 90th, 95th, and 99th percentiles are 20.99, 27.28, 34.25, 44.57, 55.33, and 103.76 days, respectively. Any other desired quantity is also obtained easily from the predictive density.

Analysis from the flowgraph model gives an entire waiting time distribution as well as the system reliability (survivor) and hazard functions for any total or partial waiting time. For decision making, this immediately provides the mean, median, or any percentile of the waiting time distribution for a variety of input distributions. The analysis from the current month along with prior information about potential delays gives a predictive distribution for the next month. This gives the project team a figure for how many days behind schedule they are likely to be for the

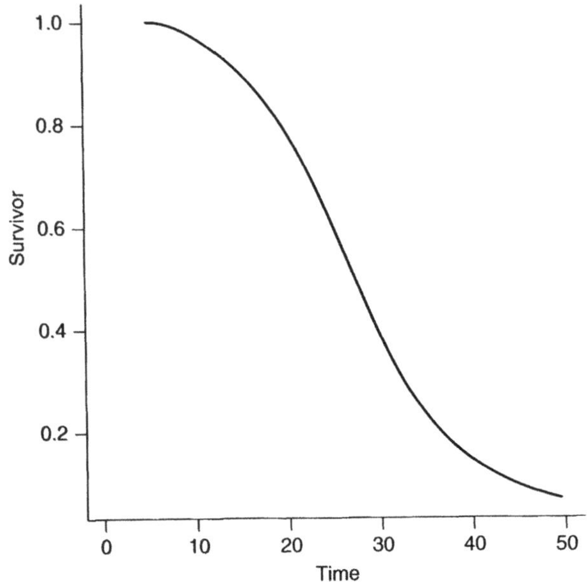

FIGURE 7.19 Bayes predictive survivor function for the completion time of the construction engineering project.

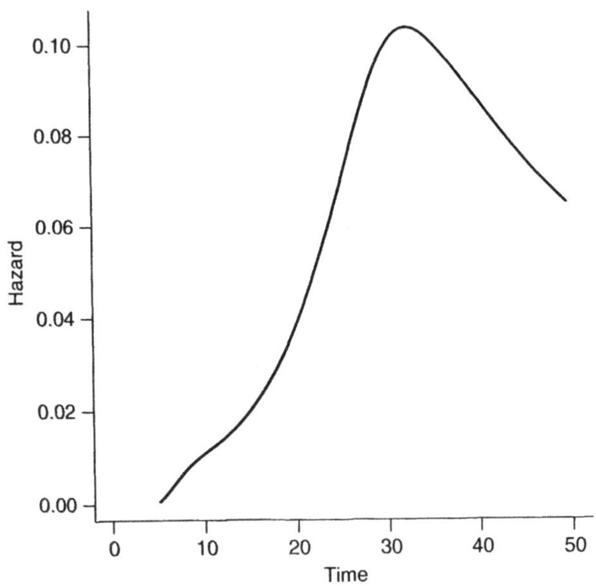

FIGURE 7.20 Bayes predictive hazard function for the completion time of the construction engineering project.

next month allowing them to make necessary modifications such as reorganizing workload or hiring additional workers. Other methods generally provide only the mean and median for a limited number of distributions. For example, current methods used (reluctantly) by these firms are the critical path method (CPM) and the project evaluation and review technique (PERT). Both of these methods are commonly used in management and engineering to identify potential problems in a project plan, but it is accepted that they have many shortcomings. AbouRizk and Wales (1997) and Pritsker et al. (1989) provide details of these methods. They are used at the beginning of a project and are not very adaptable to the project once the work has begun. They do not provide much assistance to the project manager in terms of timely updates to the project schedule (cf. Leung and Tam, 1999). CPM is also limited to analyzing deterministic systems with no feedback or failures in the project. Projects are always assumed to complete successfully. PERT is limited to using beta and normal distributions. Some current trends include the use of genetic algorithms (cf. Li et al., 1999). Genetic algorithms and other evolutionary programming methods require large amounts of data for the algorithm's learning process and a great deal of computation to implement. Flowgraphs can work with a relatively small amount data and incorporate prior information about planning, rework, possible delays, and completion times from the project team. A sequence of flowgraph models can be constructed for each monthly phase where each subsequent flowgraph uses the results from the previous month's model along with new prior information. The model can be updated as frequently as the project finds it necessary.

7.4 PHASE TYPE DISTRIBUTIONS

A nonnegative random variable T is said to have a *phase type* (PH) distribution if T is the time until absorption in a finite-state continuous time Markov chain. Neuts (1981) provides an introduction to PH distributions. Assuming that the waiting times between states are exponential, the distribution of T is a generalization of the Erlang distribution, which is gamma with an integer-valued shape parameter. Every exponential random variable is of PH, so that, in practice, exponential waiting times are used to obtain PH distributions. Even the use of mixtures of exponentials leading to hyperexponential or mixed Erlang distributions give highly complicated analytical expressions (cf. Neuts, 1981), so generally, mixtures of exponential distributions that are simple enough to be analytically tractable are assumed. Cox (1955) showed that the absorption time in a finite state Markov process has a rational Laplace transform that can be represented by a series of exponential phases. However, note that the waiting times in a PH distribution are not required to be exponential. It is only analytically convenient if they are. The pervasiveness of the exponential distribution as a basis for deriving PH distributions is so thorough that the *Encyclopedia of Statistics* entry for PH distributions makes this assumption (cf. Shaked and Shantikumar, 1983).

PH distributions have found applications in modeling survival data (cf. Aalen, 1995) and queuing (cf. Ramaswami and Lucantoni, 1979; Ramaswami and Neuts, 1980), reliability, and branching processes. For example, the progression of a disease such as cancer, AIDS, kidney disease, or diabetic retinopathy can be modeled using a PH distribution. All involve progression through several stages to an absorbing end state of death, dialysis, or blindness and may involve stages of partial recovery before the end stage. The same is true of the hydraulic pump queuing system of Example 2.8 (Figure 2.10) or of manufacturing systems such as the car stereo assembly of Example 2.15 (Figure 2.29) and the parts assembly system of Example 5.3 (Figure 5.5). Flowgraph methods analyze problems involving PH distributions because they are inherently modeling the same overall waiting time distribution. In addition, flowgraphs allow nonexponential waiting times, extending the practical application of PH distributions (cf. Huzurbazar, 1999). In this section we present three examples of models whose analysis is based on exponential PH distributions. The first involves the Coxian model. The second is a model with multiple exits. PH distributions also arise in processes that have arbitrary initial states, and these processes are also amenable to flowgraph modeling. The last example illustrates this for a birth and death process. The flowgraph analysis allows all of these to have nonexponential waiting times.

Example 7.1: COXIAN MODEL. A Coxian model (cf. Cox, 1955) is a series model with transitions to the next state as well as transitions to the end state. The flowgraph representation of a Coxian model using the progress of recovery of bone marrow transplant patients is shown in Figure 7.21. Bone marrow transplants are used for acute leukemia. A patient begins in state 1 with transplant. After transplant, the patient's platelet count may return to normal levels, state 2 (PLATELET RECOVERY), or the patient may experience relapse or death, state 4. From state 2, the patient can progress to state 3, acute or chronic graft versus host disease (ACUTE GVHD), or again experience a relapse or death, state 4. From state 3, the

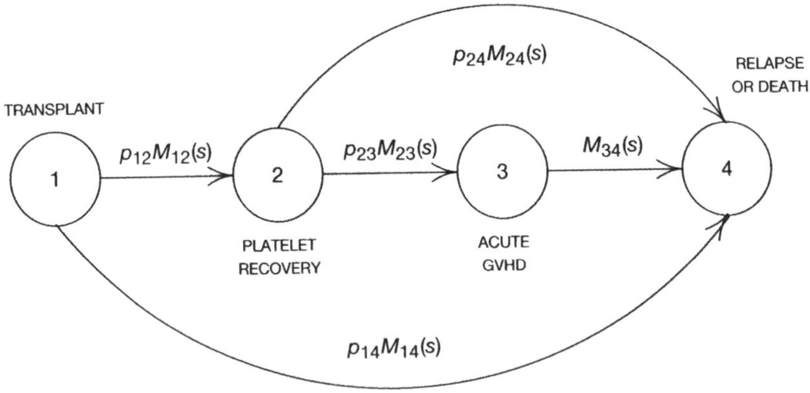

FIGURE 7.21 Flowgraph representation of a Coxian model for the recovery process from bone marrow transplant.

only passage is to state 4. The AIDS example of Section 7.2 (Figure 7.2) can also be modeled using a Coxian model where we allow a direct transition to death from each intermediate state.

Cox (1955) bases this model on the fact that any distribution having a rational Laplace transform can be represented by a sequence of exponential phases. Therefore, the standard analysis for this Coxian model is a PH analysis with exponential waiting times, often with the transition probability from $2 \to 3$ set equal to 0, as in Aalen (1995). This greatly simplifies the analysis because the resulting distribution is a finite mixture of a convolution of two exponentials with one additional exponential. Huzurbazar (1999) presents a probabilistic analysis without data using flowgraph models with general distributions for transitions. We present the framework for how an analysis using general waiting time distributions would proceed. Simulated data from the illustrative model is presented in Exercise 7.3, which asks the reader to perform a Bayesian data analysis. The branch waiting time distributions are:

- $\text{Exp}(\beta_1)$ for $1 \to 2$
- $IG(\lambda_1, \lambda_2)$ for $1 \to 4$
- $\text{Exp}(\beta_2)$ for $2 \to 3$
- $\text{Gamma}(\alpha_4, \beta_4)$ for $2 \to 4$
- $\text{Exp}(\beta_3)$ for $3 \to 4$

We use Mason's rule (2.26) to find the MGF of the overall waiting time distribution from $1 \to 4$. There are three paths but no loops. The path transmittances are

$$P_1(s) = p_{12}p_{23}M_{12}(s)M_{23}(s)M_{34}(s)$$
$$P_2(s) = p_{12}p_{24}M_{12}(s)M_{24}(s)$$
$$P_3(s) = p_{14}M_{14}(s).$$

The MGF of the overall waiting time distribution is

$$M(s) = p_{12}p_{23}M_{12}(s)M_{23}(s)M_{34}(s) + p_{12}p_{24}M_{12}(s)M_{24}(s) + p_{14}M_{14}(s), \tag{7.32}$$

where $p_{12} + p_{14} = 1$, $p_{23} + p_{24} = 1$, and

$$M_{12}(s) = \frac{\beta_1}{\beta_1 - s} \qquad \text{for } s < \beta_1$$

$$M_{14}(s) = \exp\{\lambda_1\lambda_2 - [\lambda_1^2(\lambda_2^2 - 2s)]^{1/2}\} \qquad \text{for } s < \lambda_2^2/s$$

$$M_{23}(s) = \frac{\beta_2}{\beta_2 - s} \qquad \text{for } s < \beta_2$$

$$M_{24}(s) = \left(\frac{\beta_4}{\beta_4 - s}\right)^{\alpha_4} \qquad \text{for } s < \beta_4$$

$$M_{34}(s) = \frac{\beta_3}{\beta_3 - s} \qquad \text{for } s < \beta_3.$$

The overall MGF for time from state $1 \to 4$, transplantation to death or relapse, is found using the MAPLE program code for this flowgraph given in Figure 7.22.

$$M(s) = \frac{p_{12}\, p_{23}\, \beta_1\, \beta_2\, \beta_3}{(\beta_1 - s)(\beta_2 - s)(\beta_3 - s)} + p_{12}(1 - p_{23})\,\beta_1 \left(\frac{\beta_4}{\beta_4 - s}\right)^{\alpha_4} (\beta_1 - s)^{-1}$$

$$+ (1 - p_{12}) e^{\lambda_1 \lambda_2 - \sqrt{\lambda_1^2(\lambda_2^2 - 2.0 s)}}. \qquad \square \qquad (7.33)$$

```
# Include package to generate C code
with(codegen,makeproc,C):

# Define the MGF as a procedure function of s
ms := proc(s)

# Define transmittances using w's for transmittances
local w12, w23, w34, w24, w14;

# Enter each transmittance

   w12 := p12*b1/(b1-s);
   w23 := p23*b2/(b2-s);
   w34 := b3/(b3-s);
   w24 := p24*((b4/(b4-s))^a4);
   w14 := (1-p12)*exp((l1*l2)-(sqrt(l1*l1*(l2*l2-2.0*s))));

# The next line defines M(s)
   w12*w23*w34+w12*w24+w14;
end proc;

# Make the MGF
mx := makeproc(ms(s),[s]):

# Define the CGF, K(s)
k := makeproc(ln(ms(s)),[s]):

# Write the CGF as a C function to a file called k.c
C(k,optimized,ansi,filename="c:/k.c"):

# Define K'(s) and write it as a C function
dk := makeproc(diff(k(s),s),[s]):
C(dk,optimized,ansi,filename="c:/dk.c"):

# Define K''(s) and write it as a C function
ddk := makeproc(diff(dk(s),s),[s]):
C(ddk,optimized,ansi,filename="c:/ddk.c"):
```

FIGURE 7.22 MAPLE code for the Coxian model of Example 7.1.

Phase type models are also used with systems that have multiple exits. The next example presents the flowgraph generalization of these models. We use data on the time between births collected on Norwegian women described in Aalen (1995).

Example 7.2: Flowgraph Model with Multiple Exits. Data were collected on 1779 Norwegian women who have had one live-born child and record the time until their second birth. Women in the study had their first child during the period 1967–1971 and were followed up until 1982. The object of this study is to analyze the time until their next birth. Aalen (1995) conducts an analysis using a PH model with unconditional exponential waiting times. We use Aalen's reported MLEs to present a flowgraph analysis.

Figure 7.23 gives a flowgraph model of this process. The initial state 1 represents the first birth. The end states 3 and 4 represent second birth and no second birth. State 2 represents temporary infertility and state 5 represents the use of contraception. The intermediate states 2 and 5 allow varying degrees of progression to the end state. For example, after the first birth, some women will decide against a second birth, $1 \to 4$, while others will spend time on contraceptives before the second birth, $1 \to 5 \to 3$, and some will have a temporary period of infertility before the second birth, $1 \to 2 \to 3$. This study does not allow a direct $1 \to 3$ transition and assumes that it is encompassed by the $1 \to 2 \to 3$ transition.

The survival time of interest is the probability of not yet having reached state 3. The waiting times for $1 \to 2$ and $2 \to 3$ are modeled as Exp(α). The times $1 \to 5$ and $5 \to 3$ are modeled as Exp(β), and $1 \to 4$, $2 \to 4$, and $5 \to 4$ are modeled as Exp(γ). The maximum likelihood estimates for the exponential parameters are given by Aalen (1995) as $\alpha = 0.190$, $\beta = 0.822$, and $\gamma = 0.0476$. Although a general flowgraph model can allow the exponential parameter values to vary for each

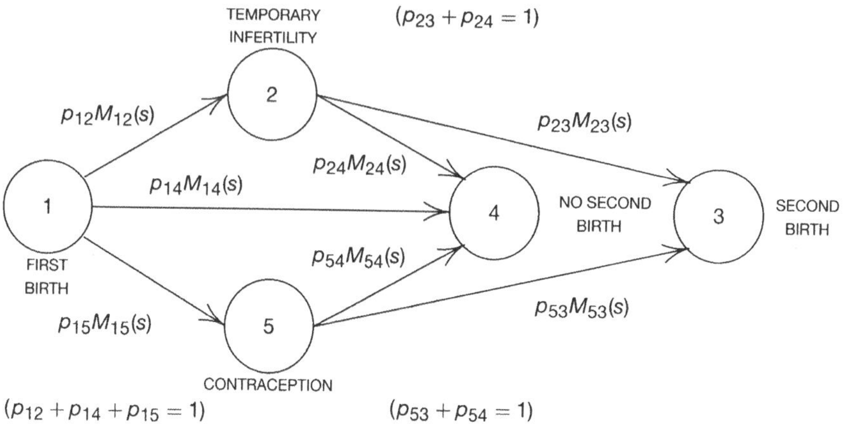

FIGURE 7.23 Multiple exits flowgraph model for the Norwegian birth interval data.

waiting time or allow for alternative distributions, we are using previously reported MLEs in this analysis, and hence we use the same parameterization. The flowgraph model is maintaining the Markov model assumptions of Aalen (1995) and is not performing a conditional analysis.

The flowgraph model allows computation of any desired waiting time distribution, including

1. The time until the second birth
2. The time until no second birth
3. The time until absorption into states 3 or 4

The last quantity is computed for illustration only. In this particular problem, it is the waiting time until a decision is made as to whether or not to have a second child.

When computing the MGF of a partial passage from a complete flowgraph, say the waiting time for the second birth ($1 \to 3$), it is important to note that one is computing a conditional distribution. This involves computing the equivalent transmittance defined in Section 2.4 (Definition 2.4.1) and normalizing it appropriately. Thus far, we have considered finite systems with a single exit, so that passage from input to output is certain to occur. In these cases, the equivalent transmittance $T(s)$ is the MGF of the overall waiting time distribution. For multiple exits, passage from input to a specific exit is not certain to occur, and therefore the equivalent transmittance $T(s)$ for the overall waiting time corresponding to a specific exit is not generally a MGF. The MGF conditional on taking the path $1 \to 3$ is

$$M_{13}(s) = \frac{T_{13}(s)}{T_{13}(0)}, \tag{7.34}$$

where

$$T_{13}(s) = p_{12}p_{23}M_{12}(s)M_{23}(s) + p_{15}p_{53}M_{15}(s)M_{53}(s) \tag{7.35}$$

and $T_{13}(0)$ is the probability of eventual passage to state 3. To compute $T_{13}(0)$, we require the corresponding probabilities from the Markov model of Aalen (1995):

$$\begin{aligned} p_{12} &= \frac{\alpha}{\alpha+\beta+\gamma}, \quad p_{23} = \frac{\alpha}{\alpha+\gamma} \\ p_{15} &= \frac{\beta}{\alpha+\beta+\gamma}, \quad p_{53} = \frac{\beta}{\beta+\gamma}. \end{aligned} \tag{7.36}$$

Using the MLEs, we find $T_{13}(0) = 0.877$, the estimated probability of eventual passage to state 3.

The MGF of the overall waiting time distribution for no second birth $1 \to 4$ is computed as follows. We denote the equivalent transmittance from $1 \to 4$ by $T_{14}(s)$,

$$T_{14}(s) = p_{14}M_{14}(s) + p_{12}p_{24}M_{12}(s)M_{24}(s) + p_{15}p_{54}M_{15}(s)M_{54}(s). \tag{7.37}$$

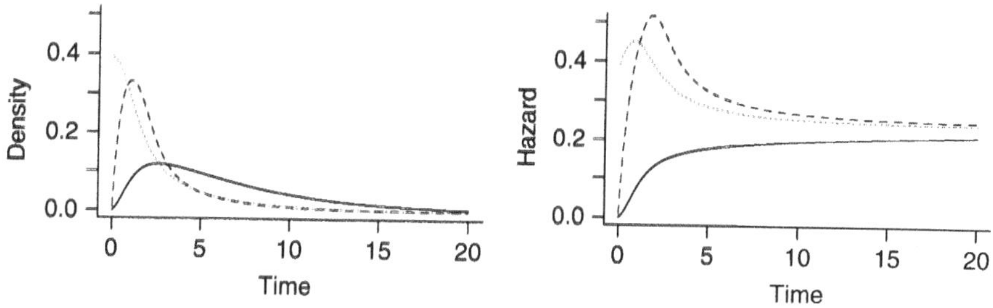

FIGURE 7.24 Birth interval data. Time to second birth (dashed line), time to no second birth (dotted line), and time to absorption into state 3 or 4 (solid line).

The MGF conditional on reaching state 4 is

$$M_{14}(s) = \frac{T_{14}(s)}{T_{14}(0)}. \tag{7.38}$$

We can estimate $T_{14}(0)$ from the MLEs of the transition probabilities (7.36) or notice that eventual passage to state 4 is complementary to absorption in state 3, and hence $T_{14}(0) = 1 - T_{13}(0) = 0.123$.

The equivalent transmittance of the overall waiting time distribution for absorption into either states 3 or 4 is the same as the MGF of the unconditional overall waiting time distribution, since passage to one of these states is certain to occur. This MGF is

$$M(s) = T_{13}(s) + T_{14}(s). \tag{7.39}$$

The densities and hazard functions for both passages, $1 \to 3$ and $1 \to 4$, and the overall time to absorption are presented in Figure 7.24. □

PH models are also used to model absorption times for situations involving arbitrary initial states. Flowgraph models can be used in the same context but allow the use of arbitrary waiting time distributions with ease. Often, we are interested in the time to absorption starting in a state other than the initial state. For example, in the HIV progression example of Section 7.2, a patient diagnosed with AIDS may already be experiencing pre-AIDS symptoms. In that case, the relevant quantities of interest would be the waiting times to full-blown AIDS or death (i.e., the waiting times $3 \to 4$ or $3 \to 5$ in the flowgraph of Figure 7.2). Similar quantities can be computed for the retinopathy flowgraph of Figure 7.4. The next example considers computing such quantities for a birth and death process over states $1, 2, \ldots, 6$ with an absorbing end state, 6.

Example 7.3: Birth and Death Process. Figure 7.25 presents a flowgraph model for a birth and death process with an absorbing end state, 6. We use this to demonstrate the effect of starting in arbitrary initial states. For illustration of PH

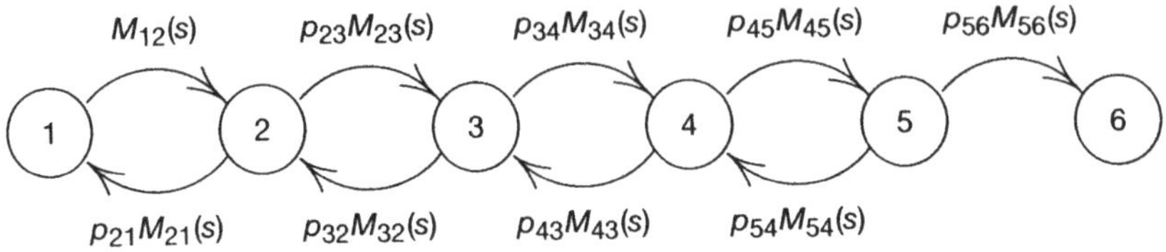

FIGURE 7.25 Flowgraph model for a birth and death process.

TABLE 7.6 Paths and Loops for the Flowgraph of a Birth and Death Process

Path	Transmittance
$1 \to 2 \to 3 \to 4 \to 5 \to 6$	$p_{23}p_{34}p_{45}p_{56}M_{12}(s)M_{23}(s)M_{34}(s)M_{45}(s)M_{56}(s)$
First order loops:	
$1 \to 2 \to 1$	$p_{21}M_{12}(s)M_{21}(s)$
$2 \to 3 \to 2$	$p_{23}p_{32}M_{23}(s)M_{32}(s)$
$3 \to 4 \to 3$	$p_{34}p_{43}M_{34}(s)M_{43}(s)$
$4 \to 5 \to 4$	$p_{45}p_{54}M_{45}(s)M_{54}(s)$
Second order loops:	
$1 \to 2 \to 1$ and $3 \to 4 \to 3$	$p_{21}p_{34}p_{43}M_{12}(s)M_{21}(s)M_{34}(s)M_{43}(s)$
$1 \to 2 \to 1$ and $4 \to 5 \to 4$	$p_{21}p_{45}p_{54}M_{12}(s)M_{21}(s)M_{45}(s)M_{54}(s)$
$2 \to 3 \to 2$ and $4 \to 5 \to 4$	$p_{23}p_{32}p_{45}p_{54}M_{23}(s)M_{32}(s)M_{45}(s)M_{54}(s)$

analysis, we assume a Markovian process with exponential waiting times. However, we perform the flowgraph computations for the case of general waiting time distributions. We compute the overall waiting time to absorption starting in states 1, 3, and 5. First we compute the MGFs of these quantities. The flowgraph has many first- and second-order loops. Table 7.6 gives the paths and loops involved for computing the overall waiting time distribution for $1 \to 6$. Using Mason's rule (2.26) along with Table 7.6, we find the equivalent transmittance for $1 \to 6$, which is the MGF of the waiting time distribution for $1 \to 6$:

$$M(s) = \frac{p_{23}p_{34}p_{45}p_{56}M_{12}(s)M_{23}(s)M_{34}(s)M_{45}(s)M_{56}(s)}{1 - L_1(s) + L_2(s)}, \tag{7.40}$$

where

$$\begin{aligned} L_1(s) &= p_{21}M_{12}(s)M_{21}(s) + p_{23}p_{32}M_{23}(s)M_{32}(s) + p_{34}p_{43}M_{34}(s)M_{43}(s) \\ &\quad + p_{45}p_{54}M_{45}(s)M_{54}(s) \\ L_2(s) &= p_{21}p_{34}p_{43}M_{12}(s)M_{21}(s)M_{34}(s)M_{43}(s) + p_{21}p_{45}p_{54}M_{12}(s)M_{21}(s) \\ &\quad \times M_{45}(s)M_{54}(s) + p_{23}p_{32}p_{45}p_{54}M_{23}(s)M_{32}(s)M_{45}(s)M_{54}(s). \end{aligned} \tag{7.41}$$

There are no first- or second-order loops that do not touch the path, so $L^i_j(s) = 0$ for all i and j.

Note that at every state, $p_{i,i+1} + p_{i,i-1} = 1$ for $i = 2, \ldots, 5$. If all the unconditional individual waiting times are exponential, the process in Figure 7.25 reduces to a Markovian birth and death process, with birth parameters λ_i and death parameters μ_i.

The equivalent transmittance for state 3 to state 6 is found by appropriate redefinition of the flowgraph of Figure 7.25 and a reapplication of Mason's rule along the main path $3 \to 4 \to 5 \to 6$. States 1 and 2 are only involved in defining loops associated with this path. The equivalent transmittance of the waiting time to reach state 6 starting in state 3 is given by

$$M_{36}(s) = \frac{p_{34}p_{45}p_{56}M_{34}(s)M_{45}(s)M_{56}(s)[1 - p_{21}M_{12}(s)M_{21}(s)]}{1 - L_1(s) + L_2(s)},$$

where $L_1(s)$ and $L_2(s)$ are the sum of the first- and second-order loop transmittances given by (7.41).

Another alternative is to begin the process in state 5. This involves using Mason's rule with the main path $5 \to 6$ giving the equivalent transmittance:

$$T_{56}(s) = \frac{p_{56}M_{56}(s)[1 - L^1_1(s) + L^1_2(s)]}{1 - L_1(s) + L_2(s)}, \tag{7.42}$$

where

$$L^1_1(s) = p_{21}M_{12}(s)M_{21}(s) + p_{23}p_{32}M_{23}(s)M_{32}(s) + p_{34}p_{43}M_{34}(s)M_{43}(s)$$
$$L^1_2(s) = p_{21}p_{34}p_{43}M_{12}(s)M_{21}(s)M_{34}(s)M_{43}(s)$$

and $L_1(s)$ and $L_2(s)$ are as in (7.41). Note that we denote this MGF as $T_{56}(s)$ to distinguish it from $M_{56}(s)$, the MGF of the waiting time distribution for direct passage from $5 \to 6$. In this calculation, transition all the way back to state 1 is allowed in the passage from $5 \to 6$.

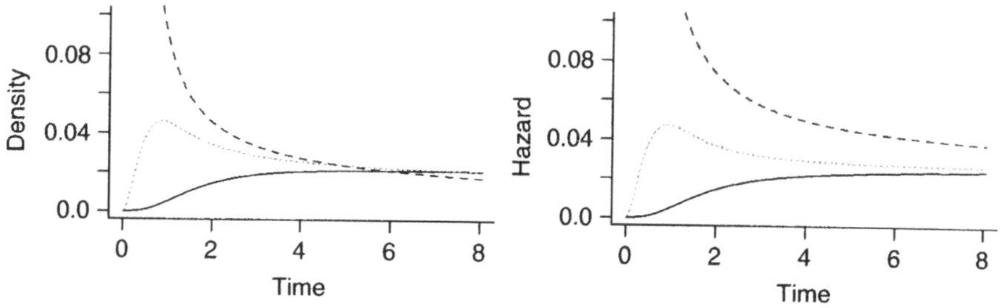

FIGURE 7.26 Density and hazard functions for the waiting time until absorption for a birth and death process with different initial states: 1 (solid line), 3 (dotted line), 5 (dashed line).

Figure 7.26 presents the density and hazard functions for the waiting time to absorption into state 6 beginning in states 1, 3, and 5, assuming a common birth parameter $\lambda_1 = \cdots = \lambda_5 = 1$ and common death parameter $\mu_1 = \cdots = \mu_5 = 2$. □

EXERCISES

7.1 Consider data on the construction engineering flowgraph of Figure 7.15. Perform a flowgraph analysis using the data provided in Table P7.1. Construct censored data histograms with overlaid distributions. Compute the Bayes predictive density based on these data.

TABLE P7.1 Data for Construction Engineering Flowgraph of Figure 7.15

Obs.	$2 \to 3$	$3 \to 4$	$3 \to 5$	$4 \to 5$
1	4.5	2		21
2	4.0	5		18
3	3.5	5		18
4	3.5	5		18
5	4.0	3		20
6	4.5	3		20
7	4.0	15		
8	4.0	5		18
9	3.5	6		17
10	4.0	20		
11	4.5	34		
12	4.5	28		
13	4.0	5		18
14	4.0	5		18
15	4.0		5	
16	3.5		40	
17	4.0		15	
18	4.5		5	
19	4.0		25	
20	4.0		20	

7.2 Refer to Example 7.2 and answer the following:

(a) Verify that the overall transmittance computed in (7.39) is a MGF.

(b) Prove that the probability of passage from $1 \to 3$, $P(1 \to 3)$, is $T_{13}(0)$.

(c) Prove that $M_{13}(s)$, the MGF of the waiting time distribution from $1 \to 3$, is

$$M_{13}(s) = \frac{T_{13}(s)}{T_{13}(0)}.$$

7.3 Consider the Coxian model of Example 7.1. The data below are simulated using an Exp(0.3) for the $1 \to 2$ branch, $IG(2, 3)$ for the $1 \to 4$ branch, Exp(3) for the $2 \to 3$ branch, Gamma(5, 7) for the $2 \to 4$ branch, and Exp(1.5) for the $3 \to 4$ branch. The probabilities used were $p_{12} = 0.7$ and $p_{23} = 0.5$. A total of 50 observations were simulated. The data are given in Tables P7.2 and P7.3. The censored observations are listed by the state in which they are censored.

TABLE P7.2 Censored Data from the Coxian Model for Exercise 7.3[a]

State 1	State 2	State 3
6.04272	0.23148	1.00592
7.62828		
7.26209		
7.46458		

[a]Column headings are the state in which censoring occurred.

(a) Compute censored data histograms of these data. Are there parametric models that appear to fit better than those used for simulating the data?

(b) Compute both a Bayes predictive density and a maximum likelihood estimated density for these data. Use appropriate parametric models (i.e., not necessarily the models used for the data simulation).

TABLE P7.3 Uncensored Data from the Coxian Model for Exercise 7.3

$1 \to 2$	$1 \to 4$	$2 \to 3$	$2 \to 4$	$3 \to 4$
5.87851	0.55606	0.122810	0.64666	0.07599
2.76795	0.44276	0.155258	0.43216	0.03188
3.47374	0.62386	0.306455	0.49533	0.05896
3.36958	0.90834	0.104286	0.38785	0.38252
5.53786	0.94529	0.158914	0.78057	0.21213
2.54920	0.59530	0.453173	0.95955	0.05746
0.42444	0.71333	0.073219	0.76170	0.03271
7.02077	0.52530	0.024199	0.11520	0.21136
0.13216	0.60365	0.048729	1.02099	1.90332
0.46382	0.68176	0.093590	0.94276	0.21709
4.07049	0.59134	0.187996	0.73397	0.40648
2.23654	1.12013	0.213918	0.59847	0.04641
5.99006	0.79887	0.392623	0.52195	0.16877
0.17531		0.633503	0.48523	2.55231

(*Continued*)

TABLE P7.3 (*Continued*)

$1 \rightarrow 2$	$1 \rightarrow 4$	$2 \rightarrow 3$	$2 \rightarrow 4$	$3 \rightarrow 4$
3.64898		0.051018	0.62957	0.01790
3.74522		0.075643		1.32935
1.28670		0.271855		
1.11451				
2.91260				
1.43268				
0.69087				
5.42923				
3.49993				
1.90993				
2.67206				
5.92624				
0.13400				
0.91085				
1.51784				
0.60293				
6.56427				
3.05279				
1.40729				

CHAPTER 8

Incomplete Data

Thus far, in our analysis of time-to-event data using flowgraph models, we have assumed that we have essentially complete information on the transitions and waiting times. We have allowed for censoring but assumed that a censored observation would only have transited to states connected by direct branches. Complete data on the flowgraph model consist of following every person through every intermediate transition. Incomplete data occur when not all of the transitions are observed. This happens in several ways; for example, consider the flowgraph model for the progression of diabetic retinopathy given in Figure 7.4. Suppose that we observe a patient only in states $1 \rightarrow 3 \rightarrow 4$. We know that the patient must have transited through state 2 but we do not know when or how often i.e., the patient's path may have been $1 \rightarrow 2 \rightarrow 3 \rightarrow 4$, but another possibility is $1 \rightarrow 2 \rightarrow 3 \rightarrow 2 \rightarrow 1 \rightarrow 2 \rightarrow 3 \rightarrow 4$. Furthermore, the patient's path may be subject to censoring. Although censored data is incomplete in that the final transition is not observed to occur, when dealing with flowgraphs we make the distinction between censored data and what we will refer to as incomplete data. The term *complete data* refers to complete transition information and includes right-censored observations. For example, in Section 7.3, the retinopathy data had many right-censored observations. If an observation were censored in state 2, the implicit assumption is that the person could transit to either state 1 or state 3, as both states are connected by direct branches from state 2, and hence the transition information is complete although the waiting time is censored.

Definition 8.0.1. *Incomplete data* consists of data that have complete information on observed waiting times but incomplete information on the associated transitions.

The contribution of the censored observation is a finite mixture distribution weighted by the respective transition probabilities. The main problem when dealing with incomplete data having missing transitions is the difficulty in calculating the contribution of the incomplete data terms to the likelihood. In this chapter

Flowgraph Models for Multistate Time-to-Event Data, By Aparna V. Huzurbazar
ISBN 0-471-26514-4

we discuss two types of incomplete data: incomplete data that are known to be incomplete and incomplete data that are not known to be incomplete, called *unrecognized incomplete data.*

8.1 CONSTRUCTED LIKELIHOOD

The first step in dealing with any type of incomplete data in a flowgraph model is to construct an approximate likelihood contribution of that data. The approximate likelihood is constructed with the saddlepoint approximation using the underlying flowgraph model. This methodology can be used in either the Bayesian or frequentist framework, and in this section we present steps for implementing both methods of analysis. Suppose that $T_1, \ldots, T_n$ are n independent and identically distributed random variables representing the total waiting time in a flowgraph model. For example, in the retinopathy flowgraph of Figure 7.4, our data would consist only of observations on the waiting time $1 \to 4$. We do not observe any intermediate transitions that the patients may have made. Since we do not observe the intermediate transitions, we cannot write down a likelihood function involving the associated waiting time distributions. However, we can construct an approximate likelihood

$$\tilde{\mathcal{L}}(\boldsymbol{\theta}|\mathcal{D}) = \prod_{j=1}^{n} \tilde{f}_T(t_j|\boldsymbol{\theta}), \tag{8.1}$$

where $\mathcal{D}$ is the data and $\tilde{f}_T(t_j|\boldsymbol{\theta})$ is the normalized saddlepoint density approximation (3.19). The flowgraph model gives the MGF of the total waiting time. The saddlepoint approximation converts this MGF to a density, giving the approximate likelihood. Thus, we are using the saddlepoint approximation to construct an approximate sampling distribution. Note that this likelihood construction depends on the entire flowgraph parameterization.

In general, observations will contribute different amounts of data. Some observations may be complete, allowing us to write down their full likelihood, whereas others may be incomplete, requiring likelihood construction. The overall likelihood function is given by

$$\mathcal{L}(\boldsymbol{\theta}|\mathcal{D}) = \mathcal{L}(\boldsymbol{\theta}|\mathcal{D}_{\text{complete}})\tilde{\mathcal{L}}(\boldsymbol{\theta}|\mathcal{D}_{\text{incomplete}}). \tag{8.2}$$

The likelihood can involve different flowgraph MGFs associated with the incomplete data cases. If each observation has different amounts of data, we must construct the likelihood in pieces. Suppose that in the retinopathy flowgraph of Figure 7.4, the first observation provides complete data on $1 \to 2 \to 3$ and the second observation provides incomplete data on $2 \to 1 \to 3$. The contribution to the likelihood from the first observation is

$$f_{12}(x_{112}|\boldsymbol{\theta})(1 - p_{21})f_{23}(x_{123}|\boldsymbol{\theta}),$$

where x_{igh} denotes the observation on patient i of the $g \to h$ transition. The second patient has a complete observation from $2 \to 1$ followed by an incomplete observation from $1 \to 3$. The MGF required to construct the likelihood from $1 \to 3$ is

$M_{13}(s)$. We assume that the $1 \to 3$ transition time reflects the first time that state 3 was reached from state 1. Note that in this discussion we do not allow for the case of unrecognized incomplete data where the true transition could have been

$$1 \to 2 \to 3 \to 2 \to 3,$$

and we observe the second time that state 3 is reached. The MGF $M_{13}(s)$ is computed by removing the $3 \to 4$ and $3 \to 2$ states of the original flowgraph so that

$$M_{13}(s) = \frac{(1 - p_{21})M_{12}(s)M_{23}(s)}{1 - p_{21}M_{12}(s)M_{21}(s)}. \tag{8.3}$$

The contribution of the second patient to the likelihood is

$$p_{21} f_{21}(x_{221}|\boldsymbol{\theta})\tilde{f}_{13}(x_{213}|\boldsymbol{\theta}).$$

The term $\tilde{f}_{13}(x_{213}|\boldsymbol{\theta})$ is constructed from $M_{13}(s)$ using the saddlepoint approximation (3.19) for given $\boldsymbol{\theta}$.

We can implement likelihood construction in the Bayesian framework as follows. Suppose that we have a flowgraph model whose parameter vector is $\boldsymbol{\theta}$. Let Λ_{id} denote the collection of transitions for which incomplete data are observed. For $j \in \Lambda_{id}$, let t_{jk}, $k = 1, \ldots, n_j$, denote the n_j waiting times observed on transition j. The posterior distribution of $\boldsymbol{\theta}$ is given by

$$\pi(\boldsymbol{\theta}|\mathcal{D}) \propto \mathcal{L}(\boldsymbol{\theta}|\mathcal{D})\pi(\boldsymbol{\theta}), \tag{8.4}$$

where $\mathcal{L}(\boldsymbol{\theta}|\mathcal{D})$ is given by (8.2). We generate posterior samples of $\boldsymbol{\theta}$ using an MCMC algorithm such as slice sampling, where $\tilde{\mathcal{L}}(\boldsymbol{\theta}|\mathcal{D}_{\text{incomplete}})$ is calculated as

$$\tilde{\mathcal{L}}(\boldsymbol{\theta}|\mathcal{D}_{\text{incomplete}}) = \prod_{j \in \Lambda_{id}} \prod_{k=1}^{n_j} \tilde{f}_j(t_{jk}|\boldsymbol{\theta}). \tag{8.5}$$

The predictive density $f_Z(z|\mathcal{D})$ is estimated by averaging the approximate overall waiting time density $\tilde{f}_Z(z|\boldsymbol{\theta})$ over the posterior samples $\boldsymbol{\theta}_1, \ldots, \boldsymbol{\theta}_m$:

$$\hat{f}_Z(z|\mathcal{D}) = \frac{1}{m}\sum_{i=1}^{m} \tilde{f}_Z(z|\boldsymbol{\theta}_i). \tag{8.6}$$

The density $\tilde{f}_Z(z|\boldsymbol{\theta})$ is calculated by inverting the MGF of Z using the saddlepoint approximation.

To estimate $\boldsymbol{\theta}$ in the frequentist setting, maximize $\mathcal{L}(\boldsymbol{\theta}|\mathcal{D})$ using any appropriate multidimensional optimization algorithm, such as Nelder–Mead simplex (cf. Nelder and Mead, 1965), simulated annealing (cf. Bohachevsky et al., 1986) or a genetic algorithm (cf. Chatterjee et al., 1996). Use this MLE in the subsequent saddlepoint computation to convert the MGF of Z to a maximum likelihood estimated density.

Example 8.1: KIDNEY EXAMPLE LIKELIHOOD CONSTRUCTION. We return to the two-state model for kidney failure first presented in Example 2.1. The flowgraph is duplicated in Figure 8.1. Example 3.1 presented the exact inversion of the MGF for the waiting time distribution for failure of both kidneys. Example 5.5 presented the Bayes predictive density for the time to failure of both kidneys, and Example 3.8 presented the maximum likelihood estimated density. Recall that state 0 represents initial diagnosis of kidney disease where the patient has two functioning kidneys. State 1 represents one failed kidney, and state 2 represents two failed kidneys, an absorbing state. Y_0 is the random waiting time in state 0 until state 1 is reached, and Y_1 is the random waiting time in state 1 until state 2 is reached. We assume that $Y_0 \sim \text{Exp}(2\lambda_0)$ is independent of $Y_1 \sim \text{Exp}(\lambda_1)$. Let $T = Y_1 + Y_2$ be the total waiting time from $0 \to 2$. Suppose that we have observations on T from n kidney patients i.e., we have incomplete data such that we observe each patient in state 0 and next in state 2. Our quantity of interest is the survival time of the patient's kidneys, that is, the waiting time from $0 \to 2$. We compute both the exact likelihood and the approximate likelihood.

We found the exact density of T in (3.6) to be

$$f_T(t|\lambda_0, \lambda_1) = \frac{2\lambda_0\lambda_1}{2\lambda_0 - \lambda_1}(e^{-\lambda_1 t} - e^{-2\lambda_0 t}) \qquad \text{for } t > 0, \lambda_0 > 0, \lambda_1 > 0, 2\lambda_0 \neq \lambda_1. \tag{8.7}$$

We simulate $n = 10$ observations using the MLEs reported in Gross et al. (1971) as the true parameter values. For the purpose of this illustration, we drop the constraint $\lambda_0 \leq \lambda_1$ and assume that $0 < \lambda_0 < 1$ and $0 < \lambda_1 < 3$. The exact log-likelihood for each (λ_0, λ_1) value is computed as

$$\sum_{j=1}^{n} \log f_T(t_j|\lambda_0, \lambda_1). \tag{8.8}$$

This is shown in Figure 8.2. The constructed log-likelihood uses the flowgraph MGF (3.3) along with the saddlepoint approximation on the same data of 10 observations. The constructed likelihood based on (8.1) is shown in Figure 8.3. As seen in the figures, the constructed likelihood compares favorably to the exact. In general, the constructed likelihood tends to be more diffuse. In these two-dimensional plots, we can see that the shape of the likelihood reflects the deconvolution problem. The gently sloping ridges along λ_0 for fixed $\lambda_1 \approx 0.25$ and along

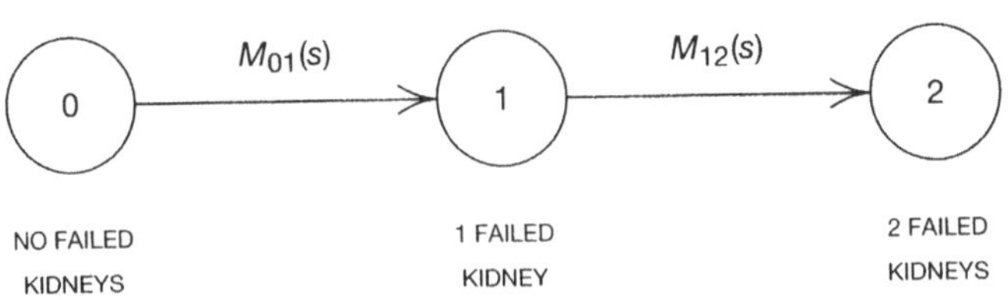

FIGURE 8.1 Series flowgraph for kidney failure.

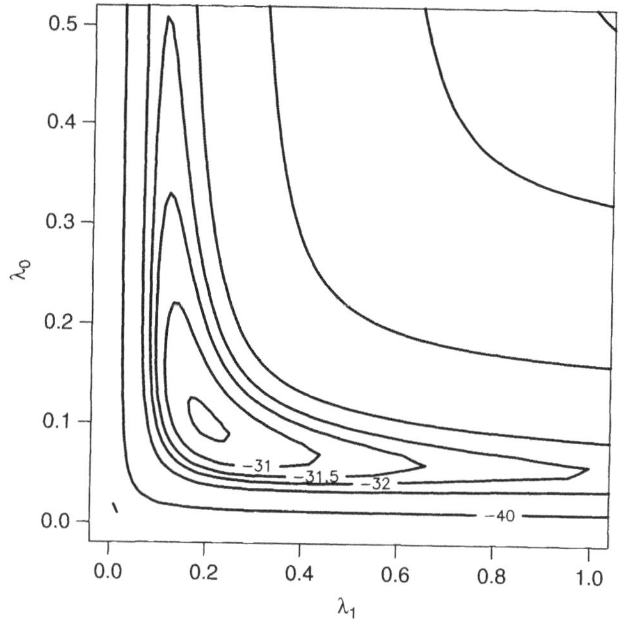

FIGURE 8.2 Exact log-likelihood for Example 8.1.

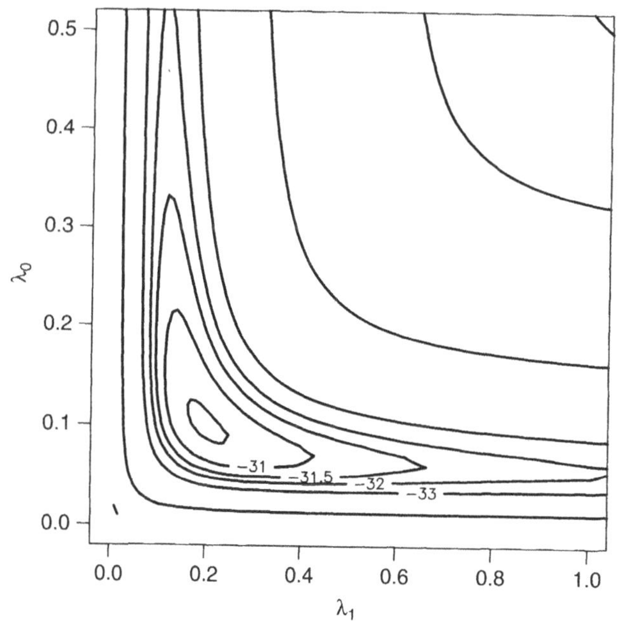

FIGURE 8.3 Saddlepoint constructed log-likelihood for Example 8.1.

λ_1 for $\lambda_0 \approx 0.075$ demonstrate the lack of identifiability in λ_0 and λ_1. We have information on the sum and are trying to deconvolve information in the likelihood for the separate parameters of the component distributions. □

Example 8.2: RETINOPATHY DATA WITH INCOMPLETE OBSERVATIONS. Recall the diabetic retinopathy example of Section 7.3.1. The full data, given in Table 7.2, consist of 11 incomplete observations which we did not include in our previous analysis. In the data, nine people were observed in state 1 and then in state 3. We do not know when transitions to state 2 were made. Two people were observed in state 2 and then in state 4, but not in between. This example extends the retinopathy model to include these 11 observations. We use the same parametric model as in Section 7.3.1 and the same priors. The likelihood function is given by (8.2), where the incomplete data portion of the likelihood is given by

$$\tilde{\mathcal{L}}(\boldsymbol{\theta}|\mathcal{D}_{\text{incomplete}}) = \prod_{i=1}^{9} \tilde{f}_{13}(x_{i13}|\boldsymbol{\theta}) \times \prod_{i=1}^{2} \tilde{f}_{24}(x_{i24}|\boldsymbol{\theta}). \quad (8.9)$$

In this likelihood, $\tilde{f}_{13}(\cdot)$ is constructed from the MGF $M_{13}(s)$, given in (8.3), for each of the nine people, and $\tilde{f}_{24}(\cdot)$ is constructed from the MGF $M_{24}(s)$, given in (7.28), for each of the two patients. The computation is performed using (8.6) for the Bayes predictive density and also by maximizing the combined likelihoods of the complete and incomplete data. Figure 8.4 gives resulting density estimates. Figures 8.5 and 8.6 give the resulting predictive survivor and hazard functions.

Figure 8.7 compares the complete data density with the density incorporating the incomplete data for both the MLE and Bayesian estimation techniques. Since there are only 11 incomplete observations, it seems surprising that the effect on the density is so dramatic. This is especially obvious when examining the quantile values for the incomplete data case, Table 8.1, and the complete data case, Table 7.3. The mean of the Bayes predictive distribution of the complete data in Table 7.3 is 639.46, the mean of the data with just the uncensored observations is 501.42. The mean of the predictive density for the complete data with the incomplete data is 475.91, lower than either of the other cases.

To investigate this further, we compute estimates of the branch waiting time means for the combined complete and incomplete data with censoring (see Table 8.2). These branch waiting time means are in line with those of the complete data given in Table 7.4. The overall mean of the predictive distribution for the data of Table 8.2 is 383.01. Recall the discussion of the data in Table 7.4. Although the overall mean increases as a consequence of the individual transition means being generally much larger with censored data, there appears to be some compensation due to lower values of $\hat{p}_{21}$ and $\hat{p}_{32}$. Figure 8.8 gives a plot of the posterior distribution of p_{32} for the combined complete and incomplete data, the complete data and the complete data with just the uncensored observations. In comparing the complete data and the complete data with just the uncensored observations, the distribution of p_{32} is similar. There is slightly more mass at lower values of p_{32} for the complete data; nonetheless, the overall mean is larger for the complete data

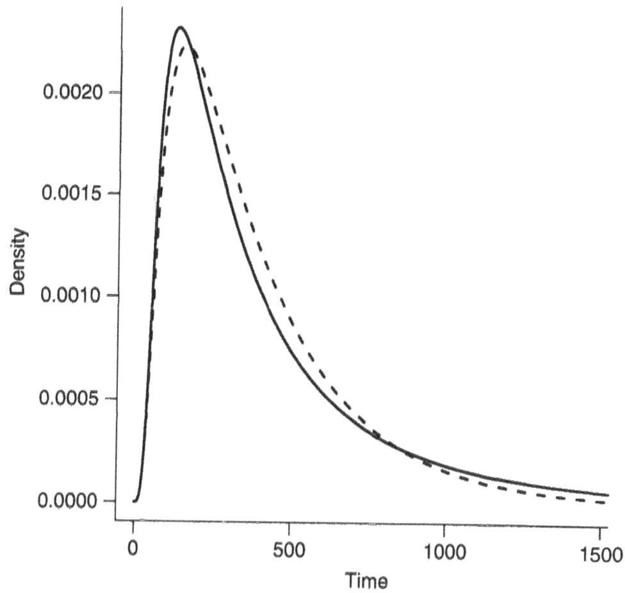

FIGURE 8.4 Bayes predictive (solid line) and maximum likelihood (dashed line) estimated densities for the retinopathy data based on a constructed likelihood using the combined complete and incomplete observations.

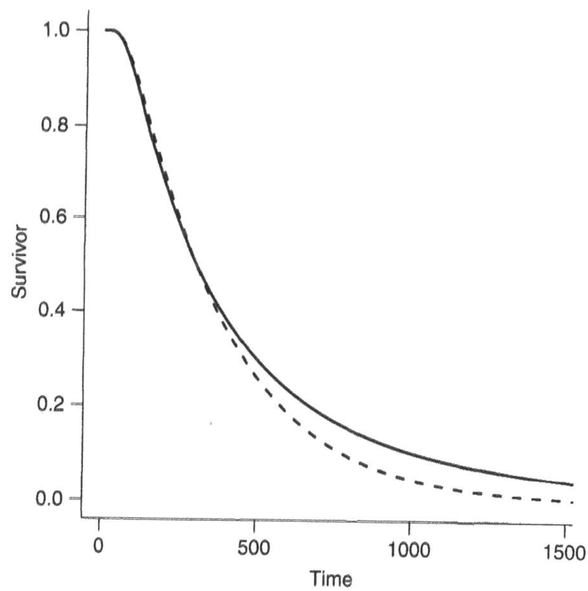

FIGURE 8.5 Bayes predictive (solid line) and maximum likelihood (dashed line) estimated survivor functions for the retinopathy data based on a constructed likelihood using the combined complete and incomplete observations.

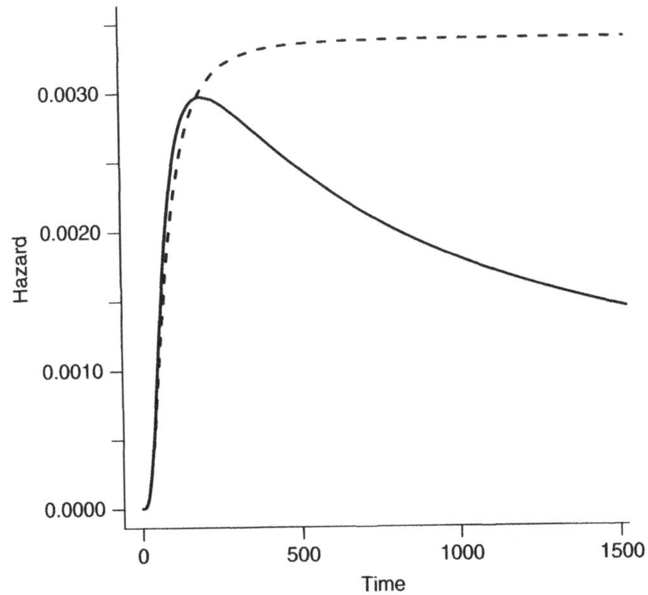

FIGURE 8.6 Bayes predictive (solid line) and maximum likelihood (dashed line) estimated hazard functions for the retinopathy data based on a constructed likelihood using the combined complete and incomplete observations.

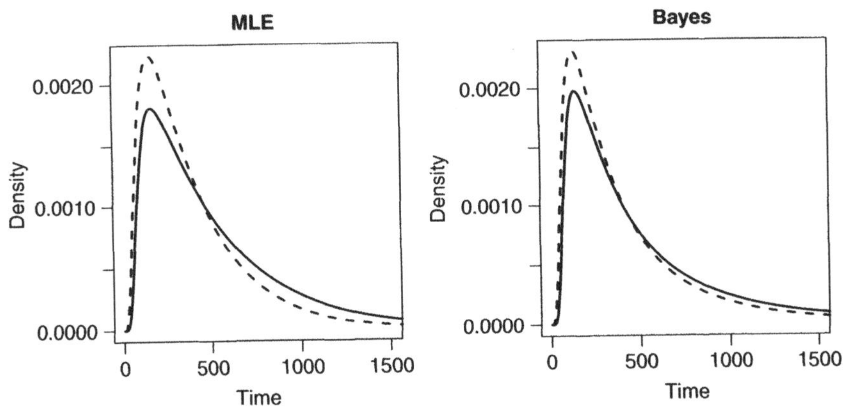

FIGURE 8.7 ML estimated densities and Bayes predictive densities for the retinopathy example with (1) complete data only (solid) and (2) incomplete data in addition to complete data (dashed).

due to the larger waiting time means for the individual transitions in the complete data. For the combined complete and incomplete data, the posterior distribution of p_{32} is shifted to the left relative to the complete data. Since the individual branches have approximately the same mean waiting times, this leads to a lower predictive

TABLE 8.1 Selected Quantiles of the Density Functions of Figure 8.4 for the Retinopathy Data[a]

Percentile	Maximum Likelihood	Bayes Predictive Density
0.25	175.1	166.48
0.50	304.35	304.95
0.75	513.2	574.77
0.90	785.11	1027.63
0.95	990.02	1455.89
0.99	1465.07	2792.08
Mean	383.01	475.91
s.d.	300.58	606.61

[a] Based on a constructed likelihood using the combined complete and incomplete observations.

TABLE 8.2 Means of Branch Waiting Time Distributions for the Combined Complete and Incomplete Retinopathy Data with Censoring

Transition	Combined Data Branch Means
$1 \to 2$	30.94
$2 \to 1$	20.03
$2 \to 3$	51.49
$3 \to 2$	34.08
$3 \to 4$	56.14
$\hat{p}_{21}$	0.265
$\hat{p}_{32}$	0.685

distribution mean because there is less of a chance of returning to earlier states in the process. This may also help to explain the lower mean for the combined complete and incomplete data relative to the mean for the complete data with uncensored data only (501.43). Even though the latter data have smaller branch waiting time means, the posterior distribution of p_{32} is shifted toward higher values. □

Example 8.3: DEGENERATIVE STAGES OF DEMENTIA. Dementia is a progressive brain dysfunction that leads to a gradually increasing restriction of daily activities. Alzheimer's disease is the best known type of dementia. Although the disease can strike at any age, it becomes more frequent with age. The prevalence is approximately 5 to 10% in people aged over 65 and approximately 20% in people aged over 80 (cf. Saunders, 1993). Alzheimer's disease is a slowly unfolding progressive

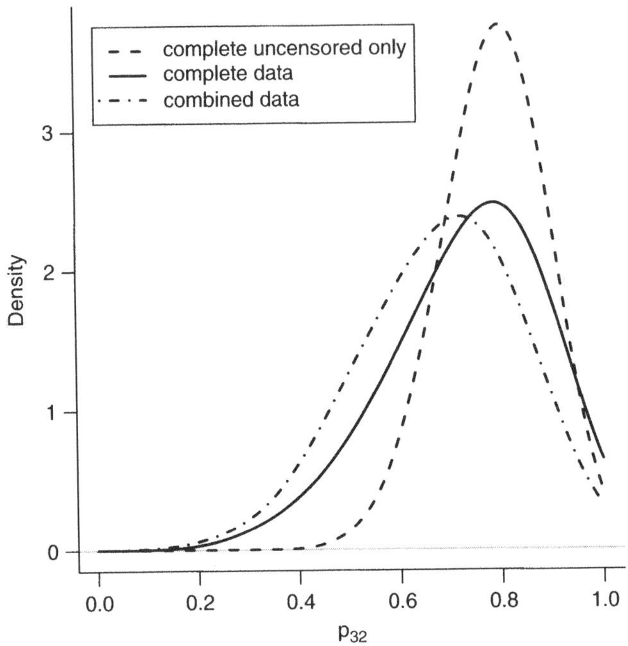

FIGURE 8.8 Posterior distribution of p_{32}. Complete uncensored only refers to the complete data with just the uncensored observations and combined full data refers to the complete data with the incomplete observations.

condition that is typically divided into three major stages: early-stage, midstage, and late-stage Alzheimer's disease (cf. Chatterjee et al., 1992). The early stage is a mild stage that is typically characterized by impairments of mental abilities as well as mood swings. The midstage can be categorized into several moderate degenerative stages. In these stages, behavioral disturbances become increasingly common, memory loss for recent events becomes more pronounced, intellectual functions begin to deteriorate, and motor skills begin to deteriorate. Late-stage Alzheimer's is an advanced stage where physical problems become dominant and patients are unable to care for themselves. There is substantial variability in the individual course of the disease. Figure 8.9 gives a flowgraph model for the degenerative stages of Alzheimer's dementia. State 0 is the early stage and states 1, 2, and 3 are degenerative conditions representing varying degrees of severity in the moderate stage of the disease. From state 0, the patient can progress to any one of these three states. Once in states 1, 2, or 3, the patient can improve or worsen, indicated by branches between these states. Eventually, the disease progresses to the advanced stage, 4, and ultimately, the patient dies from the disease, state 5. This characterization of stages was proposed by Commenges (1986), and Butler and Huzurbazar (1997) present a flowgraph representation. Commenges et al. (1998) analyze data on Alzheimer's patients using a simple three-state model. The real data are not available to us; thus we simulate data on the more complicated model

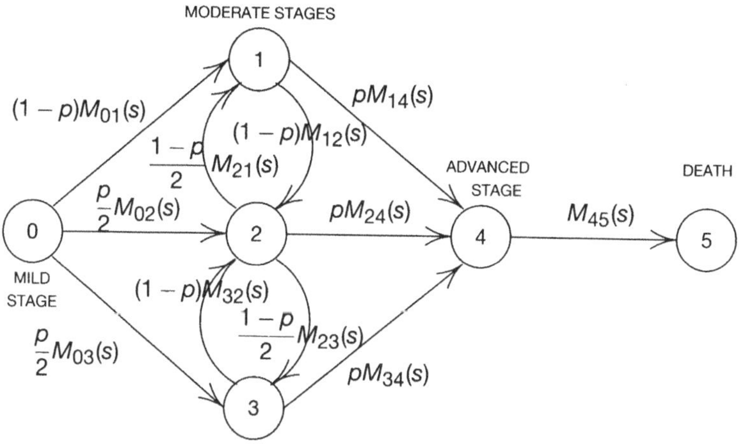

FIGURE 8.9 Flowgraph model for the degenerative stages of Alzheimer's dementia.

of Figure 8.9. We follow the analysis of Williams and Huzurbazar (2004) using a combination of complete and incomplete data.

A patient beginning in state 0 is in a competing risks situation for making a transition to one of the moderately degenerative conditions: state 1, 2, or 3. We simulate data from the flowgraph model assuming that the waiting time distributions from state 0 to states 1, 2, and 3 are Exp(λ_1), Exp(λ_2), and IG(η_1, η_2). The corresponding branch probabilities are $(1-p)$, $p/2$, and $p/2$. The waiting time distributions for states $1 \to 2$ and $2 \to 3$ are both Exp(μ_1) with branch probabilities $p_{12} = (1-p)$ and $p_{23} = (1-p)/2$. The waiting time distributions for states $2 \to 1$ and $3 \to 2$ are both Exp(μ_2) with $p_{21} = (1-p)/2$ and $p_{32} = (1-p)$. The waiting time distributions for states $1 \to 4$, $2 \to 4$, and $3 \to 4$ are each modeled as Gamma(α, β) with probabilities $p_{14} = p_{24} = p_{34} = p$. The waiting time distribution for state $4 \to 5$ is IG(δ_1, δ_2). The model parameter vector is $\boldsymbol{\theta} = (p, \lambda_1, \lambda_2, \eta_1, \eta_2, \mu_1, \mu_2, \alpha, \beta, \delta_1, \delta_2)$. We simulate complete data on 100 patients without censoring using $\boldsymbol{\theta} = (0.4, 0.6, 0.3, 6, 2, 0.5, 0.3, 4, 0.25, 4, 7)$. We make the complete observations incomplete by recording only the time it takes for a patient to reach state 5 (death). Thus, we have incomplete observations on states $0 \to 5$ for each patient. In particular, n = 20 of the complete data observations are randomly selected and recorded in full, while the remaining 80 have only their total waiting time from $0 \to 5$ recorded. We do not keep any information on the intermediate transitions or the associated waiting times for these incomplete observations.

We assume a Jeffreys prior (5.5) prior on θ. We use the constructed likelihood of (8.1) for the incomplete observations in computation of the predictive density in (8.6). Note that Butler and Huzurbazar (1997) perform the computation in a slightly different manner. They sample $\theta_1, \ldots, 0_m$ from the prior distribution and use those values of θ in (8.7). Their "predictive density" is therefore not a predictive density because it converges to the prior mean of $f_Z(z|\boldsymbol{\theta})$, not the posterior mean. The

MGF of the flowgraph used in the likelihood construction involves nine paths and two loops. Using (2.26), the MGF of the distribution of the waiting time until death due to dementia ($0 \to 5$) is

$$M(s) = \left\{ P_1(s)\left[1 - \frac{1}{2}(1-p)^2 M_{23}(s)M_{32}(s)\right] + \sum_{i=2}^{8} P_i(s) \right.$$
$$\left. + P_9(s)\left[1 - \frac{1}{2}(1-p)^2 M_{12}(s)M_{21}(s)\right]\right\}$$
$$\times \left[1 - \frac{1}{2}(1-p)^2 M_{12}(s)M_{21}(s) - \frac{1}{2}(1-p)^2 M_{23}(s)M_{32}(s)\right]^{-1}, \quad (8.10)$$

where $P_1, \ldots, P_9$ are the path transmittances given by

$$\begin{aligned}
P_1(s) &= (1-p)pM_{01}(s)M_{14}(s)M_{45}(s) \\
P_2(s) &= (1-p)^2 pM_{01}(s)M_{12}(s)M_{24}(s)M_{45}(s) \\
P_3(s) &= \frac{1}{2}(1-p)^3 pM_{01}(s)M_{12}(s)M_{23}(s)M_{34}(s)M_{45}(s) \\
P_4(s) &= \frac{1}{4}p^2(1-p)M_{02}(s)M_{21}(s)M_{14}(s)M_{45}(s) \\
P_5(s) &= \frac{1}{2}p^2 M_{02}(s)M_{24}(s)M_{45}(s) \\
P_6(s) &= \frac{1}{4}p^2(1-p)M_{02}(s)M_{23}(s)M_{34}(s)M_{45}(s) \qquad (8.11) \\
P_7(s) &= \frac{1}{4}p^2(1-p)^2 M_{03}(s)M_{32}(s)M_{21}(s)M_{14}(s)M_{45}(s) \\
P_8(s) &= \frac{1}{2}p^2(1-p)M_{03}(s)M_{32}(s)M_{24}(s)M_{45}(s) \\
P_9(s) &= \frac{1}{2}p^2 M_{03}(s)M_{34}(s)M_{45}(s).
\end{aligned}$$

Path 1 shares no common nodes with the first order loop $2 \to 3 \to 2$, so $L_1^1(s) = \frac{1}{2}(1-p)^2 M_{23}(s)M_{32}(s)$ and similarly path 9 has the nontouching loop $1 \to 2 \to 1$, so $L_1^9(s) = \frac{1}{2}(1-p)^2 M_{12}(s)M_{21}(s)$. Again, while the MGF may appear complicated, it is only the corresponding MAPLE-generated C functions for $K(s)$, $K'(s)$, and $K''(s)$ that are necessary for computation. Exercise 8.2 asks the reader to perform the MAPLE computations.

Figure 8.10 gives the Bayes predictive density and hazard functions based on the simulated complete data on the flowgraph, the modified dataset with 80% incomplete cases, and the dataset containing only the 20 complete cases from this modified dataset. The slice sampling algorithm used 2000 posterior samples to estimate the predictive distributions based on each of these datasets, after discarding 1000

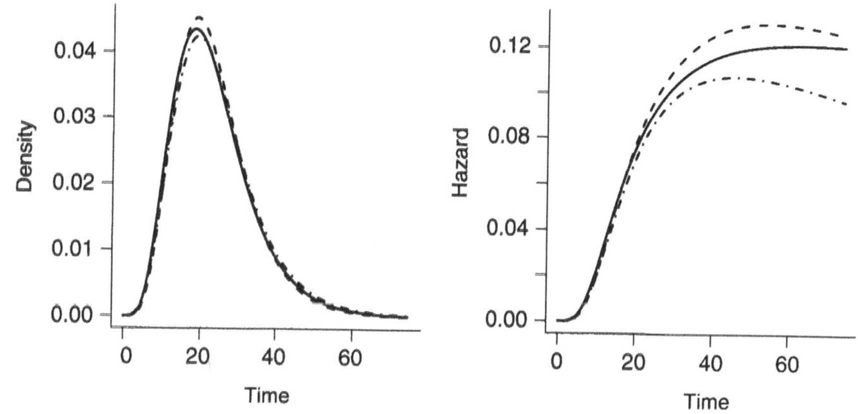

FIGURE 8.10 Bayes predictive density and hazard function based on complete data (solid line), 80% incomplete data (dashed line), and deleted incomplete data (dotted and dashed line) for the Example 8.3.

burn-in samples. With multiple paths from $0 \rightarrow 5$, the presence of incomplete data causes a deconvolution problem. This is indicated by additional diffuseness in the posterior distributions of each parameter relative to the 100% complete data case. However, it is apparent that substantial information is lost by deleting the incomplete data from the analysis. Even in the presence of deconvolution, the 80% incomplete dataset leads to a predictive distribution very similar to the 100% complete data predictive distribution. However, the estimated predictive distribution is much more diffuse when the incomplete data are discarded. Table 8.3 gives the quantile values of the complete and incomplete data predictive densities. Note that the standard deviation of the predictive distribution based on the deleted incomplete data, containing only the 20 complete cases, is 5.5% larger than the complete data

TABLE 8.3 Quantiles of the Predictive Distributions Based on Complete, 80% Incomplete, and Deleted Incomplete Data for Example 8.3

Percentile	Complete Data ($n = 100$)	80% Incomplete Data	Complete Data ($n = 20$)
.25	15.79	16.21	16.35
.50	21.65	21.84	22.36
.75	29.15	28.94	30.09
.90	37.63	36.86	39.00
.95	43.63	42.42	45.49
.99	57.08	54.85	60.68
Mean	23.55	23.53	24.36
s.d.	10.73	10.16	11.32

($n = 100$) and 11.4% larger than the 80% incomplete data predictive distribution standard deviations.

The hazard function for the 80% incomplete data is close to the complete data hazard up to the median of the predictive distribution but subsequently diverges. This divergence becomes much more extreme when the incomplete data are discarded. In the context of this simulation exercise, it is not surprising that the complete and incomplete datasets of the same size produce similar results, as both were generated from the same underlying model. This fact also provides context for the result that information is lost when the incomplete observations are discarded. In practice, incorporation of incomplete data has the potential of adding useful information, particularly when it involves transitions for which only limited amounts of complete data are present. On the other hand, incomplete data can introduce biases if it is not the result of the same underlying process governing generation of complete data. In most applications, relatively small portions of the data will be incomplete, so that the deconvolution problem will generally be under control. □

8.2 SIMULATIONS OF UNRECOGNIZED INCOMPLETE DATA

One potential problem with any sampling scheme is having unrecognized incomplete data i.e., data that are missing but unknown to the investigator. Consider the diabetic retinopathy flowgraph of Figure 7.4, the full retinopathy model. In this figure, if we observe a patient making a transition $1 \to 3$, we recognize that the data are incomplete. However, if in a well-designed study we observe a transition $1 \to 2 \to 3$, we typically assume that it is complete data. Unfortunately, that need not be the case. The patient could have transited as $1 \to 2 \to 1 \to 2 \to 3$ without our observing the second transition to state 1 and one of the transitions to state 2 but not knowing which transition to state 2 was actually observed. For example, in $1 \to 2 \to 1 \to 2 \to 3$, the initial loop $1 \to 2 \to 1$ could be missed and reported as all being time in state 1, or the loop $2 \to 1 \to 2$ could be missed and treated as time only in state 2. In good studies, this should not occur, but with anything less than continuous monitoring, it is possible.

Definition 8.2.1. *Unrecognized incomplete data* are data that appear to have complete information on the transitions and waiting times but in reality are incomplete with respect to transition information.

Yau and Huzurbazar (2002) investigate the effect of such unrecognized incomplete data on flowgraph models. A comparison to a Markovian birth and death process model with exponential waiting times is also made because previous analyzes of the retinopathy data have used this model. Our analysis follows the maximum likelihood based analysis of Yau and Huzurbazar (2002).

Data are simulated on the subset of the original model given in Figure 8.11. The simulated data are generated to be similar to the observed data. Yau and Huzurbazar (2002) use their reported MLEs to simulate 277 patient records. Unrecognized

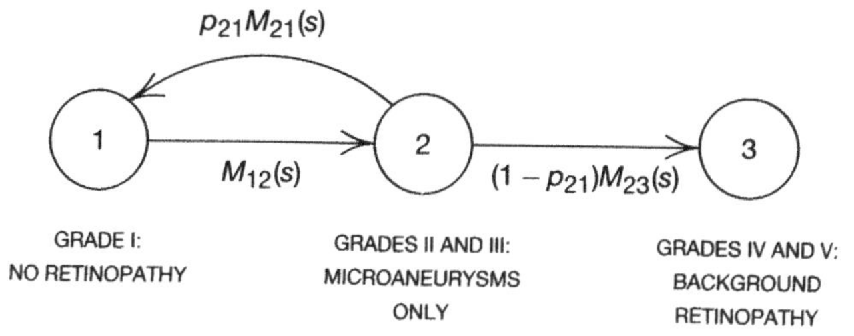

FIGURE 8.11 Subflowgraph model for the progression of diabetic retinopathy.

TABLE 8.4 Number of Transitions for One Run of the First Simulation

Transition	Complete Data	Unrecognized Incomplete Data
$1 \to 2$	448	371
$2 \to 1$	171	94
$2 \to 3$	277	277
Total	896	742

incomplete data are created around the $1 \to 2 \to 1$ loop transition. There are two simulation cases considered. Each simulation consists of 10 realizations. The first simulation is constructed so that about half of the $2 \to 1$ transitions are unrecognized incomplete and the second is set up so that a higher percentage, two-thirds, are unrecognized incomplete. For the first simulation, the probability of transition from state 2 back to state 1 is set to be 0.40. One realization of the first simulation is given in Table 8.4. All of the 277 patients make a $1 \to 2 \to 3$ transition; however, a subset first make $1 \to 2 \to 1$ loops and some patients make multiple $1 \to 2 \to 1$ loops. Altogether there were 171 such loops, giving a total of $277 + 171 = 448$ $1 \to 2$ transitions. This gives $896 (= 448 + 171 + 277)$ transitions in the complete simulated data. To create unrecognized incomplete data, as with the actual retinopathy data, 21% of the 448 transitions from state 2 (94 transitions) are assumed to be observed as returning to state 1. The remaining $171 - 94 = 77$ transitions that return to state 1 are taken as unrecognized incomplete; that is, 77 transitions from $2 \to 1$ are taken as part of an unobserved (unrecorded) loop. This gives a total of $277 + 94 = 371$, $1 \to 2$ transitions in the observed (incomplete) simulated data. Thus, the unrecognized incomplete data has 742 transitions made up of 371, $1 \to 2$; 94, $2 \to 1$; and 277, $2 \to 3$ transitions. The unrecognized incomplete data are generated so as to have numbers of transitions that are very similar to the actual retinopathy data. This was done by going into the complete simulated data and "losing the records" for 77 loops. In practice, unrecognized incomplete data are more likely to occur because patients are not observed frequently enough rather than because data are actually lost. To form the unrecognized incomplete data, we consider the 277 simulated individual complete data "patient" records, and randomly removed the records for some $1 \to 2 \to 1$ and $2 \to 1 \to 2$ loops. In particular, this was done by randomly selecting 77 of the 171 $2 \to 1$ transitions for elimination and then determining whether a $1 \to 2 \to 1$ or $2 \to 1 \to 2$ loop is being missed by flipping a coin. If a $1 \to 2 \to 1$ loop is missing, the $1 \to 2$ and $2 \to 1$ waiting times are added to the $1 \to 2$ waiting time. If a $2 \to 1 \to 2$ loop is missing, the $2 \to 1$ and $1 \to 2$ waiting times are added to the $2 \to 3$ waiting time. Yau and Huzurbazar (2002) simulate data using the following distributions: IG(8.05, 0.23) for transition $1 \to 2$, Gamma(6, 3.33) for transition $2 \to 1$, and IG(7.14, 0.086) for transition $2 \to 3$. This simulation was run 10 times, getting numbers of transitions similar to those of Table 8.4 for the other nine simulations.

Both the flowgraph and Markov model with exponential waiting times are fitted to this simulated incomplete data. The flowgraph model fits an inverse Gaussian distribution with parameters λ_1 and λ_2 for transition $1 \to 2$, a gamma distribution with parameters α and β for the transition from $2 \to 1$, which occurs with probability p_{21}, and an inverse Gaussian distribution with parameters μ_1 and μ_2 for transition $2 \to 3$. These are the distributional assumptions made in Yau and Huzurbazar (2002). The MGF of this flowgraph model is given by

$$M_{13}(s) = \frac{(1 - p_{21})M_{12}(s)M_{23}(s)}{1 - p_{21}M_{12}(s)M_{21}(s)}, \tag{8.12}$$

where $M_{12}(s)$, $M_{21}(s)$, and $M_{23}(s)$ are the MGFs of the IG(λ_1, λ_2), Gamma(α, β), and IG(μ_1, μ_2) distributions. The likelihood is

$$L(\lambda_1, \lambda_2, \alpha, \beta, \mu_1, \mu_2 | \mathcal{D}) = \prod_{i=1}^{n_{12}} f_{12}(x_{i12}|\lambda_1, \lambda_2) \prod_{i=1}^{n_{21}} p_{21} f_{21}(x_{i21}|\alpha, \beta) \prod_{i=1}^{n_{23}} (1 - p_{21}) f_{23}(x_{i23}|\mu_1, \mu_2), \tag{8.13}$$

where n_{12}, n_{21}, and n_{23} are the number of unrecognized incomplete data transitions from $1 \to 2$, $2 \to 1$, and $2 \to 3$, and $f_{12}(\cdot)$, $f_{21}(\cdot)$, and $f_{23}(\cdot)$ are the corresponding densities. We converted the MGF of (8.12) to a density using the saddlepoint approximation of (3.19). Table 8.5 gives the MLEs, averaged over 10 simulations, of the unrecognized incomplete data along with their standard errors.

The Marshall and Jones (1995) Markov model with exponential waiting times is also fitted to these simulated data. The Markov model MLEs $(\hat{\lambda}_{12}, \hat{\lambda}_{21}, \hat{\lambda}_{23})$, along with their standard errors for the incomplete data, are [0.21(0.006), 0.06(0.004), 0.01(0.0002)].

Figure 8.12 shows a plot of the true density from state 1 to state 3, and flowgraph and Markov model densities based on unrecognized incomplete data averaged over 10 simulated data sets. The true density is based on the parameter values for the simulation. The Markov model density is quite different from the true density.

TABLE 8.5 MLEs of the Incomplete Data Averaged Over 10 Simulations

Transition	Distribution	MLE (Unrecognized Incomplete Data)[a]
$1 \to 2$	Inverse Gaussian	$\hat{\lambda}_1 = 7.94$ (0.011), $\hat{\lambda}_2 = 0.17$ (0.003)
$2 \to 1$	Gamma	$\hat{\alpha} = 2.28$ (0.139), $\hat{\beta} = 12.99$ (1.138)
$2 \to 3$	Inverse Gaussian	$\hat{\mu}_1 = 7.36$ (0.082), $\hat{\mu}_2 = 0.06$ (0.003)
	p_{21}	$\hat{p}_{21} = 0.26$ (0.003)

[a] Standard errors in parentheses.

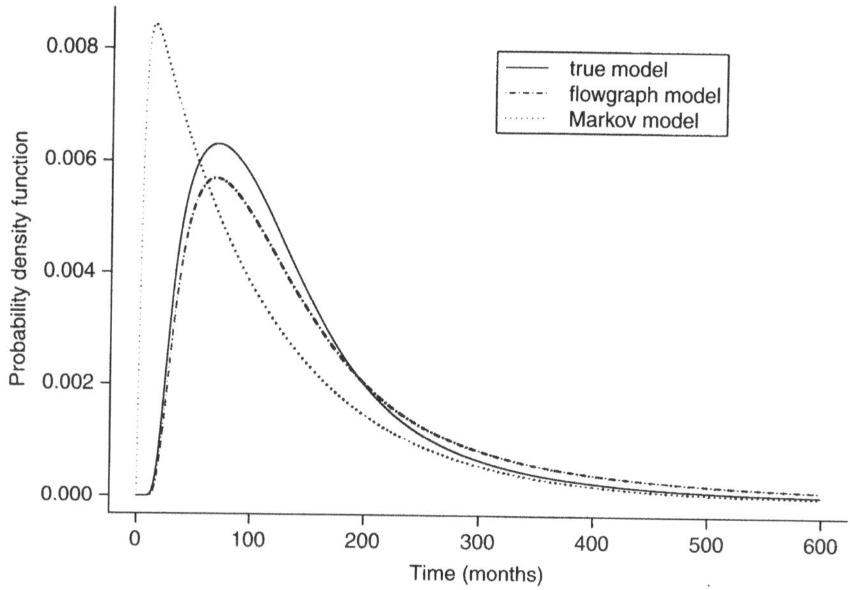

FIGURE 8.12 Unrecognized incomplete data simulation with $p_{21} = 0.4$ and approximately half of the $2 \to 1$ observations missing.

The incomplete data flowgraph model is quite close to the true density. In theory the exponential-based Markov model does not change very much under incomplete data. This is because the proportion of incomplete data is small compared to complete data and the exponential model is not as sensitive. In theory, under complete data the exponential-based Markov model handles feedback loops automatically due to the memoryless property of the exponential but this is not true for incomplete data where the waiting times for those loop transitions are biased so as to be longer. With more incomplete data, this would be reflected in the resulting density. Note also that the flowgraph is fitting richer parametric models than the exponential. Table 8.6 presents selected values of the survivor function from state 1 to state 3. The last column gives a range from 0.15 to 0.05 because the densities

TABLE 8.6 Selected Quantiles of Survivor Function (in Months) for the Incomplete Data Simulation with $p_{21} = 0.4^a$

Survivor Function	0.75	0.50	0.25	(0.15, 0.05)
True quantile	71	113	178	(230, 362)
Flowgraph	76 (1.3)	126 (2.3)	216 (4.6)	[297 (7.8), 528 (19.3)]
Markov model	34 (0.5)	77 (1.1)	149 (2.2)	[203 (3.0), 318 (4.8)]

[a] Standard errors are in parentheses.

TABLE 8.7 Selected Quantiles of the Survivor Function (in Months) for the Incomplete Data Simulation with $p_{21} = 0.6^a$

Survivor Function	0.75	0.50	0.25	(0.15, 0.05)
True quantile	97	157	236	(291, 420)
Flowgraph	94 (1.6)	160 (2.7)	263 (3.8)	[346 (5.0), 550 (17.8)]
Markov model	47 (0.4)	107 (1.1)	213 (3.0)	[288 (3.0), 453 (4.8)]

[a] Standard errors are in parentheses.

are very flat this far out in the right tail and the range gives an idea of the extent of this flatness. The Markov model seriously underestimates the true quantiles in the middle of the distribution but does better in the tails. The flowgraph model based on incomplete data tends to overestimate the true quantiles, but it is quite close to the truth for the more important higher survival values.

The second simulation is analogous to the first except a much higher percentage of data are unrecognized as incomplete. For this case, the probability of transitions from state 2 to state 1 is set to 0.60, so that approximately two-thirds of the $2 \to 1$ transitions are missed. Selected values of the survivor function are presented in Table 8.7. Figure 8.13 gives a plot of the true density, and the incomplete data densities from the flowgraph and Markov models averaged over 10 sets of simulated data. The Markov model density behaves similar to the first case with one-half of the $2 \to 1$ transitions missing ($p_{21} = 0.4$) and, again, is quite different from the true density except in the right tail. The incomplete data flowgraph density is now a little further off the target; however, it is still preferable to the Markov model density. Yau and Huzurbazar (2002) also performed the simulation with 70% and 80% probabilities for the $2 \to 1$ transition, and the results were similar to the 60% case.

The survivor function values show that up to the median, the incomplete data flowgraph model does quite well. In this context, reaching state 3 is a bad outcome. For example, for the 60% incomplete data simulation, half the people will get to state 3 in 157 months. The Markov model indicates that half the people will get there in 107 months, and the flowgraph model indicates that half the people will get there in 160 months. In this context, it is not good to underestimate by so large an amount. At $\hat{S}^{-1}(0.25)$ and further in the tail, the incomplete data flowgraph model overestimates the truth. This is a consequence of the fact that we are using very

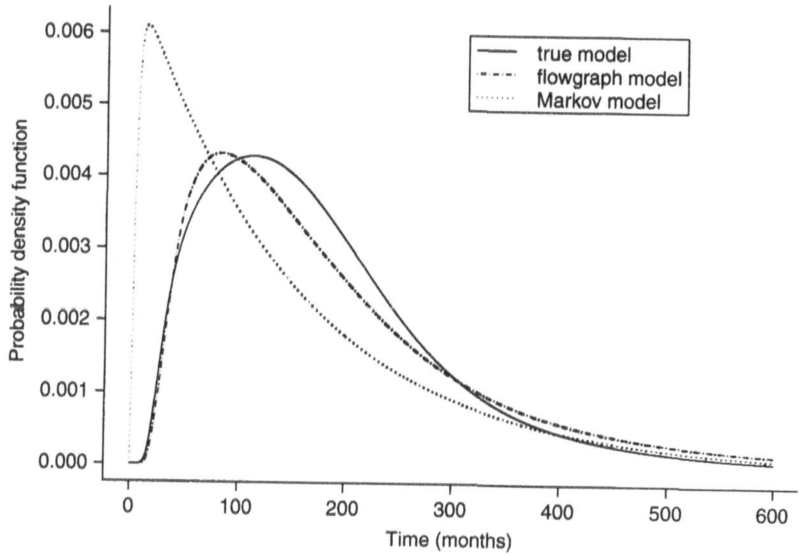

FIGURE 8.13 Unrecognized incomplete data with $p_{21} = 0.6$ and 2/3 of the $2 \rightarrow 1$ observations missing.

heavy tailed densities, inverse Gaussians and gamma, in the flowgraph model. If we know that the transitions are truly missing, we should use the flowgraph likelihood construction methods. But, in practice, we may not know when loops are missing. For data that do not follow an exponential-based Markov model, the flexibility of flowgraph models can give results that are superior to fitting an inappropriate Markov model, even with unrecognized incomplete data.

EXERCISES

8.1 Consider the diabetic retinopathy flowgraph of Figure 7.4. For each case of known incomplete data, write down the relevant MGF and the contribution to the likelihood of the observation:

(a) An incomplete observation from $3 \rightarrow 1$.

(b) An incomplete observation from $3 \rightarrow 2$ where you are told that the person transited through state 1 but you do not know when.

(c) An incomplete observation from $3 \rightarrow 3$.

8.2 Write MAPLE code to construct the MGF of the waiting time from $0 \rightarrow 5$ for the dementia flowgraph of Figure 8.9 using the parametric assumptions of Example 8.3.

8.3 Derive the MGF of the waiting time distribution from $2 \rightarrow 5$ for the dementia flowgraph of Figure 8.9. Then repeat Exercise 8.2.

CHAPTER 9

Flowgraph Models for Queuing Systems

Queuing systems are an important topic of study in statistics and probability theory with many applications in areas as diverse as (but not limited to) manufacturing, computer systems and networks, engineering systems, resource allocation in medicine, and operations research. Yet most people, within and outside of statistics, appear thoroughly bored at even the mention of the word *queue*. Like it or not, queues are part of our daily lives whether or not we are aware of it. Examples include waiting on hold on the phone with a customer service representative to discuss billing questions, waiting in line at the grocery store and changing lines only to find that the line just left is finally moving, or ordering pizza on Superbowl Sunday and discovering that it is a four-hour wait. These are all examples of queues that affect us if only as inconveniences. More serious examples of queuing systems include hospital emergency rooms or organ transplant lists. If one is waiting with a possibly broken foot in the emergency room, chances are that the gunshot victim wheeled in later will get treated first, a *last in, first out* queue. Similarly, although we would like organ transplant waiting lists to be fair, a "*first come, first served*" queue, factors such as blood work and donor compatibility can result in treatment being received in a different order than indicated by the transplant list. Many times, queuing systems affect us in secondary ways. For example, as a result of the west coast dockworkers strike of 2002, consumers were short of certain goods. Some companies planned ahead for the strike using tools from inventory modeling or supply chain management, an application of queuing. For example, Boeing's Wichita plant made arrangements to ship inventory by air rather than by sea so as to have minimal impact on their production lines (cf. *Wichita Business Journal*, 2002). The production lines themselves are also queuing systems. During the strike, ships could be seen "queued" up in the ocean off the Los Angeles coast. Even after the strike was settled, it took weeks for the backlog to clear. With such a variety of interesting applications, why are queuing systems

Flowgraph Models for Multistate Time-to-Event Data, By Aparna V. Huzurbazar
ISBN 0-471-26514-4

considered, well, boring? This author believes that it is due to an overemphasis on the classical machine–repairman problem, cumbersome mathematics for queues, and the time spent deriving limiting distributions. The general focus has been on stationary Markovian queues while many real-world queuing systems tend to be non-Markovian and nonstationary. The requirement that the arrival process to a queue be Poisson or that the service time distributions be exponential imposes restrictions that may not hold for the real problem being considered. For most applications, what is needed is often a short term-solution which may be transient. Although there is interest in deriving long-term characteristics of a queuing system, such as whether the system reaches steady state, the mean queue length at that time, the mean time to failure, and so on, often this information is not useful in the short term. For example, if we put a new manufacturing system on-line today, our immediate interest is in how it is behaving presently and in the near future. Initially, we are not concerned with how it will behave a year from today. In fact, a year from now, the parameters of this system will have been adjusted in ways that may hardly resemble the system that we put into operation today.

In this chapter we focus on the applications of flowgraph modeling to queuing systems. The flowgraph approach differs from the traditional approach in that it models the observable waiting time distributions and makes no assumptions about the arrival process. We present the standard analysis for queues briefly and refer the reader to vast amounts of literature for additional detail. There are thousands of papers on this subject. A search using SciSearch, ISI's Science Citation Index electronic database, yielded 18,460 records. Portions of the SciSearch database go back only to the year 1970. Therefore, an additional search, prior to 1970, was performed using *JSTOR*, which brought up an additional 307 items. Even these did not contain many earlier works cited in papers such as Kendall (1951), who cites Johannsen (1907). Kendall (1951, 1953) and Lindley (1952) are early works that present queuing theory in the format that statisticians are familiar with today. Queues are generally covered in a later chapter of books on stochastic processes. We refer the reader to the references cited at the beginning of Chapter 7. In addition, some books specifically devoted to queues are Bolch et al. (1998), Cox and Smith (1961), Gross and Harris (1998), and Kleinrock (1975), to name a few. There is also a great deal of literature in computer science, industrial engineering, and operations research devoted to queues, particularly the simulation of queuing networks. For example, Bolch et al. (1998) is directed to computer scientists, Harrington and Tumay (2000) is a reference for industrial and process engineers, and Kulkarni (1999) is an operations research text. This chapter begins with a brief review of queuing terminology.

9.1 REVIEW OF QUEUING TERMINOLOGY

A queue is characterized by a flow of *customers* requiring *service* performed by *servers*. The words *customer* and *service* are used generically. For example, in the dockworker strike described above, if our interest is in unloading the ships, the customer is the ship and the service is the unloading of the ship. The dockworkers are the *servers*. If our interest were an aircraft production line, the customer would be an aircraft that is being built, the service would be the parts required to build the aircraft, and the servers would be both the people and the machines involved in the building of the aircraft. In the emergency room example above, the customers are the patients, the service is medical treatment, and the servers are emergency room personnel. We are interested in how customers enter the system, the order in which they are served, and the amount of time they spend waiting for service. The *arrival* of customers forms the arrival or input distribution of the queue. Erlang (1909) showed that the Poisson process applies to random telephone traffic, and since then, Poisson arrivals have been a staple in queuing theory. The distribution of the waiting times required to serve a customer is called the *service time distribution.* A notation used with queues is A/B/c, where A is an abbreviation for the arrival distribution, B is an abbreviation for the service time distribution, and c refers to the number of servers. The notation is due to Kendall (1953). Exponential distributions are denoted by M for "memoryless"; deterministic times are denoted by D; gamma distributions with integer-valued shape parameters (k), called *Erlang distributions*, are denoted by Ek; and general waiting time distributions are denoted by G. General waiting time distributions are anything other than an exponential, Erlang, or deterministic distribution. Thus, an M/M/2 queue has an exponential arrival time distribution, an exponential service time distribution, and two servers. Note that the parameters of the exponential arrivals and service times can be different, and hence some authors prefer the notation M(λ)/M(μ)/2 when the arrival distribution is Exp(λ) and the service time distribution is Exp(μ). A queue with an exponential arrival time distribution and a general service time distribution with 1 server is denoted by M/G/1.

We will assume that an arriving customer joins the end of the queue and is served on a first come, first served basis. There are many other queue disciplines. For example, queues with *balking* allow an arriving customer to decide the line is too long and leave the queue; queues with *group service* where waiting customers are served in groups such as at a subway station or a bus stop; or what the author likes to call *Indian queuing* (named after the country of India) wherein there is a complete absence of queue discipline and customers are pushing their way toward any available server, such as at a ticket counter. Such different disciplines can make queues quite difficult to analyze. For the moment, we will assume orderly customers who wait patiently (i.e., they do not leave) to be served on a first come, first served basis. We define some additional terms:

- *Queuing time:* The time from when the customer enters the system to the instant at which the customer's service begins.
- *Waiting time:* The time from when the customer enters the system up to the instant at which the customer's service is complete.
- *Busy period:* Begins when a customer arrives to find the server free and lasts until the next instant at which the server is free i.e., it is the period of time during which the server is continuously occupied.

The busy period is distinct from service time in that service time refers to one customer whereas the busy period can involve several customers. These definitions imply the following relationship,

$$\text{Waiting time} = \text{Queuing time} + \text{Service time}.$$

In some of the queuing literature, especially in computer science, queuing time is called *waiting time* and waiting time as defined here is called *response time*.

9.2 M/M/1 QUEUE

In earlier chapters of this book we have used queuing models but not referred to them as such. For example, the birth and death processes of Chapter 7 are examples of queues. The M/M/1 queue assumes that arrivals occur according to a Poisson process with rate parameter λ so that the time between arrivals is exponentially distributed, Exp(λ). The service times are also exponentially distributed as Exp(μ). The size of the queue forms a birth and death process. The probability that an arrival occurs in a small interval $[t, t+h)$ is

$$\lambda h + o(h),$$

and the complementary probability of no arrivals in the interval $[t, t+h)$ is

$$1 - \lambda h + o(h).$$

The probability that service is completed in $[t, t+h)$ given that a customer is being served is $\mu h + o(h)$, and the complementary probability is $1 - \mu h + o(h)$. The service rate μ applies to one server. If there are c servers, the service rate is $c\mu$. Let $X(t)$ denote the number of customers in the queue at time t. An arriving customer adds one to the state $X(t)$ and a departing customer subtracts one from $X(t)$. Let k be the number of customers at time t, $X(t) = k$. The probability that another customer is added in $[t, t+h)$ is calculated from the equivalent event that a new customer arrived and no customer being served finished in that time period, that is,

$$\begin{aligned} P[X(t+h) = k+1 | X(t) = k] &= [\lambda h + o(h)][1 - \mu h + o(h)] \\ &= \lambda h + o(h) \qquad \text{for } k = 0, 1, 2, \ldots. \end{aligned} \tag{9.1}$$

Similarly,

$$\begin{aligned} P[X(t+h) = k-1 | X(t) = k] &= [\mu h + o(h)][1 - \lambda h + o(h)] \\ &= \mu h + o(h) \qquad \text{for } k = 1, 2, \ldots. \end{aligned} \tag{9.2}$$

Notice that $X(t)$ is a birth and death process, discussed in Section 7.1, and notice that (9.1) and (9.2) satisfy (7.8) with $\lambda_i = \lambda$ for $i = 0, 1, 2, \ldots$ and $\mu_i = \mu$ for

$i = 1, 2, 3, \ldots$ and $\mu_0 = 0$. The probability that an arrival occurs before a service is complete is just

$$\frac{\lambda}{\lambda + \mu} \tag{9.3}$$

and the probability that a service is complete before a customer arrives is

$$\frac{\mu}{\lambda + \mu}. \tag{9.4}$$

The stationary probabilities, or the limiting distribution of the queue length, are defined as

$$\pi_k = \lim_{t \to \infty} P(X(t) = k) \qquad \text{for } k = 0, 1, \ldots. \tag{9.5}$$

When they exist, these quantities are computed as

$$\begin{aligned} \eta_0 = 1 \quad &\text{and} \quad \eta_j = \frac{\lambda_0 \lambda_1 \cdots \lambda_{j-1}}{\mu_1 \mu_2 \cdots \mu_j} \\ \pi_0 = \frac{1}{\sum_{j=0}^{\infty} \eta_j} \quad &\text{and} \quad \pi_k = \eta_k \pi_0 = \frac{\eta_k}{\sum_{j=0}^{\infty} \eta_j}, \end{aligned} \tag{9.6}$$

where the denominator of π_k is simply a normalizing constant. The *traffic intensity* is defined as

$$\rho = \frac{\text{mean service time of a customer}}{\text{mean interarrival time of customers}}. \tag{9.7}$$

In some of the literature, this is also called the *utilization*. For our M/M/1 queue, this is

$$\rho = \frac{\lambda}{\mu}.$$

The unit of traffic intensity is the ***erlang***, named after A. K. Erlang, a pioneer in congestion theory (cf. Cox and Smith, 1961, p. 40). The stationary probabilities and other quantities of interest can be computed for $\rho < 1$. As ρ approaches 1, the mean queue length becomes infinite. For the M/M/1 queue, we have

$$\pi_k = \left(1 - \frac{\lambda}{\mu}\right)\left(\frac{\lambda}{\mu}\right)^k \qquad \text{for } k = 0, 1, \ldots, \text{ and } \rho < 1. \tag{9.8}$$

Further results for birth and death processes are obtained by using Little's law (cf. Little, 1961), which states that the average number of customers in a stable queue is equal to the average arrival rate multiplied by the average time spent waiting by a customer in the queue.

Theorem 9.2.1: LITTLE'S LAW. Suppose that L is the expected number of customers in the queue at steady state, λ the arrival rate of the customers, and W the expected waiting time for a customer in the queue at steady state. Then Little's law states that

$$L = \lambda W. \tag{9.9}$$

Notice that these computations all depend on the limiting distribution and provide relations based on long-term means. The next example illustrates an application of a flowgraph model for an M/M/1 queue. Although the flowgraph equivalent of the limiting distribution is presented here, flowgraph models are not restricted to the limiting case.

Example 9.1: FINITE STATE M/M/1 QUEUE: HYDRAULIC PUMP SYSTEM WITH FEEDBACK. We consider the hydraulic pump system with feedback of Figure 2.10 discussed in Example 2.8. The original Markovian analysis is due to Limnios (1992), and the original flowgraph analysis is due to Huzurbazar (2000b). The system has two pumps that are physically in parallel and is designed to work with one or two pumps. The states represent the number of failed pumps. We allow for a failed pump to be repaired, which makes this a series flowgraph model with feedback. We assume that each pump operates independently and fails according to an Exp(λ) waiting time distribution. Once failed, a pump is repaired according to an Exp(μ) waiting time distribution. In state 0 we have two functioning pumps and waiting time to failure is the minimum of two independent Exp(λ) random variables, Exp(2λ). In state 1 we have one working pump and one failed pump. The transition from state 1 to state 2 occurs if the working pump fails before the failed pump is fixed, and transition occurs to state 0 if the failed pump is repaired before the good pump fails. Therefore, we observe the minimum of an Exp(λ) with an Exp(μ), which is Exp($\lambda + \mu$). The corresponding transition probabilities and waiting time MGFs for the flowgraph are

$$\begin{aligned} p_{01} &= 1, & M_{01}(s) &= \frac{2\lambda}{2\lambda - s}, \quad \text{for } s < 2\lambda, \\ p_{10} &= \frac{\mu}{\lambda + \mu}, & M_{10}(s) &= \frac{\mu + \lambda}{\mu + \lambda - s}, \quad \text{for } s < \mu + \lambda, \text{ and} \\ p_{12} &= \frac{\lambda}{\lambda + \mu}, & M_{12}(s) &= \frac{\mu + \lambda}{\mu + \lambda - s}, \quad \text{for } s < \mu + \lambda. \end{aligned} \tag{9.10}$$

Note that the probabilities p_{12} and p_{10} are just (9.3) and (9.4), where an arrival is a failed pump and service is a repaired pump.

The flowgraph analysis can be used to give not only the mean time to total failure (MTTF) but also the entire waiting time distribution of the time to total failure, $0 \to 2$. We found the flowgraph MGF of the waiting time to total failure (TTF), in (2.20):

$$M(s) = \frac{p_{12} M_{01}(s) M_{12}(s)}{1 - p_{10} M_{01}(s) M_{10}(s)} \qquad \text{where} \quad p_{12} + p_{10} = 1. \tag{9.11}$$

Substituting (9.10) into (9.11) gives

$$M(s) = \frac{\left(\frac{\lambda}{\lambda+\mu}\right)\left(\frac{2\lambda}{2\lambda-s}\right)\left(\frac{\lambda+\mu}{\lambda+\mu-s}\right)}{1-\left(\frac{\mu}{\lambda+\mu}\right)\left(\frac{\mu+\lambda}{\mu+\lambda-s}\right)\left(\frac{2\lambda}{2\lambda-s}\right)}, s < \min(2\lambda, \mu+\lambda). \qquad (9.12)$$

Limnios (1992) performed the original Markovian queue analysis and found system parameter estimates of $\hat{\lambda} = 1/1000$ and $\hat{\mu} = 1/100$, and after much detailed algebra, the MTTF. The flowgraph analysis of the same model is much simpler and gives the MTTF, $M'(0) = K'(0) = 6500$ with considerably less effort. While the MTTF is a quantity of interest in systems analysis and derivable from well-known methods for Markovian systems, the flowgraph model provides it in a very simple form and gives much more than the MTTF. Flowgraph models provide the *entire* waiting time distribution of the TTF as well as the system reliability and hazard functions for this M/M/1. We rewrite (9.12) as

$$M(s) = \frac{2\lambda^2}{s^2 - (3\lambda+\mu)s + 2\lambda^2} \qquad (9.13)$$

$$= \frac{2\lambda^2}{\alpha-\beta}\left(\frac{1}{\beta-s} - \frac{1}{\alpha-s}\right) \quad \text{for } s < \min(\alpha,\beta), \alpha \neq \beta, \qquad (9.14)$$

where $\alpha = \alpha(\lambda, \mu)$ and $\beta = \beta(\lambda, \mu)$ are the factors of the denominator of (9.13). The estimated parameter values give $\hat{\alpha} = 0.012844$ and $\hat{\beta} = 1.5571 \times 10^{-4}$. Each term in (9.14) is the MGF of an exponential distribution. Inverting gives the density of the time to total failure as the mixture of exponentials,

$$f(t) = \frac{2\lambda^2}{\alpha-\beta}(e^{-\beta t} - e^{-\alpha t}) \quad \text{for } \lambda > 0, \mu > 0, t > 0. \qquad (9.15)$$

Figure 9.1 gives the maximum likelihood estimated density, $\hat{f}(t)$, and hazard function, $\hat{h}(t)$. If branch waiting times are not exponential, a situation in which methodology for Markovian systems does not give answers easily, the flowgraph model is still applicable, as the next example illustrates. □

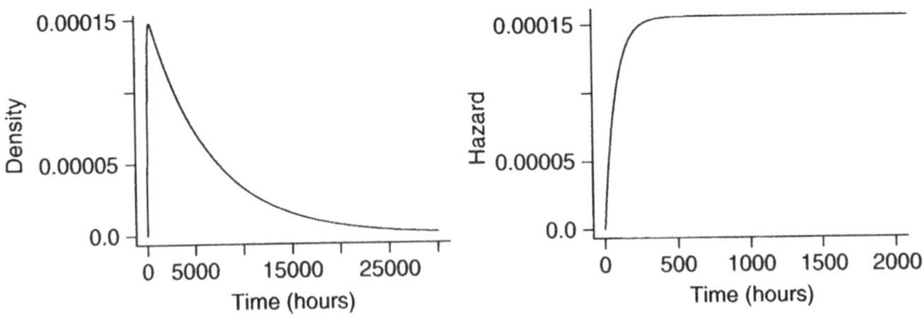

FIGURE 9.1 Maximum likelihood estimated density and hazard function for hydraulic pump application.

Example 9.2: HYDRAULIC PUMP SYSTEM WITH FEEDBACK GENERAL WAITING TIME DISTRIBUTIONS. In Example 4.5 we considered the hydraulic pump system with feedback of Figure 2.10 with nonexponential waiting times. We simulated data on this system with $p_{10} = 0.7$ using an Exp(0.01) distribution for the $0 \to 1$ branch, IG(22.36,2.236) for $1 \to 0$, and a Gamma(1000,5) distribution for the $1 \to 2$ branch. We simulated 100 realizations of the process, 27 of which were censored. The censored data histograms for the state-to-state waiting times are given in Figures 4.10, 4.11, and 4.12. This is a queue with general waiting time distributions. Note that in the traditional queuing analysis, the parametric models are made on the unconditional arrival and service time distributions. The flowgraph makes no assumptions about these unconditional distributions. The flowgraph models the observable conditional waiting times and transition probabilities. Figure 4.10 indicates that the Weibull, gamma, and exponential models all fit equally well. Since the exponential is a special case of both distributions, this makes sense for exponential data. Figures 4.11 and 4.12 indicate that both the inverse Gaussian and gamma models fit the data of each transition well. We model the flowgraph with an Exp(μ) for the $0 \to 1$ branch, an IG(λ_1, λ_2) for the $1 \to 0$ branch, a Gamma(α, β) for the $1 \to 2$ branch, and $p = p_{12}$. Note that we could just as easily have reversed the models for the $1 \to 0$ and $1 \to 2$ transition. The MGF of the time to total failure is found by substituting these probabilities and MGFs into (9.11):

$$M(s) = \mu p \left(\frac{\beta}{\beta - s}\right)^{\alpha} (\mu - s)^{-1} \left[1 - \frac{\mu(1-p)e^{\lambda_1\lambda_2 - [\lambda_1^2(\lambda_2^2 - 2s)]^{0.5}}}{\mu - s}\right]^{-1}. \quad (9.16)$$

□

The MAPLE program code to compute the MGF, the CGF, and its derivatives is given in Figure 9.2. With diffuse priors on $\boldsymbol{\theta} = (p, \mu, \lambda_1, \lambda_2, \alpha, \beta)$, this example shows what can go wrong with any MCMC algorithm. It turns out that for (λ_1, λ_2) and (α, β), there is long-term dependence in the posterior samples, giving bad convergence diagnostics for these parameters. One way around this problem is to reparameterize using the mean and variance of the gamma and inverse Gaussian. Let μ_{IG} and σ^2_{IG} be the mean and variance of the inverse Gaussian and let μ_{G} and σ^2_{G} be the mean and variance of the gamma. Then, for the inverse Gaussian we have

$$\lambda_1 = \lambda_2 \mu_{\text{IG}} \quad \text{and} \quad \lambda_2 = \sqrt{\frac{\mu_{\text{IG}}}{\sigma^2_{\text{IG}}}}, \quad (9.17)$$

and for the gamma we have

$$\alpha = \beta \mu_{\text{G}} \quad \text{and} \quad \beta = \frac{\mu_{\text{G}}}{\sigma^2_{\text{G}}}. \quad (9.18)$$

The transformed parameters $(\mu_{\text{IG}}, \sigma^2_{\text{IG}})$ and $(\mu_{\text{G}}, \sigma^2_{\text{G}})$ have much better convergence properties. After taking samples on the transformed parameters, we transform back

```
# Include package to generate C code
with(codegen,makeproc,C):

# Define the MGF as a procedure function of s
ms := proc(s)

# Define transmittances using w's for transmittances
local w01,w10,w12;

# Enter each transmittance

   w01 := mu/(mu-s)
   w10 := (1-p)*exp(l1*l2-((l1^2)*((l2^2)-2*s))^(.5));
   w12 := p*((b/(b-s))^a);

# Calculate M(s)
   (w01*w12)/(1-(w01*w10));

end proc;

# Make the MGF, CGF, and its derivatives
mx := makeproc(ms(s),[s]):

k := makeproc(ln(ms(s)),[s]):
C(k,optimized,ansi,filename="c:/fg/hydraulicpump/k.c"):

dk := makeproc(diff(k(s),s),[s]):
C(dk,optimized,ansi,filename="c:/fg/hydraulicpump/dk.c"):

ddk := makeproc(diff(dk(s),s),[s]):
C(ddk,optimized,ansi,filename="c:/fg/hydraulicpump/ddk.c"):
```

FIGURE 9.2 MAPLE code for Example 9.2.

to $\boldsymbol{\theta}$ and proceed with the analysis. Figure 9.3 shows the convergence properties of iterations of β for the original and the transformed case. We can see that sampling in the transformed parameter space leads to better coverage of the original parameter space. The solid gray horizontal line is the posterior mode. The dashed gray vertical line separates the burn-in samples from the samples used for prediction. Figure 9.4 gives the sample autocorrelation function (ACF) of the samples used for prediction. We can see that samples of β in the original case show long-term dependence. This shows the importance of making a plot of the MCMC iterations for all problems. The figures are given for the parameter β only; however, the behavior of the other three parameters is similar.

For illustration we also create some incomplete data samples from these data. We do this by taking 10 of the uncensored complete data realizations and using

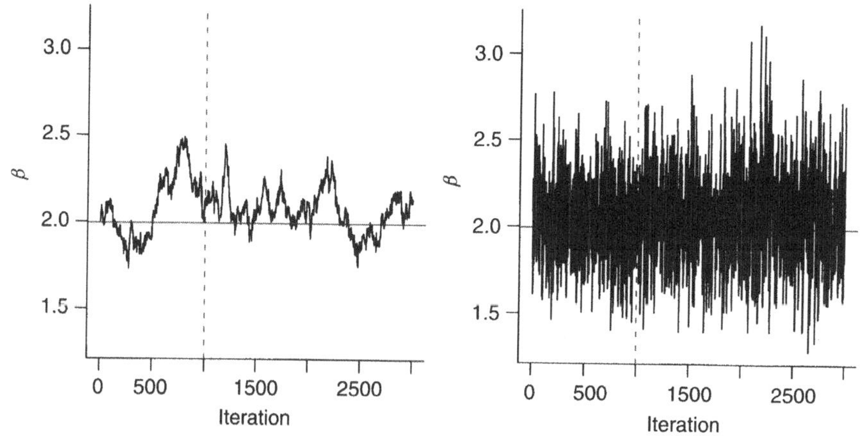

FIGURE 9.3 Convergence properties of iterations of β. The left-hand panel is the original case and the right-hand panel is the transformed case. The posterior mode is indicated by the solid gray horizontal line, and the dashed gray vertical line divides the burn-in samples from the samples used.

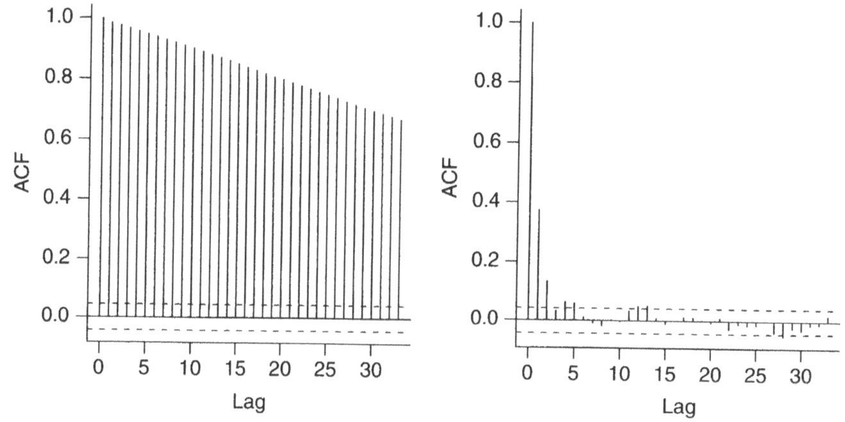

FIGURE 9.4 ACF for samples of β (without burn-in samples). The left-hand panel is the original case and the right-hand panel is the transformed case.

only their total time from $0 \rightarrow 2$. The incomplete data consist of the same 100 realizations as the complete data except that the breakdown is 63 realizations that are complete and do not have censoring, 27 realizations that have censoring, and 10 realizations that are incomplete with only the total $0 \rightarrow 2$ time recorded. Table 9.1 presents quantile values of the Bayes predictive density for the complete and incomplete data case. Figure 9.5 gives a plot of the complete and incomplete data densities. Since the incomplete data are created from the corresponding complete data and the flowgraph and saddlepoint do a good job of constructing the likelihood, the results for the complete and incomplete data are very close. In both cases, there is sufficient information on each of the branches to accurately

TABLE 9.1 Summary of Predictive Distribution Results for Example 9.2 for Both the Complete and Incomplete Data Case for the Flowgraph with General Waiting Time Distributions

Percentile	Complete Data	Incomplete Data
0.25	294.65	293.93
0.50	419.88	418.65
0.75	634.72	632.60
0.90	921.28	918.06
0.95	1140.06	1136.13
0.99	1654.79	1649.75
Mean	510.74	509.00
s.d.	313.43	312.36

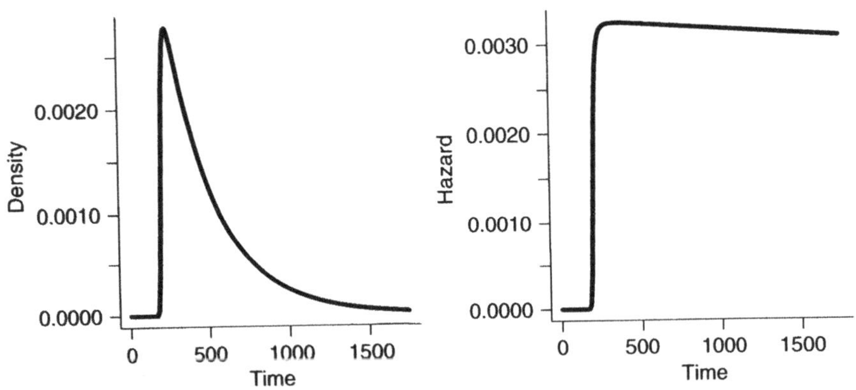

FIGURE 9.5 Bayes predictive density for complete and incomplete data for the hydraulic pumps of Example 9.2. Thick black line is based on the complete data and dashed gray overlay is based on the incomplete data. The two lines overlap.

estimate or predict all of the parameters. There is not any loss of information due to the incomplete data. This was not so for the retinopathy data of Example 8.2, where the incomplete observations added information to the analysis that resulted in modification of the complete data predictive distribution.

9.3 CELLULAR TELEPHONE NETWORK: M/M/c QUEUE

An M/M/c queue extends an M/M/1 queue by assuming that a fixed number c of servers are available. We assume that a server is never idle if customers are waiting. This is a birth and death process with $\lambda_i = \lambda$ and $\mu_i = k\mu$ for $k = \min\{i, c\}$ and $\mu_i = c\mu$ for $i > c$. The state of the system is $X(t)$, the number of customers in the queue at time t. The number of customers receiving service is then $\min\{X(t), c\}$ and the number of customers waiting for service is $\max\{X(t) - c, 0\}$. Again standard computations for limiting distributions and stationary probabilities are available. We present a flowgraph analysis of an M/M/3 queue arising from a cellular telephone network. The original analysis is due to Huzurbazar (2000a).

A cellular telephone network consists of cells that are located throughout a service area. We focus our attention on one cell of such a network. A cell is essentially a big box that consists of radios that handle phone calls. Radio malfunctions are frequent and reduce the system capacity dramatically. They also cost the company money, as the bad service results in lost customers. The number of radios in a cell can vary from 2 to 15 depending on the location of the cell in a rural or urban area. Although it is desirable to have a large number of radios, too many radios cause interference with one another, and since the power to a cell is divided among the radios, their coverage radius is reduced. Having fewer radios means that the radios can transmit greater distances but they can handle fewer calls. For an introduction to cellular telephone networks, we refer the reader to Lee (1993, 1995). We consider an urban cell with a total of eight radios: six of these are online and two are reserve radios that can be switched online from a remote location. The cell has coverage from three cellular technicians who attempt first to fix remotely small problems within the cell and then visit the cell physically if necessary to make repairs. The flowgraph model for the number of failed radios in the cell is given in Figure 9.6.

State 0 refers to the state where all radios are functioning. Failure of a radio means a forward transition to a higher state in Figure 9.6. Repair of radio means a backward transition to a lower state. States 1 through 8 refer to the number of failed radios. When all radios have failed, the cell has broken down completely and the process is in state 8. Radio malfunctions often begin as small problems that can quickly compound into larger problems. Our interest is in modeling the behavior of these radios and predicting time to failure of a cell or time to partial failure of a cell. Failure of a cell means that the cell can not handle any calls. Partial failure of a cell means that it cannot handle calls to capacity.

In Figure 9.6 at state 2, for example, p_{23} is the probability of moving to three failed radios, and $p_{21} = 1 - p_{23}$ is the probability of having a radio repaired and moving to one failed radio. $M_{23}(s)$ is the MGF of the waiting time distribution until reaching state 3. If we are interested in the time to total failure of our cell, the MGF is $M_{08}(s)$. Our interest is in the waiting time from state 0 to state 4. State 4 is considered a "critical state" because once the process reaches state 4, it can quickly deteriorate toward total system breakdown. Standard practice is that once state 4 is reached, rather than working on the cell remotely, a technician is sent to the physical site for making repairs to the cell.

There is only one path from $0 \to 4$: namely, $0 \to 1 \to 2 \to 3 \to 4$. Its transmittance and all other necessary transmittances are given in Table 9.2. Figure 9.6 contains seven first-order loops. These are $0 \to 1 \to 0$ through $6 \to 7 \to 6$. However, our interest is in the waiting time from states 0 to 4, so we only need to account for the loops $0 \to 1 \to 0$, $1 \to 2 \to 1$, and $2 \to 3 \to 2$. Table 9.2 gives

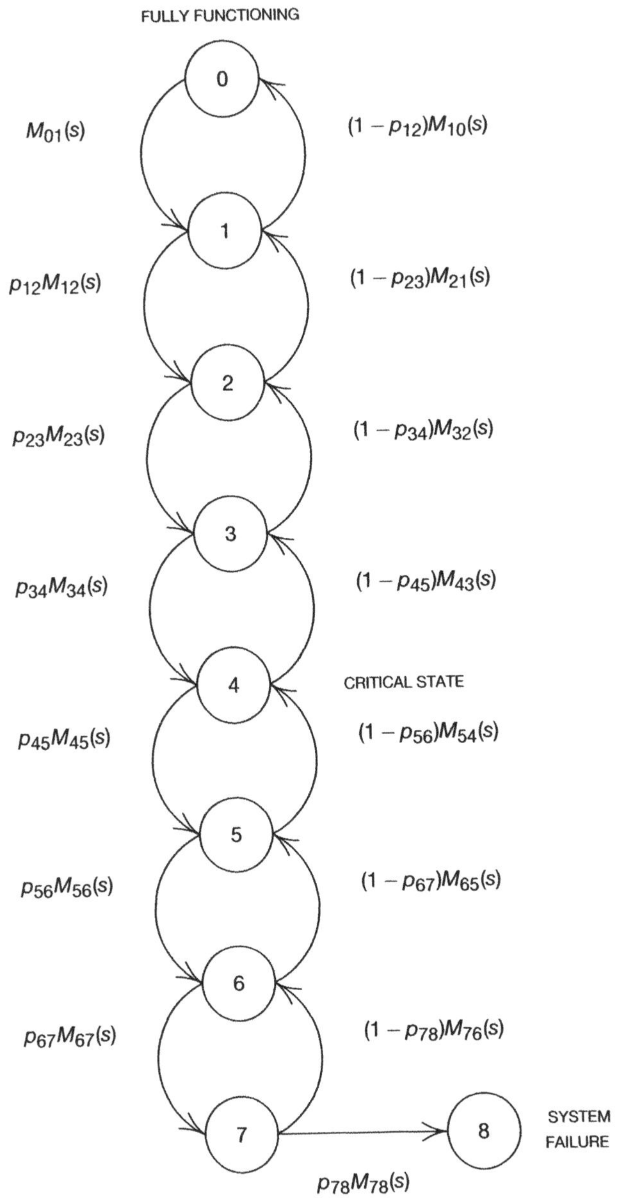

FIGURE 9.6 Flowgraph model for one cell from a cellular telephone network.

the transmittances for these loops. Note that in waiting for state 4 there is no possibility of moving from 4 back to 3. However, if we were waiting for any state greater than 4, we would have to consider the possibility of the system returning to state 3. In computing the transmittance $0 \to 4$, there is one second-order loop because $0 \to 1 \to 0$ and $2 \to 3 \to 2$ are first-order loops that do not touch. There

TABLE 9.2 Summary of Paths and Loops for Solving the Flowgraph

Path	Transmittance
$0 \to 1 \to 2 \to 3 \to 4$	$p_{12}p_{23}p_{34}M_{01}(s)M_{12}(s)M_{23}(s)M_{34}(s)$
First-order loops:	
$0 \to 1 \to 0$	$p_{10}M_{01}(s)M_{10}(s)$
$1 \to 2 \to 1$	$p_{12}p_{21}M_{12}(s)M_{21}(s)$
$2 \to 3 \to 2$	$p_{23}p_{32}M_{23}(s)M_{32}(s)$
Second-order loops:	
$0 \to 1 \to 0$ and $2 \to 3 \to 2$	$p_{10}p_{23}p_{32}M_{01}(s)M_{10}(s)M_{23}(s)M_{32}(s)$

are no loops of order three or more. For our partial waiting time from state 0 to state 4, this is

$$M_{04}(s) = \frac{P_1(s)}{1 - L_1(s) + L_2(s)}, \tag{9.19}$$

where

$$P_1(s) = p_{12}p_{23}p_{34}M_{01}(s)M_{12}(s)M_{23}(s)M_{34}(s)$$

$$L_1(s) = p_{10}M_{01}(s)M_{10}(s) + p_{12}p_{21}M_{12}(s)M_{21}(s) + p_{23}p_{32}M_{23}(s)M_{32}(s)$$

$$L_2(s) = p_{10}p_{23}p_{32}M_{01}(s)M_{10}(s)M_{23}(s)M_{32}(s).$$

Note that there are no loops that do not touch the path, so $L_1^1(s) \equiv 0 \equiv L_2^1(s)$. Note also, for illustration, that if we were computing the waiting time from state 0 to state 8, there would be one fourth-order loop, as well as various second- and third-order loops. One can imagine how much more complicated $M_{08}(s)$ is than $M_{04}(s)$.

9.3.1 Cellular Telephone Network: Complete Data

The waiting time distribution for the transition $0 \to 4$ can be predicted from the MGF given in (9.19) using the saddlepoint approximation of (3.19) and the Bayesian prediction method of Chapter 5. For this example, our data consist of one *total lifetime* of the cell i.e., all transitions and waiting times that state 8 was reached. This took about two weeks (20,910 minutes). There are a total of 171,486 complete observations on various transitions and their waiting times. The transition data are summarized in Table 9.3. Note that backward transitions are omitted because they are always one less than the corresponding forward transition. For example, there were 30,678 transitions from $0 \to 1$ and 30,677 transitions from $1 \to 0$. We have data on the complete system $0 \to 8$; however, our problem is to estimate the waiting time distribution for the passage $0 \to 4$ given these data on the entire system.

Histograms of the data indicate that exponential distributions for failure times and service times are adequate. Historically, most stochastic networks have been

TABLE 9.3 Data Summary for Sample Cell Site

Transition	No. Obs.	Dist.	Transition	No. Obs.	Dist.
$0 \to 1$	30,678	Exp(6λ)	$4 \to 5$	1,076	Exp($4\lambda + 3\mu$)
$1 \to 2$	32,100	Exp($6\lambda + \mu$)	$5 \to 6$	192	Exp($3\lambda + 3\mu$)
$2 \to 3$	16,859	Exp($6\lambda + 2\mu$)	$6 \to 7$	20	Exp($2\lambda + 3\mu$)
$3 \to 4$	4,821	Exp($5\lambda + 3\mu$)	$7 \to 8$	1	Exp($\lambda + 3\mu$)

analyzed using exponential distributions because they are easy to work with mathematically. We are using them only because the data suggest that they are appropriate. The flowgraph methods employed here apply generally to the branch waiting time distributions, which have tractable MGFs as described in Chapter 2.

To model this problem as a flowgraph equivalent of an M/M/3 queue, we suppose that each radio fails independently according to an Exp(λ) distribution (mean $1/\lambda$) and is repaired according to an Exp(μ) distribution (mean $1/\mu$). The transmittances in Figure 9.6 are obtained as follows. Recall that the cell has six radios online and two backup (spare) radios. In state 0 we are waiting for the minimum of six independent Exp(λ) distributions to occur; therefore, the distribution of the waiting time for the $0 \to 1$ transition is Exp(6λ). Once in state 1, there are still six working radios because one of the backups is now online and one failed radio is under repair. In state 1 we observe the minimum of seven independent distributions, of which six are Exp(λ) and one is Exp(μ); hence the $1 \to 2$ and $1 \to 0$ transitions have the Exp($6\lambda + \mu$) distribution. Transition is made to state 2 if a radio fails before one is repaired, and the probability that this occurs is $p_{12} = 6\lambda/(6\lambda + \mu)$, the probability that the Exp(6λ) random waiting time occurs before the Exp(μ) random waiting time. The complementary event, that a radio is repaired before another one fails, has probability $p_{10} = \mu/(6\lambda + \mu)$. Table 9.3 gives the waiting time distributions for the forward transitions. The waiting time distribution for state i to $i-1$ is the same as that for state i to $i+1$. We can summarize this process with the following notation. Let $n_i = \max(2-i, 0)$ be the number of spare radios available in state $i, m_i = \min(6, 8-i)$ be the number of working radios in state i, and $q_i = \min(3, i)$ be the number of radios under repair in state i, where $q_i \leq 3$ because there are only three cellular technicians available to make repairs. We assume that each technician fixes one radio at a time and returns it online as soon as it is fixed. Although not always realistic, this assumption is deemed reasonable for repair from a remote location. For example, in state 5, there are $m_5 = 3$ working radios, $q_5 = 3$ radios under repair, two radios waiting for repair, and $n_5 = 0$ spare radios.

Let $T_{i,i+1}(s)$ denote the transmittance for the transition from state i to state $i+1$, so that

$$\begin{aligned} T_{i,i+1}(s) &= p_{i,i+1} M_{i,i+1}(s) \\ &= \frac{m_i\lambda}{m_i\lambda + q_i\mu}\left(1 - \frac{s}{m_i\lambda + q_i\mu}\right)^{-1} \quad \text{for } i = 0, \ldots, 7. \end{aligned} \tag{9.20}$$

Let $T_{i,i-1}(s)$ denote the transmittance for the transition from state i to state $i-1$, so that

$$T_{i,i-1}(s) = p_{i,i-1} M_{i,i-1}(s)$$
$$= \frac{q_i \mu}{m_i \lambda + q_i \mu}\left(1 - \frac{s}{m_i \lambda + q_i \mu}\right)^{-1} \qquad \text{for } i = 1, \ldots, 7. \quad (9.21)$$

Using MAPLE gives the $0 \rightarrow 4$ MGF using equation (9.19) with the transmittances given in (9.20) and (9.21). The MAPLE code is given in Figure 9.7. The MGF is given by

$$M(s) = 1080\lambda^4 (6\lambda - s)^{-1} (6\lambda + \mu - s)^{-1} (6\lambda + 2\mu - s)^{-1} (5\lambda + 3\mu - s)^{-1}$$
$$\times \left[1 - 6\frac{\lambda \mu}{(6\lambda - s)(6\lambda + \mu - s)}\right.$$
$$- 12\frac{\lambda \mu}{(6\lambda + \mu - s)(6\lambda + 2\mu - s)} - 18\frac{\lambda \mu}{(6\lambda + 2\mu - s)(5\lambda + 3\mu - s)}$$
$$\left. + 108\frac{\lambda^2 \mu^2}{(6\lambda - s)(6\lambda + \mu - s)(6\lambda + 2\mu - s)(5\lambda + 3\mu - s)}\right]^{-1}.$$

Note that our assumptions make the flowgraph model equivalent to a Markovian M/M/3 queue (cf. Taylor and Karlin, 1984). However, the analysis here gives the actual transient distribution, not the limiting distribution derived from queuing theory. If the data indicated that the waiting times were not exponential, the queuing model would not be tractable, but the flowgraph model would still be valid and useful. In this case we would model each branch probability and waiting time distribution separately for the flowgraph based on examining histograms of the data to choose parametric models for branch waiting time distributions. Note that this approach models the observable random variables, which are conditional on each branch. It makes no assumption of independence for unconditional (unobservable) branch waiting times, only a much weaker conditional independence assumption.

The likelihood function for the M/M/3 queue has a contribution of a probability and a waiting time density from each branch. For example, the $1 \rightarrow 2$ transition contributes 32,100 data points. The probability of the transition is $6\lambda/(6\lambda + \mu)$ and the density is $\text{Exp}(6\lambda + \mu)$, as indicated in Table 9.3. The overall contribution to the likelihood of this transition is

$$\left(\frac{6\lambda}{6\lambda + \mu}\right)^{32{,}100} \times (6\lambda + \mu)^{32{,}100} \exp\left[-(6\lambda + \mu)\sum_{i=1}^{32{,}100} t_{i12}\right],$$

where t_{i12} represents the ith waiting time for $1 \rightarrow 2$. The backward transition $1 \rightarrow 0$ occurs with probability $\mu/(6\lambda + \mu)$ and also has density $\text{Exp}(6\lambda + \mu)$. The contribution to the likelihood for $1 \rightarrow 0$ is

$$\left(\frac{\mu}{6\lambda + \mu}\right)^{30{,}677} \times (6\lambda + \mu)^{30{,}677} \exp\left[-(6\lambda + \mu)\sum_{i=1}^{30{,}677} t_{i10}\right],$$

```
# Include package to generate C code
with(codegen,makeproc,C):

# Define the MGF as a procedure function of s
ms := proc(s)

# Define local variables, w's for transmittances
  local w01,w10,w12,w21,w23,w32,w34,num,l1,l2,den;

  w01 := 6*lb/(6*lb-s);
  w12 := 6*lb/(6*lb+mu-s);
  w10 := mu/(6*lb+mu-s);
  w23 := 6*lb/(6*lb+2*mu-s);
  w21 := 2*mu/(6*lb+2*mu-s);
  w34 := 5*lb/(5*lb+3*mu-s);
  w32 := 3*mu/(5*lb+3*mu-s);

# num for numerator of Mason's rule
  num:=w01*w12*w23*w34:

# define loops
  l1:=w01*w10+w12*w21+w23*w32:
  l2:=w01*w10*w23*w32:

# den for denominator of Mason's rule
  den:=1-l1+l2:    num/den;
end proc:

mx := makeproc(ms(s),[s]):

# Make C functions for the CGF and its derivatives
k := makeproc(ln(ms(s)),[s]):
C(k,optimized,ansi,filename="c:/k04.c"):

dk := makeproc(diff(k(s),s),[s]):
C(dk,optimized,ansi,filename="c:/dk04.c"):

ddk := makeproc(diff(dk(s),s),[s]):
C(ddk,optimized,ansi,filename="c:/ddk04.c"):
```

FIGURE 9.7 MAPLE code for M/M/3 queue, $0 \to 4$.

where t_{i10} represents the ith waiting time for $1 \to 0$. The likelihood is the product of similar terms for each branch.

With the large amount of data available on two parameters, (μ, λ), we use maximum likelihood (ML) methods for the analysis. Results from a Bayesian analysis were indistinguishable from the maximum likelihood methods. The ML estimates

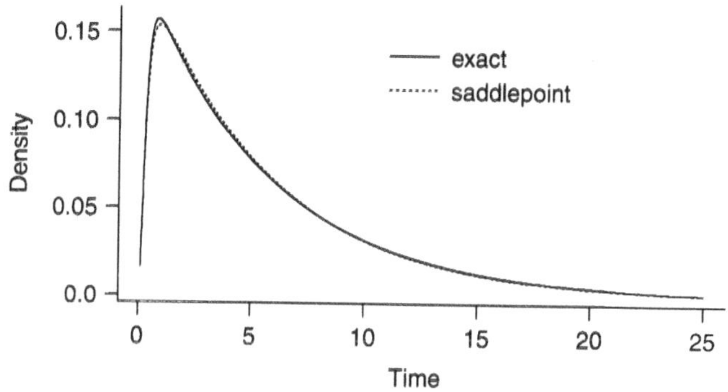

FIGURE 9.8 Exact (solid) and saddlepoint estimated (dashed) densities for the time to partial failure ($0 \to 4$) of the cell of Figure 9.6.

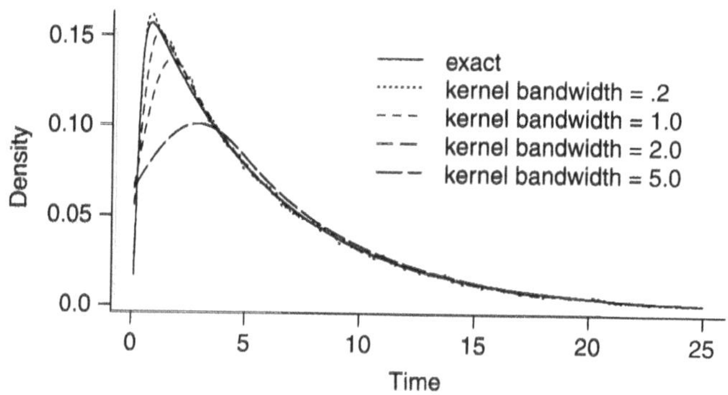

FIGURE 9.9 Exact density (solid) with kernel density estimates for the time to partial failure ($0 \to 4$) of the cell of Figure 9.6.

are $\hat{\mu} = 3.9892$ and $\hat{\lambda} = 0.7052$. Figures 9.8 and 9.9 give the waiting time density for time to partial failure ($0 \to 4$) using three methods:

1. Exact inversion of the MGF from the flowgraph at the MLEs
2. Approximate inversion of the MGF using saddlepoint methods at the MLEs
3. Kernel density estimation based on direct simulation of the network

Exact inversion is possible only in rare cases, and it is therefore important to consider alternatives such as saddlepoints and simulation. In this problem exact inversion is possible only because all of the waiting times are exponential. If any of the waiting times were not exponential, we would have to use an approximate inversion or simulation. The solid line in Figure 9.8 is the exact inversion.

The dotted line is the saddlepoint inversion of the flowgraph model. From Figure 9.8 we see that the saddlepoint inversion provides an excellent approximation to the exact density. Note that with these data, the density puts most of the probability in the first few minutes. This is a very poorly functioning cell, which is why it was being analyzed. Since someone must be sent out every time a cell reaches state 4, it makes sense to have a person permanently assigned to stay at this cell.

With complete data to give the MLEs of all the parameters, we can also simulate data from the network using the MLEs as the system parameter values. Figure 9.9 gives several kernel density estimates (cf. Silverman, 1986) based on simulated data.

Each simulated data set contained 100,000 observations. We use a biweight kernel with varying bandwidths, four of which are shown: 0.2, 1.0, 2.0, and 5.0. The 1.0 bandwidth seems to do best, 2.0 and 5.0 oversmooth, and 0.2 undersmooths and is quite close to the 1.0 bandwidth. Note the variability in the kernel density estimates. If an exact density were not available for comparison, it would be very difficult to choose a good kernel density estimate with confidence. On the other hand, the saddlepoint density approximation is very close to the exact.

Simulation, the preferred method of analysis for computer scientists working on queues, is feasible here because we have complete data on the transitions and waiting times. If we did not observe individual waiting times and transitions, for example, we had incomplete observations from state 0 to state 8, the M/M/3 queue likelihood would not apply, so that direct simulation would be very difficult. For the flowgraph, the incomplete data methods of Chapter 8 can be applied.

Figure 9.10 gives the hazard function for the time to first failure. The hazard function is that of a finite mixture of exponentials, a hyperexponential distribution (see Balakrishnan and Basu, 1995, p. 300). The increasing hazard flattens out to the equilibrium behavior of the process for time to partial failure, $0 \to 4$. Such flattening of the hazard occurs because we have a Markovian process with exponential waiting times. The kernel hazard is based on converting the kernel density to a CDF and then a hazard function.

A hazard analysis is especially useful when designing new cells. When a new cell is put online, its hazard function is initially increasing. Ideally, we want a hazard that is either decreasing over time or one that is relatively flat at some small

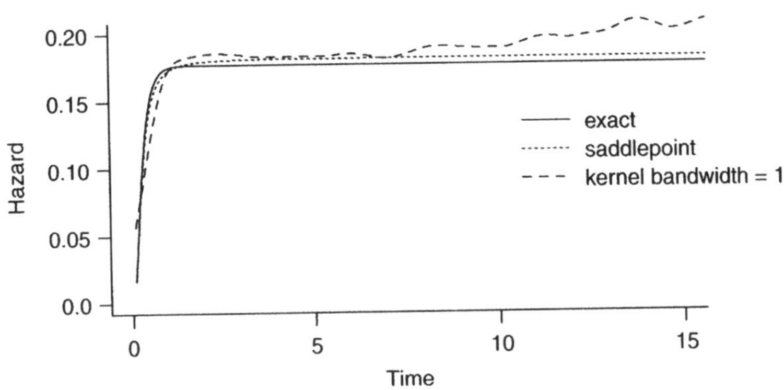

FIGURE 9.10 Exact (solid line), saddlepoint estimated (dotted line), and kernel based (dashed line, bandwidth = 1.0) hazard functions for the time to partial failure ($0 \to 4$) of the cell of Figure 9.6.

constant. Decisions about how many radios to put in each cell have traditionally been made using the experience of field engineers. The most quantitative tool currently used is based on Erlang charts (Lee, 1993), which give *average* cell use [see Yeung and Nanda (1996) for further details on channel management]. Erlang charts are crude approximations of system performance. They are based on averages such as the average number of calls and average calling time, and give the average cell size required to meet performance. In situations such as cell phone usage during rush hour on an interstate highway, such approximations are inappropriate. Decisions based on experience about a particular cell, its location, and its similarity to other cells of its type are far better than chart projections. The flowgraph methodology is not based on averages and is useful for investigating the behavior of a cell under various situations, especially for constructing startup cells.

9.4 STARTUP CELLS AND INCOMPLETE DATA

Incomplete data arise in a network when new cells, called *startup cells*, are brought online. Suppose that a startup cell in a remote location is being initialized. Startup cells are expected to malfunction frequently. The malfunction information is available to the operations manager but not to the technicians. In this case, we have data only on the time to total failure, not on any of the intermediate transitions, i.e. we observe the cell in state 0 and in state 8 (total failure) but not in any intermediate state. Our aim is still to predict the time to partial failure $0 \to 4$ so that total failure may be averted, but our procedure is based only on data for complete system failure, $0 \to 8$.

A solution to this problem cannot be simulated effectively, except in special cases when the $0 \to 8$ exact density is available, because there is no good way to estimate the parameters of the model. However, flowgraph methods with saddlepoint approximations can be used for likelihood construction as discussed in Chapter 8. Let T denote the waiting time $0 \to 8$. Flowgraph methods give the MGF, $M_{08}(s)$ (see Figures 9.11 and 9.12 for MAPLE code) for fixed $\boldsymbol{\theta}$. Either exact inversion of the MGF or the approximate inversion of saddlepoint approximation (3.19) gives an analytical formula for the density of T. Let Z represent the waiting time for partial failure, $0 \to 4$. The likelihood construction for this problem is based on $n = 50$ observations of the total waiting time, $0 \to 8$.

Samples are generated from the posterior using slice sampling on the constructed likelihood to compute the approximate Bayes predictive density of (8.6). Maximum likelihood estimates are derived from the constructed likelihood as well. The approach we use is that of Williams and Huzurbazar (2004). The constructed likelihood (8.1) is calculated from inverting the flowgraph MGF for $0 \to 8$. Let Z represent the waiting time for partial failure, $0 \to 4$. The saddlepoint density approximation $\tilde{f}_Z(z|\boldsymbol{\theta}_i)$ is computed by inverting the MGF of the time to partial failure $0 \to 4$. The priors used are independent Uniform(0, 10) distributions on λ and μ. The resulting density and hazard approximations for $0 \to 4$ given in Figures 9.13 and 9.14 are based on 2000 slice samples after discarding 1000 burn-in samples. Initially, the hazard function increases very steeply, as expected

```
# Include package to generate C code
with(codegen,makeproc,C):

# Define the MGF as a procedure function of s
ms := proc(s)

# Define local variables, w's for transmittances
  local w01,w10,w12,w21,w23,w32,w34,w45,w43,w56,w54,
    w65,w67,w78,num,l1,l2,l3,l4,den,w1,w2,w3,w4,w5,w6,w7;

  w01 := 6*lb/(6*lb-s);
  w12 := 6*lb/(6*lb+mu-s);
  w10 := mu/(6*lb+mu-s);
  w23 := 6*lb/(6*lb+2*mu-s);
  w21 := 2*mu/(6*lb+2*mu-s);
  w34 := 5*lb/(5*lb+3*mu-s);
  w32 := 3*mu/(5*lb+3*mu-s);
  w45 := 4*lb/(4*lb+3*mu-s);
  w43 := 3*mu/(4*lb+3*mu-s);
  w56 := 3*lb/(3*lb+3*mu-s);
  w54 := 3*mu/(3*lb+3*mu-s);
  w67 := 2*lb/(2*lb+3*mu-s);
  w65 := 3*mu/(2*lb+3*mu-s);
  w78 := lb/(lb+3*mu-s);

# num for numerator of Mason's rule
  num := w01*w12*w23*w34*w45*w56*w67*w78:

# define loops
  w1:=w10*w01:
  w2:=w12*w21:
  w3:=w23*w32:
  w4:=w34*w43:
  w5:=w45*w54:
  w6:=w56*w65:
  w7:=w67*w76:

# l1: first order, l2: second order
  l1 := w1+w2+w3+w4+w5+w6+w7:
  l2 := w1*(w3+w4+w5+w6+w7)+w2*(w4+w5+w6+w7)+
          w3*(w5+w6+w7)+w4*(w6+w7)+w5*w7:

# Continued in Figure 9.12
```

FIGURE 9.11 MAPLE code for M/M/3 queue, $0 \rightarrow 8$.

```
# l3: third order, l4: fourth order

   l3 := w1*w3*(w5+w6+w7)+w1*w4*(w6+w7)+w1*w5*w7+
         w2*w4*(w6+w7)+w2*w5*w7+w3*w5*w7:
   l4 := w1*w3*w5*w7;

# den for denominator of Mason's rule
   den := 1.0-l1+l2-l3+l4:
   num/den;

end proc:

mx := makeproc(ms(s),[s]):

# Make C functions for the CGF and its derivatives
k := makeproc(ln(ms(s)),[s]):
C(k,optimized,ansi,filename="c:/k08.c"):

dk := makeproc(diff(k(s),s),[s]):
C(dk,optimized,ansi,filename="c:/dk08.c"):

ddk := makeproc(diff(dk(s),s),[s]):
C(ddk,optimized,ansi,filename="c:/ddk08.c"):
```

FIGURE 9.12 MAPLE code for M/M/3 queue, $0 \to 8$ (*continued*).

for this poorly performing cell, and then drops slightly and levels off. Hazards that exhibit both increasing and decreasing behavior over disjoint regions of the time domain are common with flowgraph networks as they parametrically model the observable waiting times and calculate the hazard based on these branch models. They do not make any direct assumptions about hazard functions themselves, which is often done in proportional hazards modeling, for example.

The constructed likelihood surface is given in Figure 9.15. The left-panel shows the log-likelihood function. The right-hand panel shows the slice sampling iterations overlaid on the log-likelihood surface. The gray areas are the burn-in samples; samples used for prediction are shown in black. Notice that the likelihood maximum is a ridge at approximately -335, the value of the log-likelihood evaluated at a maximum likelihood estimate. The slice sampling method initially gets stuck near the origin but then manages to thoroughly cover the ridge. The ridge is related to the traffic intensity of this queue. The estimation problem is very difficult: we have information on the total waiting time from $0 \to 8$ from which we are attempting to deconvolve information about μ and λ. Quantile values for the predictive distribution corresponding to the are 0.53, 1.02, 1.87, 3.08, 4.06, and 6.52 for the 25th, 50th, 75th, 90th, 95th, and 99th percentiles, respectively. The predictive distribution mean is 1.4 hours and the standard deviation is 1.33 hours. This startup cell performs poorly as it typically reaches partial failure in under half a workday. This

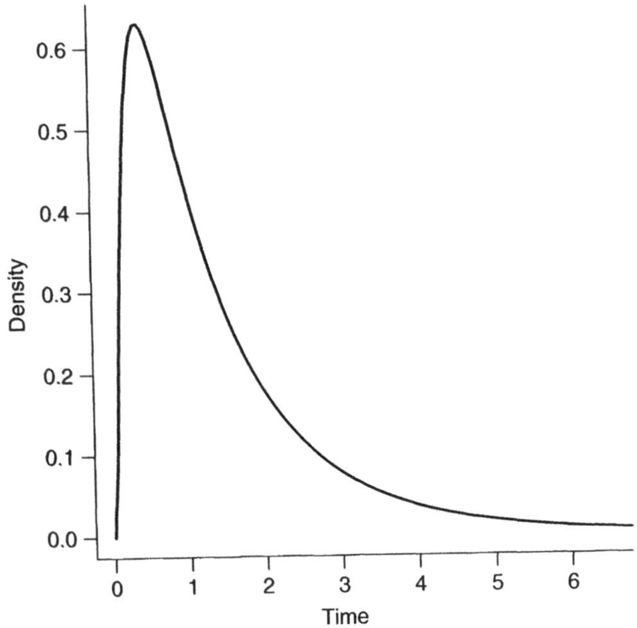

FIGURE 9.13 Incomplete data density for the time (in hours) to partial failure, $0 \to 4$, of the cell of Figure 9.6.

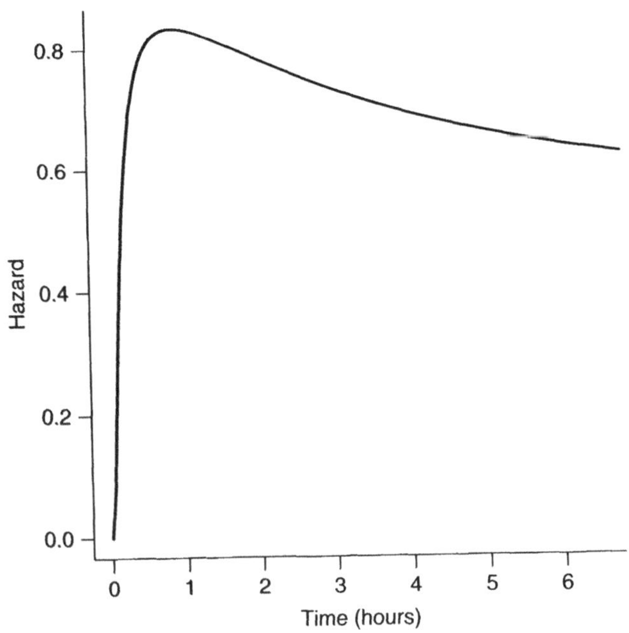

FIGURE 9.14 Incomplete data hazard for the time to partial failure ($0 \to 4$) of the cell of Figure 9.6.

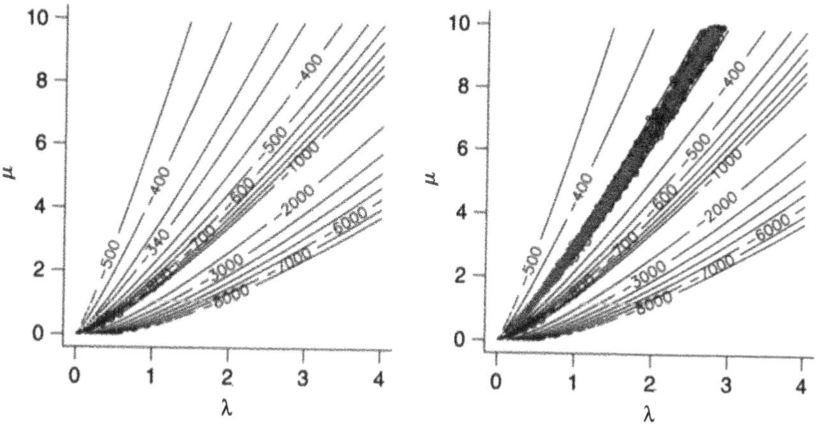

FIGURE 9.15 Contours of the incomplete data constructed log-likelihood for the cell of Figure 9.6. The figure on the left shows the likelihood function. The figure on the right shows the slice sampling iterations overlaid on the likelihood: gray areas are the burn-in samples; samples used for prediction are shown in black.

analysis informed the development of a protocol to automatically and remotely check cells before they reach total failure. A cell can reach partial failure and be even closer to total failure many times and recover before eventually reaching total failure. The MLEs for μ and λ are $\hat{\lambda} = 0.175$ and $\hat{\mu} = 0.333$.

Likelihood construction is model dependent. The waiting time distribution $0 \rightarrow 8$ depends heavily on all of its constituent branch models. In this cellular telephone example, we have substantial evidence from other cells that the M/M/3 queue model is reasonable. Likelihood construction is a powerful tool that can be used in a wide variety of incomplete data situations. It is most effective when one is combining complete and incomplete data, because the complete data can inform on appropriate branch models. Otherwise, as in this example, one has only theory or other similar examples to rely upon for branch transition time modeling.

9.5 CELLULAR TELEPHONE NETWORK: M/G/1 QUEUE

We review briefly the standard analysis for the M/G/1 queue before discussing the alternative flowgraph analysis. An M/G/1 queue has an exponential arrival time distribution, a general service time distribution, and one server. We assume that the service times do not depend on the number of customers waiting. In this case, writing down the equilibrium equations is not straightforward. Recall that our derivation of (9.1) and (9.2) for what occurs in the time interval $[t, t + dt)$ used the memoryless property of the exponential service time distribution. When the service times are not exponential, we cannot easily compute the probability associated with the time spent by the server waiting on a customer. One way around this difficulty is to create an embedded Markov chain within $X(t)$, the number of customers in the queue at time t. Note that this Markov chain is a discrete-time process, whereas $X(t)$ is a continuous-time process. We define random variables as follows:

- Let D_n be the time at which the nth customer leaves the system.
- Let $Q(D_n)$ be the number of customers who are left behind. This does not include the customer who just left.
- Let U_n be the number of customers who arrive during the service time of customer n.

We focus on the increment $Q(D_{n+1}) - Q(D_n)$. If there is at least one customer waiting [i.e., $Q(D_n) > 0$], the next customer, i.e., the $(n+1)$st customer, receives service immediately. The $(n+1)$st customer's service time begins at time D_n. During this random service time, a random number of customers arrive to join the queue. When customer $(n+1)$ leaves the system after receiving service, the number of customers left behind are

$$U_{n+1} + Q(D_n) - 1.$$

This says that the queue left behind by customer $n+1$ is one customer less than the queue left behind by customer n plus any additional arrivals during the service time of customer $n+1$. Now, if $Q(D_n) = 0$, i.e. there is no one waiting in line, then the server must wait for the next, $(n+1)$st, customer's arrival. When this customer arrives, receives service, and subsequently leaves, the number of customers left behind is exactly

$$U_{n+1}.$$

We can summarize this as

$$Q(D_{n+1}) = \begin{cases} U_{n+1} + Q(D_n) - 1 & \text{if } Q(D_n) > 0 \\ U_{n+1} & \text{if } Q(D_n) = 0. \end{cases} \tag{9.22}$$

Equation (9.22) holds for any queue in general. However, for an M/G/1 queue, U_{n+1} depends only on the service time of the $(n+1)$st customer, and since we have Poisson arrivals, is independent of $Q(D_n)$ This means that the sequence of random variables $\{Q(D_n); n > 1\}$ forms a Markov chain. Practically, to use (9.22), we need the distribution of U_n. If the random service time Y_n of customer n is y, then U_n has a Poisson distribution with mean (λy):

$$P(U_n = k | Y_n = y) = \frac{\exp(-\lambda y)(\lambda y)^k}{k!}. \tag{9.23}$$

But Y has an assumed general waiting time distribution with CDF $G(y)$, so that

$$P(U_n = k) = \int_0^{\infty} \frac{\exp(-\lambda y)(\lambda y)^k}{k!} \, dG(y). \tag{9.24}$$

Since $U_n \sim \text{Poisson}(\lambda Y_n)$ we can write down the form of the transition matrix for the Markov chain formed by $\{Q(D_n); n > 1\}$. Let δ_j be the probability that exactly j customers join the queue during a typical service time Y_n. Then

$$\delta_j = E_Y\left[\frac{\exp(-\lambda Y)(\lambda Y)^j}{j!}\right]. \tag{9.25}$$

The transition matrix of the Markov chain formed by $\{Q(D_n); n > 1\}$ has entries p_{ij} where

$$p_{ij} = \delta_j \qquad \text{for } i = 0, \tag{9.26}$$

and for $i \geq 1$,

$$p_{ij} = \begin{cases} \delta_{j-i+1} & \text{for } j - i + 1 \geq 0 \\ 0 & \text{for } j - i + 1 < 0. \end{cases} \tag{9.27}$$

At this point, the standard analysis focuses on the asymptotic behavior of the embedded Markov chain. Notice that the p_{ij} entries of the transition matrix are averages over the service time. For the M/G/1 queue, the traffic intensity ρ, defined in (9.7), is

$$\rho = \lambda E(Y).$$

The Markov chain is stationary if $\rho < 1$, transient for $\rho > 1$, and null persistent for $\rho = 1$. For the case where $\rho < 1$, the standard analysis can find quantities such as the busy period and the equilibrium distribution for the waiting time in the queue. We refer the reader to the many books on stochastic processes and queuing theory mentioned at the beginning of this chapter for further details and technical proofs of the results presented here. This is also the point at which the flowgraph analysis diverges from the standard analysis in that it does not focus on asymptotic behavior.

We illustrate the application for a flowgraph model equivalent of an M/G/1 queue by using a cell from the cellular telephone network. We suppose that our cell has seven radios, five of which are active and two of which are backup, and that there is only one cellular technician available for service. We assume that the radios fail independently according to an Exp(μ) distribution and we assume that the service time is IG(λ_1, λ_2). Such a heavy-tailed distribution is appropriate for cellular technicians, who often have to drive very long distances to reach a cell site to perform repairs.

To remain consistent with the standard analysis, the state of the system is taken to be the last observed queue length where queue lengths are observed only at the end of each service time. Radios that have failed are brought into the queue, and required spare radios are brought online only at the end of each service time. Although this assumption seems unrealistic because spares would be brought online immediately in a real system, it is necessary for the M/G/1 model to work. Note that with a flowgraph analysis, we do not require such an assumption and we can set-up and perform a correct analysis based on the actual process. When all radios have failed, one radio is in service, so the queue length is 6. Our interest is in predicting the time to total failure, the waiting time from state 0 to state 6, denoted by Z. Starting in state 0, the waiting time for one radio to fail is the minimum of five independent Exp(μ) distributions, which is Exp(5μ). By the assumptions for the M/G/1 queue, from state 1 the queue can proceed directly to any of the states $0, \ldots, 5$. From any state $i \in \{2, \ldots, 5\}$ the queue can go to states $i-1, \ldots, 6$. The branch transmittances among states are summarized as

$$T(s) = \begin{bmatrix} 0 & T_{01}(s) & 0 & 0 & 0 & 0 & 0 \\ T_{10}(s) & T_{11}(s) & T_{12}(s) & T_{13}(s) & T_{14}(s) & T_{15}(s) & 0 \\ 0 & T_{21}(s) & T_{22}(s) & T_{23}(s) & T_{24}(s) & T_{25}(s) & T_{26}(s) \\ 0 & 0 & T_{32}(s) & T_{33}(s) & T_{34}(s) & T_{35}(s) & T_{36}(s) \\ 0 & 0 & 0 & T_{43}(s) & T_{44}(s) & T_{45}(s) & T_{46}(s) \\ 0 & 0 & 0 & 0 & T_{54}(s) & T_{55}(s) & T_{56}(s) \\ 0 & 0 & 0 & 0 & 0 & 0 & 0 \end{bmatrix},$$

where $T_{01}(s) = (1 - s/5\mu)^{-1}$. Beyond T_{01}, calculating the branch transition probabilities and MGFs requires more work. The other individual transmittances $T_{ij}(s)$ are derived as follows. The probability of transition out of state $i \geq 1$ in time $[t, t+dt)$, given that transition to state i was made at time 0, is $f(t|\lambda_1, \lambda_2)dt$. The conditional probability of passing to state j in time $[t, t+dt)$ is the chance of $j-i+1$ out of m_i radios failing during the service time t; since each radio fails with probability $1 - e^{-\mu t}$, this probability is

$$q_{ij}(t) = \binom{m_i}{j-i+1}(1 - e^{-\mu t})^{j-i+1}(e^{-\mu t})^{m_i-(j-i+1)} \tag{9.28}$$

where $i - 1 \leq j \leq m_i + i - 1$. Therefore, with the service time distribution $T \sim f(t|\lambda_1, \lambda_2)$, the transmittance T_{ij} is

$$\begin{aligned} T_{ij}(s) &= E[q_{ij}(T)e^{sT}] \\ &= \int_0^\infty \binom{m_i}{j-i+1}(1 - e^{-\mu t})^{j-i+1}(e^{-\mu t})^{m_i-(j-i+1)} e^{st} f(t|\lambda_1, \lambda_2)\, dt. \end{aligned}$$

Note that since the probabilities themselves depend on the service time T, a characteristic of the semi-Markov process, this transmittance must include the probability function $q_{ij}(t)$ within the integral for the MGF. Expanding the powers of $1 - e^{-\mu t}$ using the binomial theorem and integrating termwise gives

$$\begin{aligned} T_{ij}(s) = \binom{m_i}{j-i+1} \sum_{k=0}^{j-i+1} \binom{j-i+1}{k} (-1)^k \\ \times \exp\{\lambda_1\lambda_2 - \lambda_1[2(s_0 - s + \mu k)]^{1/2}\} \end{aligned} \tag{9.29}$$

defined for $s \le s_0 = \frac{1}{2}[\lambda_2^2 + 2\mu(m_i - j + i - 1)] > 0$. Note that to obtain the MGF of the waiting time distribution specific to branch $i \to j$, we must divide by $T_{ij}(0)$, so that the MGF of the waiting time distribution conditional on passage $i \to j$ is

$$\frac{T_{ij}(s)}{T_{ij}(0)} \qquad \text{for } s \le s_0 = \frac{1}{2}[\lambda_2^2 + 2\mu(m_i - j + i - 1)] > 0.$$

Note also that the convergence of this MGF at the endpoint s_0 means that it generates a tilted exponential family that is not regular. However, the tilted family is steep (cf. Barndorff-Nielsen, 1978), so that despite this nonregularity, saddlepoint methods can still be used for its inversion. Steepness of $\log[T_{ij}(s)/T_{ij}(0)] = K_{ij}(s)$ holds if $K'_{ij}(s)$ maps from $(-\infty, s_0)$ onto $(0, \infty)$. As $s \uparrow s_0$, all the terms in $K'_{ij}(s)$ stay bounded except for the $k = 0$ term. After some algebra it can be shown that $K'_{ij}(s) \sim (s_0 - s)^{-1/2}$, so $K'_{ij}(s) \uparrow \infty$ as $s \uparrow s_0$. The main point is that we can still use saddlepoint methods to convert $T_{ij}(s)/T_{ij}(0)$ into a density.

Alternatively, we can develop these transmittances using a *Bayesian argument*. A priori, the service time while in state i is $\text{IG}(\lambda_1, \lambda_2)$. Now suppose that we have additional information that the system has made the transition to state j in the time interval $[t, t + dt)$. Due to the semi-Markov nature of the process, the state transition conveys information about the value of the service time T through the binomial likelihood term in (9.28). The distribution of the number of failed radios given the service time t and state i is $\text{Bin}(m_i, 1 - e^{-\mu t})$. Thus, the posterior distribution of T given the transition to state j is

$$f(t|\{i \to j\}, \lambda_1, \lambda_2) \propto \binom{m_i}{j - i + 1}(1 - e^{-\mu t})^{j-i+1}(e^{-\mu t})^{m_i-(j-i+1)} f(t|\lambda_1, \lambda_2) \tag{9.30}$$

by an application of Bayes theorem (5.2). For the purposes of this analysis, we treated t, an observed service time, as a parameter and found a posterior distribution for it based on additional transition information.

These transmittances are the branch transmittances that are used in Mason's rule. The flowgraph for this queue is quite cluttered and so is not drawn. The root flowgraph is much like the that of Figure 9.6 except that there are seven states, 0 to 6, and many additional branches for direct transition between states $i \to j$ for $i \in \{2, \ldots, 5\}$ and $j \in i - 1, \ldots, 6$. In addition, there are self-loops at each state with transmittances $T_{ii}(s)$. To find the overall MGF from, say, any state i to any state j, first we use Mason's rule to find the equivalent transmittance for $i \to j$. To avoid confusion with the branch transmittances, let us denote that as $W_{ij}(s)$. Then the MGF of the waiting time distribution from $i \to j$, conditional on taking the path from $i \to j$, is

$$M_{ij}(s) = W_{ij}(s)/W_{ij}(0), \tag{9.31}$$

where since j may not be an absorbing state, we divide by $W_{ij}(0)$, the marginal probability of passage from $i \to j$. While Mason's rule can still be applied in this situation, enumeration of the paths and loops becomes complex. One alternative that helps for this particular queue is based on work by Pyke (1961). Pyke (1961) derives distribution functions of first passage time in semi-Markov processes based on matrix renewal theory. These can be directly applied to the flowgraph transmittances.

A general finite-state semi-Markov process is characterized by the collection of transmittances linking the states of the process. For example, suppose that an $(n+1)$-state semi-Markov process has transmittances of $T_{ij}(s)$ from state i to j for $i, j = 0, \ldots, n$ with $T_{ij}(s) \equiv 0$ when direct passage is not possible. Then the $(n+1) \times (n+1)$ matrix $T(s) = ((T_{ij}(s)))$ characterizes the semi-Markov process. In our parametric Bayesian approach, each component of $T(s)$ is allowed to depend on the process parameter $\boldsymbol{\theta}$.

Let $E_{ij}(s)$ for $i, j = 0, \ldots, n$ denote the first passage transmittances from state i to state j. Using the methods of Pyke (1961), we can write down an expression for the $E_{ij}(s)$ for $i, j = 0, \ldots, n$ in matrix form. If $E(s) = (E_{ij}(s))$,

$$E(s) = T(s)[I - T(s)]^{-1}(\{[I - T(s)]^{-1}\}_d)^{-1} \qquad \text{for } s < 0, \tag{9.32}$$

where $\{\cdot\}_d$ is the matrix operator, which replaces all nondiagonal elements with zero. As written, $E(s)$ is not well defined at $s = 0$ since $I - T(0)$ has an eigenvalue of 0 and is not invertible. There are two distinct ways of modifying (9.32) to compute first-passage transmittances that converge in a neighborhood of zero. The first method inverts $[I - T(s)]$ in terms of its adjoint matrices and is discussed in Butler and Huzurbazar (2000). The second method consists of making the destination state an absorbing state. We discuss only the second method here.

Consider first passage from arbitrary state i to state $j \neq i$ in the reducible chain described above with transmittances given by (9.29). The first passage transmittance to state j is unaffected by making state j absorbing, and a benefit of this is that the determinant $|I - T(0)|$ is made positive. State j is made absorbing by modifying matrix T so that the row corresponding to the jth state consists entirely of zeros. Note that this holds in the matrix of flowgraph transmittances (9.29). The matrix $I - T(s)$ is now invertible in a neighborhood of $s = 0$, and therefore expression (9.32) is well defined in a neighborhood of $s = 0$.

For our problem, to compute the MGF of the waiting time from $0 \to 6$, we need only $E_{06}(s)$ from this formulation. We use MAPLE to compute $M(s) = E_{06}$ using the transmittances from (9.29), and to compute $K'(s) = dE_{ij}(s)/ds$ and $K''(s) = d^2E_{ij}(s)/ds^2$.

Complete data for this M/G/1 queuing system are simulated using $\mu = 0.4$, $\lambda_1 = 3$, and $\lambda_2 = 4$. The likelihood is comprised of three separate contributions:

1. Let r_1 be the number of transitions observed from $0 \to 1$. In the data, $r_1 = 5$. This contributes five independent Exp(5μ) variates with a total waiting time of 1.37.
2. Let r_2 be the observed number of radios that are repaired in total. In the data, $r_2 = 116$. This contributes 116 independent IG(λ_1, λ_2) terms as $t_1, \ldots, t_{116}$ with minimally sufficient statistics $u_1 = \sum t_i^{-1} = 166.0$ and $u_2 = \sum t_i = 86.6$.

3. There are r_2 binomial terms, each counting the number of radio failures during service times with the form

$$\prod_{i=1}^{r_2} \binom{m_i}{n_i} (1 - e^{-\mu t_i})^{n_i} (e^{-\mu t_i})^{m_i - n_i},$$

where m_i is the number of radios operating just prior to the ith service time and n_i is the number that subsequently fail during the service time.

With flat independent priors, the posterior is such that μ and (λ_1, λ_2) are independent. The posterior on μ has a Gamma$(5, 5\mu)$ term along with the binomial terms above, which essentially contribute the likelihood of a logistic regression of the number of radio failures during service times on t using the log link function. The posterior on (λ_1, λ_2) can be shown (with some algebra) to be:

- $\lambda_1^2 | \mathcal{D} \sim \text{Gamma}[\frac{1}{2} r_2 + 1, \frac{1}{2}(u_1 - r_2^2/u_2)] = \text{Gamma}(59, 5.04)$. Note that $(u_1 - r_2^2/u_2)$ is always guaranteed to be nonnegative because the arithmetic mean is always at least as large as the harmonic mean.
- $\lambda_2 | \lambda_1, \mathcal{D} \sim \text{Normal}(r_2 \lambda_1 / u_2, 1/u_2) = \text{Normal}(1.34\lambda_1, 0.11^2)$.

The MGF of the $0 \to 6$ waiting time distribution is $E_{06}(s)$. The MAPLE code for computing $K_{06}(s)$ and its derivatives is given in Figures 9.17 and 9.18. The predictive density is given in Figure 9.16 along with a kernel density estimate. The predictive density was computed using rejection sampling for μ and direct sampling for (λ_1, λ_2) with $m = 400$. The kernel density estimate is based on brute-force simulation of the queue. This involved generating inverse Gaussian variates using Atkinson's (1982) bipartite algorithm. For an alternative Bayesian analysis of an M/G/1 queue based on PH distributions using a mixture of Erlangs for the service time density, see Ausin et al. (2004).

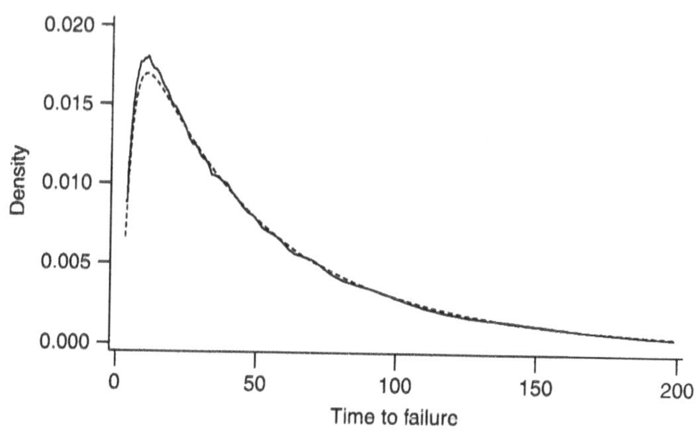

FIGURE 9.16 Bayes predictive density (dashed line) with kernel density estimate (solid line) for the time to total failure, $0 \to 6$, for the M/G/1 queue.

```
with(linalg):
with(codegen, makeproc, C):
ms := proc(s)

local m,n,i,j,k,mi,a,b,e1,p,p1,p1row,mgf1,iminv,tmat,
      tmat3,tmats;


m:=5;
n:=2;
for i from 1 to 5 do
  mi[i]:=min(m,m+n-i)
od;
for i from 1 to 5 do
 for k from (i-1) to (mi[i]+i-1) do
  a[i,k]:=0
 od;
od;

for i from 1 to 5 do
 for k from (i-1) to (mi[i]+i-1) do
  for j from 0 to k-i+1 do
    b:=binomial(k-i+1,j);
    e1:=exp(l1*(l2-((l2^2+2*(mi[i]-k+i-1+j)*mu-2*s)^.5)));
    a[i,k]:=a[i,k]+((-1)^j)*b*e1;
  od;
  a[i,k]:=a[i,k]*binomial(mi[i],k-i+1);
 od;
od;
p:=matrix(7,7):
for i from 1 to 7 do for j from 1 to 7 do p[i,j]:=0 od; od;
 for i from 1 to 5 do
  for k from (i-1) to (mi[i]+i-1) do
   p[i+1,k+1]:=a[i,k]:
  od; od;
p[1,2]:=1:
mgf1:=1/(1-(s/(5*mu)));
p1:=matrix(7,7):

# Continued in Figure 9.18.
```

FIGURE 9.17 MAPLE code for M/G/1 queue.

Suppose that we wish to analyze the long-run behavior of the queue when state 6 is not absorbing so that the queue can pass out of 6 to either 5 or 6. Allowing passage out of state 6 makes the queue stationary, and the posterior stationary probabilities can be computed using the standard techniques for the M/G/1 queue as well as with flowgraph methods. Viewing the queue as a semi-Markov process in

```
for j from 1 to 7 do p1[1,j]:=mgf1*p[1,j] od;
for j from 1 to 7 do p1[7,j]:=0 od;
for i from 2 to 6 do
   for j from 1 to 7 do
      p1[i,j]:=p[i,j]
   od;
od;

a:=array(identity,1..7,1..7):
iminv:=inverse(a-p1):
p1row:=transpose(row(p1,1)):
tmat:=multiply(p1row,col(iminv,7)):
tmat3:=tmat/iminv[7,7]:
tmats:=simplify(tmat3):
tmats;

end proc;

mx:=makeproc(ms(s),[s]);

k:=makeproc(ln(ms(s)),[s]):
C(k,optimized,ansi,filename="c:/k.c");
dk:=makeproc(diff(k(s),s),[s]):
C(dk,optimized,ansi,filename="c:/dk.c");
ddk:=makeproc(diff(dk(s),s),[s]):
C(ddk,optimized,ansi,filename="c:/ddk.c");
```

FIGURE 9.18 MAPLE code for M/G/1 queue (*continued*).

continuous time, the posterior stationary probabilities are computed by combining MAPLE computation with these simulation-based techniques. The stationary probability of state i is the mean waiting time in state i divided by the mean first return time. For our purposes, these are averaged over the posterior samples of $\boldsymbol{\theta}$. Using the flowgraph, these are

$$p_i(\boldsymbol{\theta}) = \sum_{j=\max(i-1,0)}^{6} T'_{ij}(0|\boldsymbol{\theta})/E'_{ii}(0|\boldsymbol{\theta}), i = 0, \ldots, 6. \tag{9.33}$$

The expressions for $p_0, \ldots, p_6$ above are computed using MAPLE and then mixed over the posterior samples of $\boldsymbol{\theta}$ to give the posterior stationary distribution when state 6 is not absorbing.

State	0	1	2	3	4	5	6
Prob.	0.0136	0.0976	0.1813	0.2887	0.2703	0.1268	0.0216

Alternatively, we can perform this computation using the probabilities (9.27) for the embedded Markov chain $\{Q(D_n); n > 1\}$,

$$\pi_0 = \frac{\delta_0/5\mu}{\delta_0/5\mu + (\lambda_1/\lambda_2)\sum_{j=1}^{6}\delta_j} \quad \text{and} \quad \pi_i = \frac{\delta_i(\lambda_1/\lambda_2)}{\delta_0/5\mu + (\lambda_1/\lambda_2)\sum_{j=1}^{6}\delta_j},$$
$$i = 1, \ldots, 6. \tag{9.34}$$

Computation of these quantities in MAPLE leads to the same posterior stationary probabilities.

These stationary probability computations confirm that the flowgraph transmittances with Pyke's (1961) matrix renewal equations give the same results as the standard analysis for an M/G/1 queue in the limit. The advantage of the flowgraph is that the distribution of any waiting time can be found without passing to the limit. We conclude this section with the MAPLE program code for the M/G/1 queue, given in Figures 9.17 and 9.18.

9.6 INFINITE-STATE M/M/c QUEUE

Consider a birth and death process on states $0, 1, 2, \ldots$ with transitions only between neighboring states. Let the transmittances be defined as

$$\begin{aligned} T_{i,i+1}(s) &= p_{i,i+1}M_{i,i+1}(s) \\ T_{i,i-1}(s) &= p_{i,i-1}M_{i,i-1}(s). \end{aligned} \tag{9.35}$$

Such a process is also a one-sided random walk. A flowgraph model for this process is given in Figure 9.19. Note that the process can have infinitely many states. Our concern is the waiting time for first return to state 0 beginning in state 0 (i.e., $0 \to 0$). Suppose for the moment that $p_{12} = 0$, so that our path is simply $0 \to 1 \to 0$. The transmittance of the path is

$$T_{01}(s)T_{10}(s) = p_{01}p_{10}M_{01}(s)M_{10}(s),$$

where $p_{01} = p_{10} = 1$. Denote the transmittance of first return to state 0 by $E_{00}^{+}(s)$. The MGF of the distribution of first return conditional on this passage is $E_{00}^{+}(s)/E_{00}^{+}(0)$.

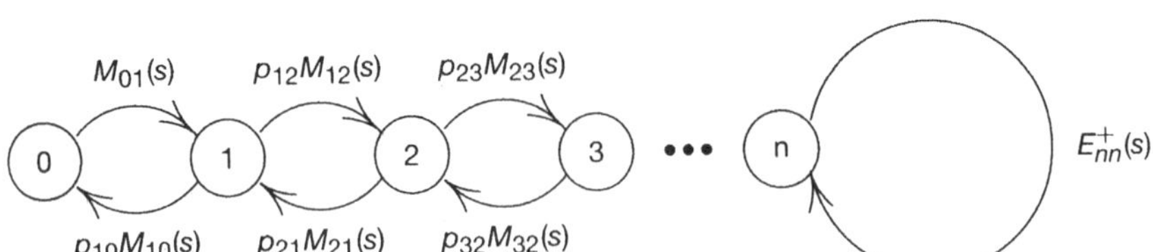

FIGURE 9.19 Flowgraph model for an infinite-state birth and death process.

Suppose that $p_{12} > 0$ but $p_{23} = 0$. Now our path is still $0 \to 1 \to 0$, but we have one loop, $1 \to 2 \to 1$. Our transmittance of interest is

$$E_{00}^{+}(s) = \frac{T_{01}(s)T_{10}(s)}{1 - p_{12}p_{21}M_{12}(s)M_{21}(s)} = \frac{T_{01}(s)T_{10}(s)}{1 - T_{12}(s)T_{21}(s)} = \frac{T_{01}(s)T_{10}(s)}{1 - E_{11}^{+}(s)}, \quad (9.36)$$

where $E_{11}^{+}(s)$ denotes the first return to state 1 from the right-hand side. If we add another loop, setting $p_{23} > 0$ and $p_{34} = 0$, we get

$$E_{11}^{+}(s) = \frac{T_{12}(s)T_{21}(s)}{1 - T_{23}(s)T_{32}(s)} = \frac{T_{12}(s)T_{21}(s)}{1 - E_{22}^{+}(s)} \quad (9.37)$$

and

$$E_{00}^{+}(s) = \frac{T_{01}(s)T_{10}(s)}{1 - \dfrac{p_{12}p_{21}M_{12}(s)M_{21}(s)}{1 - p_{23}p_{32}M_{23}(s)M_{32}(s)}} = \frac{T_{01}(s)T_{10}(s)}{1 - \dfrac{T_{12}(s)T_{21}(s)}{1 - T_{23}(s)T_{32}(s)}} = \frac{T_{01}(s)T_{10}(s)}{1 - E_{11}^{+}(s)}. \quad (9.38)$$

Iterating in this manner, we get an infinite continued fraction (cf. Olds, 1963). The recursions

$$E_{ii}^{+}(s) = \frac{T_{i,i+1}(s)T_{i+1,i}(s)}{1 - E_{i+1,i+1}^{+}(s)} \qquad \text{for } i = 0, 1, 2, \ldots \quad (9.39)$$

give the first return transmittances for each state i. In most finite systems, $E_{00}^{+}(0) = 1$, so that $E_{00}^{+}(s)$ is the MGF of the waiting time distribution for the first return to state 0. In infinite-state processes, these recursions simplify if the process is homogeneous beyond state n so that

$$T_{i,i+1}(s) = T_{\text{up}}(s) \quad \text{and} \quad T_{i+1,i}(s) = T_{\text{down}}(s) \qquad \text{for } i \geq n.$$

The recursions (9.39) apply for $i = 0, 1, \ldots, n-1$. After state n, due to the homogeneous structure,

$$E_{ii}^{+}(s) = E_{i+1,i+1}^{+}(s) \qquad \text{for } i \geq n.$$

This means that the continued fraction (9.38) converges and we can solve for $E_{nn}^{+}(s)$, picking the (smaller) negative root,

$$E_{nn}^{+}(s) = \frac{1}{2} - \frac{1}{2}\left[1 - 4\,T_{\text{up}}(s)\,T_{\text{down}}(s)\right]^{1/2}. \quad (9.40)$$

This is the first return transmittance where the embedded Markov chain is a one-sided random walk (cf. Feller, 1968, Sec. XIII.4, eq. 4.4), but now we have random transition times with MGFs

$$\frac{T_{\text{up}}(s)}{T_{\text{up}}(0)} \quad \text{and} \quad \frac{T_{\text{down}}(s)}{T_{\text{down}}(0)}.$$

The probability of returning to state 0 is now $E_{00}^+(0)$ computed recursively from (9.39) for $i = 0, \ldots, n-1$ with $E_{nn}^+(0)$ as in (9.40). A random walk over positive and negative integers or a two-sided birth and death process is handled similarly by defining a corresponding transmittance of first return to 0 from the negative side, $E_{00}^-(s)$, so that the first return to 0 is $p_{01}E_{00}^+(s) + (1 - p_{01})E_{00}^-(s)$.

We now use the flowgraph of Figure 9.19 to illustrate the infinite-state M/M/c queue. Consider the M/M/c queue, and let the number of customers $m \to \infty$ and $\mu(m) \to 0$ in such a way that $m\mu(m) \to \nu$. The parameter $\mu(m)$ of the Poisson arrival distribution is decreasing as the number of arrivals increase. The interarrival times of customers to the queue are Exp(ν) and service times are Exp(λ). The transmittances between neighboring states are

$$T_{i,i+1}(s) = \begin{cases} \dfrac{\nu}{\nu + i\lambda} M_{\nu+i\lambda}(s), & i = 0, \ldots, c \\ \dfrac{\nu}{\nu + c\lambda} M_{\nu+c\lambda}(s), & i > c \end{cases}$$
$$T_{i,i-1}(s) = \begin{cases} \dfrac{i\lambda}{\nu + i\lambda} M_{\nu+i\lambda}(s), & i = 0, \ldots, c \\ \dfrac{c\lambda}{\nu + c\lambda} M_{\nu+c\lambda}(s), & i > c, \end{cases} \tag{9.41}$$

where

$$M_{\nu+i\lambda}(s) = \frac{\nu + i\lambda}{\nu + i\lambda - s} \qquad \text{for } s < \nu + i\lambda \text{ and } i = 0, 1, \ldots.$$

The transmittances of (9.41) imply that the queue is homogeneous beyond state c. The queue can be either stationary or nonstationary, depending on whether $\nu \le c\lambda$. Recall that $E_{00}^+(0)$ is the probability of returning to state 0. The queue is nonstationary when $E_{00}^+(0) < 1$ and stationary when $E_{00}^+(0) = 1$ (cf. Grimmett and Stirzaker (1992), Section 5.3). We consider two examples, one with a stationary queue and the other with a nonstationary queue.

For an M/M/2 queue, the recursion equations (9.39) and (9.40), with the transmittances (9.41), simplify to

$$E_{00}^+(s) = \frac{T_{01}(s)T_{10}(s)}{1 - E_{11}^+(s)}$$
$$E_{11}^+(s) = \frac{T_{12}(s)T_{21}(s)}{1 - E_{22}^+(s)} \tag{9.42}$$
$$E_{22}^+(s) = \frac{1}{2} - \frac{1}{2}\left\{1 - 4\frac{\nu}{\nu + 2\lambda}\frac{2\lambda}{\nu + 2\lambda}[M_{\nu+2\lambda}(s)]^2\right\}^{1/2}.$$

Example 9.3: Infinite Queue: Stationary Case. This example considers the stationary M/M/c infinite-state queue with $c = 2$ and $E_{00}^+(0) = 1$ for $\nu \le 2\lambda$. Data were simulated from the queue with $\nu = 5.0$ and $\lambda = 5.1$ and consist of complete information on five realizations, starting in 0 and returning to 0. The posterior distribution of (ν, λ), assuming a flat prior, has ν and λ independent with $\nu|\mathcal{D} \sim$ Gamma(17, 3.08) and $\lambda|\mathcal{D} \sim$ Gamma(17, 2.58). Density approximations based on 400 and 1600 posterior samples were computed and found to be graphically indistinguishable; therefore, the density based on 400 posterior samples is used.

The MAPLE code for the time of first return to state 0 is given in Figure 9.20. For the purposes of the flowgraph, the entire $E_{22}^{+}(s)$ transmittance is treated as a loop

```
# Include package to generate C code
with(codegen,makeproc,C):

# Define the MGF as a procedure function of s
ms := proc(s)

# Define transmittances using w's for transmittances
   local w01,w10,w12,w21,tup,tdn,e22;

# Enter each transmittance

   w01 := nu/(nu-s);
   w10 := (lb/(lb+nu))*((nu+lb)/(nu+lb-s));
   w12 := (nu/(lb+nu))*((nu+ lb)/(nu+lb-s));
   w21 := (2*lb/(2*lb+nu))*(nu+2*lb)/(nu+2*lb-s));
   tup := nu/(nu+2*lb);
   tdn := 2*lb/(nu+2*lb);
   e22 := .5-.5*(1-4*tup*tdn*(((nu+2*lb)/
          (nu+2*lb-s))^2))^(.5);

# Calculate E+00(s)
   w01*w10*(1-e22)/(1-w12*w21-e22);
end proc;

# Make the MGF, CGF, and its derivatives
mx := makeproc(ms(s),[s]):

k := makeproc(ln(ms(s)),[s]):
C(k,optimized,ansi,filename="c:/fg/infq/k.c"):

dk := makeproc(diff(k(s),s),[s]):
C(dk,optimized,ansi,filename="c:/fg/infq/dk.c"):

ddk := makeproc(diff(dk(s),s),[s]):
C(ddk,optimized,ansi,filename="c:/fg/infq/ddk.c"):
```

FIGURE 9.20 MAPLE code for M/M/2 infinite-state queue.

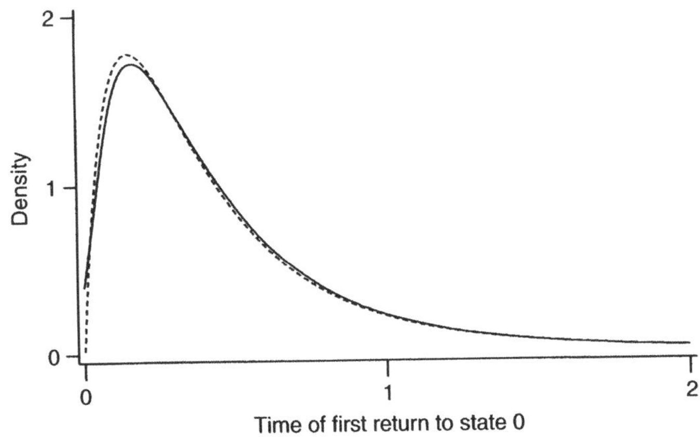

FIGURE 9.21 Bayes predictive density (dashed) with kernel density estimate (solid) for the time to first return to state 0 for the M/M/2 infinite-state queue.

that is not touching the main path $0 \to 1 \to 0$. The loop $1 \to 2 \to 1$ is treated as touching the path. The same result would be obtained by working directly with $0 \to 1 \to 0$ and $E_{11}^{+}(s)$ where $E_{11}^{+}(s)$ touches the path.

Figure 9.21 presents a plot of the Bayes predictive density for this queue. The solid line is a kernel density estimate based on simulation of this queue and the dashed line is the saddlepoint approximation.

All of the posterior samples satisfying $\nu > 2\lambda$ result in nonstationary queues. These queues must be analyzed in a different manner since they are not certain to return to 0 and some of the realization times are very long. The saddlepoint approximation is computed from the MGF

$$E_{00}^{+}(s)/E_{00}^{+}(0).$$

This estimates a density conditional on the event that the queue returns to state 0. The kernel density has the same interpretation, having been computed based only on sample paths that returned to state 0. Since some of the waiting times are extremely long, any simulated queue value that did not return by time 10,000 was declared to be unreturned. Even with this cutoff, simulation of this network took on the order of hours, whereas the flowgraph analysis took minutes. For additional details on this simulation, see Butler and Huzurbazar (2000). □

Example 9.4: M/M/C INFINITE QUEUE: MIXING OVER STATIONARITY AND NON-STATIONARITY. This example considers the infinite-state queue with $c = 2$ and a posterior that places substantial weight over both the stationary portion of the parameter space $S = \{(\nu, \lambda) : \nu \leq 2\lambda\}$ and the nonstationary portion S^c. We obtain the independent posteriors $\nu|\mathcal{D} \sim \text{Gamma}(15, 8)$ and $\lambda|\mathcal{D} \sim \text{Gamma}(15, 12)$. We obtained 400 posterior samples from these posteriors. The prediction can be summarized with the calculation of three posterior probabilities and three posterior predictive densities. Suppose that

$$R = \{\text{return to state } 0\}.$$

Then:

1. The posterior probability of stationarity is

$$P(S|\mathcal{D}) = 0.7827.$$

2. The probability that the queue returns to state 0 given nonstationarity is

$$P(R|\mathcal{D}, S^c) = E_{\boldsymbol{\theta}|\mathcal{D},S^c}[E_{00}^{+}(0)] = 0.7143.$$

Note that

$$E_{\boldsymbol{\theta}|\mathcal{D},S^c}[E_{00}^{+}(0)] = \frac{E_{\boldsymbol{\theta}|\mathcal{D}}[E_{00}^{+}(0)|S^c]}{P(S^c|\mathcal{D})}.$$

3. The unconditional probability of return is

$$P(R|\mathcal{D}) = P(S|\mathcal{D}) + P(R|\mathcal{D}, S^c)[1 - P(S|\mathcal{D})] = 0.9379.$$

This results in three densities. Figure 9.22 gives the density of time to first return based on the sample of (ν, λ) values from S. This is the conditional predictive density of time to first return to state 0 given stationarity, $\hat{f}_{00}(z|\mathcal{D}, S)$. Figure 9.23 gives the density of the time to first return based on the sample of (ν, λ) values from S^c. This is the conditional predictive density of time to first return to state 0

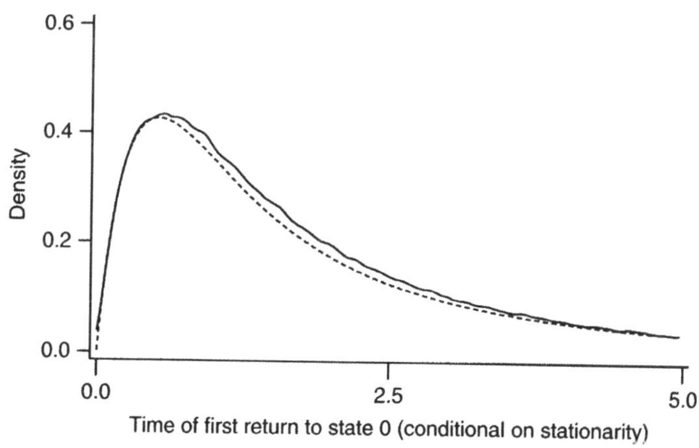

FIGURE 9.22 Conditional Bayes predictive density (dashed) with kernel density estimate (solid) given stationarity for the time to first return to state 0 for the M/M/2 infinite-state queue.

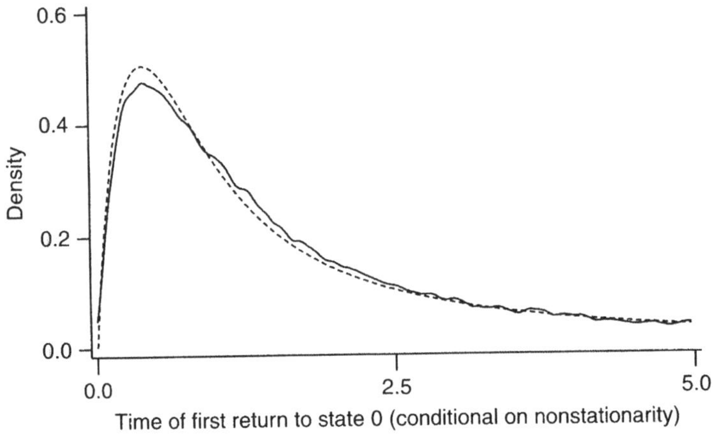

FIGURE 9.23 Conditional Bayes predictive density (dashed) with kernel density estimate (solid) given nonstationarity for the time to first return to state 0 for the M/M/2 infinite-state queue.

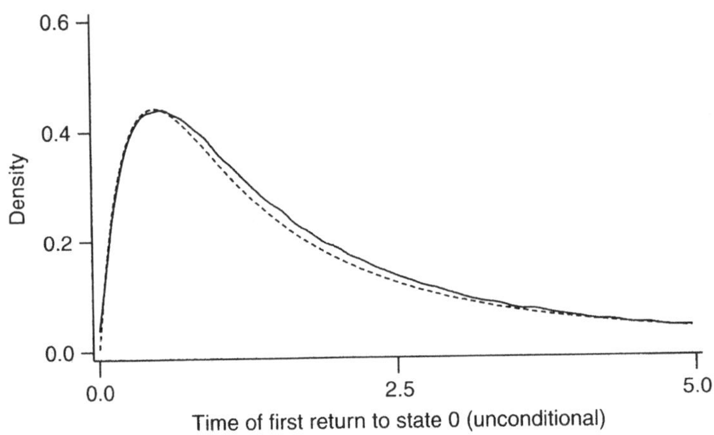

FIGURE 9.24 Unconditional Bayes predictive density (dashed) with kernel density estimate (solid) for the time to first return to 0 for the M/M/2 infinite-state queue.

under nonstationarity, $\hat{f}_{00}(z|\mathcal{D}, S^c, R)$. Figure 9.24 gives the unconditional density of time to first return to state 0 based on mixing over the samples in S and S^c. The resulting predictive density based on the posterior mixture is

$$\hat{f}_{00}(z|\mathcal{D}, R) = \hat{f}_{00}(z|\mathcal{D}, S)P(S|\mathcal{D}) + \hat{f}_{00}(z|\mathcal{D}, S^c, R)[1 - P(S|\mathcal{D})].$$

Network simulation for the corresponding kernel density estimates uses the same posteriors.

There are many interesting problems related to saddlepoint methods that arise in the context of this infinite-state queue that are beyond the scope of this book. While the saddlepoint density approximation is stable, it turns out that the Lugannani and Rice (1980) saddlepoint CDF approximation totally breaks down near the boundary of stationarity and nonstationarity which is the line $\nu = 2\lambda$ in the (ν, λ) space. It turns out that a saddlepoint CDF approximation based on the inverse Gaussian distribution (cf. Wood et al., 1993) works well under both stationarity and nonstationarity. Butler and Huzurbazar (2000) investigate the behavior of this queue under various scenarios using different saddlepoint approximations. □

9.7 CONCLUDING REMARKS

Many queuing examples are quite exotic. One will rarely run into an infinite-state queue or desire to formulate the M/G/1 queue with such a level of complexity for a real system. The discussion of the M/M/c queue and the M/G/1 queue was presented to formulate the flowgraph within the queuing framework. Now suppose that we have a finite-state queue and the number of servers is unknown or variable. In such a case, the flowgraph model adapts very nicely. The power of flowgraph models is that data are needed only on the observed transitions and their associated waiting times. For example, both the AIDS and retinopathy examples of Chapter 7, as well as the hydraulic pump system with general waiting time distributions of Example 9.2, were birth and death queuing networks. None of them had a specified number of servers. Yet the flowgraph methods are able to analyze these processes. In fact, flowgraph models bypass the tedious computation of standard queuing analysis by modeling the observable system. Of course, if one prefers to formulate the flowgraph as a queue, the methods of Sections 9.3 and 9.5 apply.

EXERCISES

9.1 Consider the hydraulic pump system of Example 9.2. Data for this example are available from the author's web site. Repeat the analysis using gamma models for the $0 \to 1$ and $1 \to 0$ branch waiting times and an inverse Gaussian model for the $1 \to 2$ branch waiting time. Compare the results to those of Example 9.2.

9.2 Formulate a flowgraph model for an M/M/4 queue as follows. Suppose that a fleet of ships consists of 10 ships. Ships arrive into a docking facility at a rate of one ship per week. It takes on average three weeks to unload and reload a ship. Four docks are available for the ships.

(a) Draw the flowgraph.

(b) Develop the transmittances by computing the appropriate probabilities and MGFs.

(c) Simulate $n = 50$ complete observations on this queue using the rates given as system parameters. An observation is a "run" that is recorded until all 10 ships are in the docking facility.

(d) Compute a Bayes predictive density and a maximum likehood estimated density for the time until seven ships are in the docking facility.

9.3 Consider an M/M/1 queue with Exp(λ) arrival distribution and Exp(μ) service time distribution. Show that the MGF of a typical busy period is

$$M(s) = \frac{\lambda + \mu - s - \sqrt{(\lambda + \mu - s)^2 - 4\lambda\mu}}{2\lambda}.$$

APPENDIX

Moment Generating Functions

A generating function is a transform of the distribution of a random variable. Let T be a random variable. Transformations are expected values of functions of T. For a general function, $g(s,t)$, the expected value is $E(g(s,T)) = \int g(s,t)f(t)\,dt$, where T has density $f(t)$. The rationale for using transforms is to allow for ease of solution of the problem at hand. There are many types of transforms. The transform used with flowgraph models is the moment generating function, which is invaluable when dealing with sums of independent random variables. MGFs are also useful for calculating moments of such random variables.

Definition .0.1. The *moment generating function* of a random variable T is defined as $M(s) = E(e^{sT})$, provided that the expectation exists for s in an open neighborhood of 0.

Often, we write $M_T(s)$ to emphasize that it is the MGF of the random variable T, or equivalently, of the distribution of T. The MGF generates *moments* of the distribution. The k-th moment of a distribution is defined as

$$\mu'_k = E(T^k),$$

Moments can be used to calculate quantities of interest about distributions. For example, the first moment μ'_1 gives us the mean of the distribution and $\mu'_2 - \mu'^2_1$ gives us the variance of the distribution. The MGF generates moments as the kth moment is the kth derivative of the MGF evaluated at $s = 0$:

$$\mu'_k = E(T^k) = \left.\frac{d^k M_T(s)}{ds^k}\right|_{s=0}.$$

The MGF, if it exists, defines the distribution of T uniquely. Moments themselves do not characterize distributions. Two distinct random variables can have identical moments of all orders and yet have different distributions (cf. Casella and Berger, 2002, p. 65).

Flowgraph Models for Multistate Time-to-Event Data, By Aparna V. Huzurbazar
ISBN 0-471-26514-4

Example .1: Gamma Distribution Moment Generating Function. Suppose that random variable T has a gamma distribution with density

$$f(t) = \frac{\beta^\alpha}{\Gamma(\alpha)} t^{\alpha-1} e^{-\beta t}, \qquad t > 0, \alpha > 0, \beta > 0.$$

The MGF of the distribution of T is

$$\begin{aligned} M(s) &= E(e^{Ts}) \\ &= \frac{\beta^\alpha}{\Gamma(\alpha)} \int_0^\infty e^{st} t^{\alpha-1} e^{-\beta t}\, dt \\ &= \frac{\beta^\alpha}{\Gamma(\alpha)} \int_0^\infty t^{\alpha-1} e^{-(\beta-s)t}\, dt. \end{aligned} \tag{.1}$$

Noting that the integrand in (.1) is the kernel of a Gamma($\alpha, \beta - s$) density, we can further simplify (.1) to give

$$\begin{aligned} M(s) &= \frac{\beta^\alpha}{(\beta - s)^\alpha} \int_0^\infty \frac{(\beta - s)^\alpha}{\Gamma(\alpha)} t^{\alpha-1} e^{-(\beta-s)t}\, dt \\ &= \left(\frac{\beta}{\beta - s}\right)^\alpha \qquad \text{for } s < \beta. \end{aligned} \tag{.2}$$

MGFs are especially useful for finding distributions of sums of independent random variables. Note that the random variables need not be identically distributed. If T_1 and T_2 are independent random variables with MGFs $M_{T_1}(s)$ and $M_{T_2}(s)$, respectively, we can find the distribution of $T_1 + T_2$ as

$$\begin{aligned} M_{(T_1+T_2)}(s) &= E\left(e^{s(T_1+T_2)}\right) \\ &= E(e^{sT_1})E(e^{sT_2}) \\ &= M_{T_1}(s)M_{T_2}(s). \end{aligned}$$

Once we have the MGF of the distribution of $T_1 + T_2$, we can convert it to the distribution of $T_1 + T_2$ either by exact or approximate inversion of the MGF. We can compute the MGF of the distribution of the sum of n independent random variables $T_1 + \cdots + T_n$ as

$$\begin{aligned} M_{(T_1+\cdots+T_n)}(s) &= E\left(e^{s(T_1+\cdots+T_n)}\right) \\ &= E(e^{sT_1}) \cdots E(e^{sT_n}) \\ &= \prod_{i=1}^n M_{T_i}(s). \end{aligned}$$

The MGF is related to the bilateral Laplace transform. The bilateral Laplace transform of a function $\phi(t)$ is defined as

$$g(s) = \int_{-\infty}^{\infty} e^{-st}\phi(t)\,dt \tag{.3}$$

if the integral converges. The function $g(-s)$, if it exists, is called the moment generating function. This is not quite the Laplace transform that is used in probability theory or in engineering systems analysis. The Laplace transform, as it is commonly known, is

$$g(s) = \int_{0}^{\infty} e^{-st}\phi(t)\,dt \equiv \mathcal{L}[\phi(t)]. \tag{.4}$$

For positive-valued random variables, this is exactly the same as the MGF with the argument $-s$ for s. We can see this if we rewrite (.3) as

$$\begin{aligned}\int_{-\infty}^{\infty} e^{-st}\phi(t)\,dt &= \int_{-\infty}^{0} e^{-st}\phi(t)\,dt + \int_{0}^{\infty} e^{-st}\phi(t)\,dt \\ &= \int_{\infty}^{0} e^{-s(-t)}\phi(-t)\,d(-t) + \int_{0}^{\infty} e^{-st}\phi(t)\,dt \\ &= \int_{0}^{\infty} e^{st}\phi(-t)\,dt + \int_{0}^{\infty} e^{-st}\phi(t)\,dt.\end{aligned}$$

It is enough to study

$$\int_{0}^{\infty} e^{-st}\phi(t)\,dt$$

and apply the results to

$$\int_{-\infty}^{0} e^{-st}\phi(t)\,dt = \int_{0}^{\infty} e^{st}\phi(-t)\,dt. \tag{.5}$$

In many real problems the random variable represents time or some other nonnegative quantity such as population size in a branching process and the left-hand integral in (.5) is 0. Thus, the bilateral Laplace transform is rarely seen in practice. □

.1 EXACT INVERSION OF MOMENT GENERATING FUNCTIONS

Moment generating functions by themselves are not useful for inference; what we require are distributions. For example, if we want to find the distribution of the sum of two independent random variables, we can use MGFs to find the MGF of the distribution of the sum as in (.3). Now we have the MGF of the distribution of the sum but we do not have the actual distribution. The most common way of converting a MGF back to a distribution is to look it up on a table of MGFs of known distributions and recognize it as the MGF of a known distribution. For example, if we had a MGF of the form (.2), we would recognize it as the MGF of a gamma distribution. Although this is done in classroom examples, it is not practical for real problems. Another method that works for a few common MGFs, such as the exponential and the gamma, with an integer-valued shape parameter, is to use the inverse Laplace transform option in a symbolic algebra package such as MAPLE.

Example .2: MAPLE INVERSION OF GAMMA DISTRIBUTION MGF. We can use the inverse Laplace transform option in MAPLE to invert the gamma MGF to a density. To do this, we must replace the argument s in the MGF with $-s$ to turn it into a Laplace transform. MAPLE does not allow a general shape parameter, an integer-valued shape parameter must be specified. The MAPLE session is as follows:

```
> with(inttrans):              # include the inverse transforms
> m(s):= (beta/(beta+s))^3;  # define the MGF for a gamma(3,beta)
                                          3
                                      beta
                        m(s) := -----------
                                             3
                                   (beta + s)
> invlaplace(m(s),s,t);        # inverse laplace transform function
                                  3   2
                     1/2 beta   t   exp(-beta t)
```

The end result from MAPLE, $f(t) = (1/2)\beta^3 t^2 e^{-\beta t}$, is the Gamma$(3, \beta)$ density. MAPLE is limited in terms of what it can invert, and for most real problems, we use approximate inversion techniques such as the saddlepoint methods of Chapter 3. □

References

Aalen, O. O. (1975). Statistical inference for a family of counting processes. Ph.D. dissertation, University of California, Berkeley, CA.

Aalen, O. O. (1978). Nonparametric inference for a family of counting processes. *Annals of Statistics*, 6, 701–726.

Aalen, O. O. (1995). Phase type distributions in survival analysis. *Scandinavian Journal of Statistics*, 22, 447–463.

Aalen, O. O., and Gjessing, H. K. (2001). Understanding the shape of the hazard rate: a point process point of view. *Statistical Science*, 16, 1–14.

Aalen, O. O., and Johansen, S. (1978). An empirical transition matrix for non-homogeneous Markov chains based on censored observations. *Scandinavian Journal of Statistics*, 5, 141–150.

AbouRizk, S., and Wales, R. (1997). Combined discrete-event/continuous simulation for project planning. *Journal of Construction Engineering and Management*, 123, 11–20.

Adler, S. (1981). Over-relaxation method for the Monte Carlo evaluation of the partition function for multiquadratic actions. *Physical Review D*, 23, 2901–2904.

Aitchison, J., and Dunsmore, I. (1975). *Statistical Prediction Analysis*. New York: Cambridge University Press.

Alioum, A., Leroy, V., Commenges, D., Dabis, F., and Salamon, R. (1998). Effect of gender, age, transmission category and antiretroviral therapy on the progression of HIV infections using multi-state Markov models: a prevalent cohort study, Bordeaux, France, 1985–1996. *Epidemiology*, 9, 605–612.

Allen, W. R. (1963). A note on conditional probability of failure when hazards are proportional. *Operations Research*, 11, 658–659.

Andersen, P. K., Borgan, O., Gill, R. D., and Keiding, N. (1993). *Statistical Models Based on Counting Processes*. New York: Springer-Verlag.

Andersen, P. K., and Keiding, N. (2002). Multi-state models for event history analysis. *Statistical Methods for Medical Research*, 11, 91–115.

Flowgraph Models for Multistate Time-to-Event Data, By Aparna V. Huzurbazar
ISBN 0-471-26514-4

Ansell, J., and Phillips, M. (1994). *Practical Methods for Reliability Data Analysis*. Oxford: Oxford University Press.

Ascher, H., and Feingold, H. (1984). *Repairable Systems Reliability: Modeling, Inference, Misconceptions and Their Causes*. New York: Marcel Dekker.

Atkinson, A. (1982). The simulation of generalized inverse Gaussian and generalized hyperbolic random variables. *SIAM Journal of Scientific Computing*, 3, 502–515.

Ausin, M., Wiper, M., and Lillo, R. (2004). Bayesian estimation for the M/G/1 queue using a phase-type approximation. *Journal of Statistical Planning and Inference*, 118, 83–101.

Balakrishnan, N., and Basu, A. (Eds.) (1995). *The Exponential Distribution: Theory, Methods and Applications*. London: Gordon & Breach.

Barlow, R. E., and Proschan, F. (1975). *Statistical Theory of Reliability and Life Testing: Probability Models*. New York: Holt, Rinehart and Winston.

Barndorff-Nielsen, O. E. (1978). *Information and Exponential Families in Statistical Theory*. New York: Wiley.

Barndorff-Nielsen, O. E. (1988). Comments on saddlepoint methods and statistical inference. *Statistical Science*, 3, 228–229.

Barndorff-Nielsen, O. E., and Cox, D. R. (1979). Edgeworth and saddlepoint approximations with statistical application (with discussion). *Journal of the Royal Statistical Society, Series B*, 41, 279–312.

Barndorff-Nielsen, O. E., and Cox, D. R. (1989). *Asymptotic Techniques for Use in Statistics*. New York: Chapman & Hall.

Barnett, O., and Cohen, A. (2000). The histogram and boxplot for the display of lifetime data. *Journal of Computational and Graphical Statistics*, 9, 759–778.

Barone, P., and Frigessi, A. (1993). Improving stochastic relaxation for Gaussian random fields. *Probability in the Engineering and Informational Sciences*, 4, 369–389.

Bartlett, M. S. (1955). *An Introduction to Stochastic Processes*. Cambridge: Cambridge University Press.

Bayes, T. (1763). An essay towards solving a problem in the doctrine of chances. *Philosophical Transactions of the Royal Society of London*, 53, 370–418. Reprinted, with an introduction by G. Barnard, in *Biometrika* in 1958, 45, 293–315.

Bernardo, J., and Smith, A. F. (2000). *Bayesian Theory*. New York: Wiley.

Berger, J. O. (1993). *Statistical Decision Theory and Bayesian Analysis*. New York: Springer-Verlag.

Berry, D., and Lindgren, B. (1996). *Statistics: Theory and Methods*, 2nd edition. Pacific Grove, CA: Duxbury.

Billingsley, P. (1986). *Probability and Measure*, 2nd edition. New York: Wiley.

Biswas, S. (1988). *Stochastic Processes in Demography and Applications*. New York: Wiley.

Blaesild, P., and Jensen, J. L. (1985). Saddlepoint formulas for reproductive exponentials. *Scandinavian Journal of Statistics*, 12, 193–202.

Bohachevsky, I., Johnson, M., and Stein, M. (1986). Generalized simulated annealing for function optimization. *Technometrics*, 28, 209–217.

Bolch, G., Greiner, S., de Meer, H., and Trivedi, K. S. (1998). *Queueing Networks and Markov Chains: Modeling and Performance Evaluation with Computer Science Applications*. New York: Wiley.

Box, G., and Tiao, G. (1973). *Bayesian Inference in Statistical Analysis*. New York: Wiley.

Brookmeyer, R., and Gail, M. (1988). A method for obtaining short-term projections and lower bounds on the size of the AIDS epidemic. *Journal of the American Statistical Association*, 83, 301–308.

Brookmeyer, R., and Liao, J. (1992). Statistical methods for reconstructing infection curves. In *AIDS Epidemiology: Methodological Issues*, ed. N. P. Jewell, K. Dietz, and V. T. Farewell. Basel: Birkhauser, pp. 39–60.

Bunday, B. (1991). *Statistical Methods in Reliability Theory and Practice*. Chichester, West Sussex, England: Ellis Horwood.

Butler, R. W., and Bronson, D. A. (2002). Bootstrapping survival times in stochastic networks using saddlepoint approximations. *Journal of the Royal Statistical Society, Series B*, 31–49.

Butler, R. W., and Huzurbazar, A. V. (1993). Prediction in stochastic networks. In *Bulletin of the International Statistical Institute, Proceedings of the 49th Session*, Book 3, Topic 23.

Butler, R. W., and Huzurbazar, A. V. (1997). Stochastic network models for survival analysis. *Journal of the American Statistical Association*, 92, 246–257.

Butler, R. W., and Huzurbazar, A. V. (2000). Bayesian prediction of waiting times in stochastic models. *The Canadian Journal of Statistics*, 28, 311–325.

Carlin, B., and Louis, T. (1996). *Bayes and Empirical Bayes Methods for Data Analysis*. New York: Chapman & Hall.

Casella, G., and Berger, R. (2002). *Statistical Inference*, 2nd edition. Pacific Grove, CA: Duxbury.

Casella, G., and George, E. (1992). Explaining the Gibbs sampler. *American Statistician*, 46, 167–174.

Chatterjee, A., Strauss, M. E., Smyth, K. A., and Whitehouse, P. J. (1992). Personality changes in Alzheimer's disease. *Archives of Neurology*, 49, 486–491.

Chatterjee, S., Laudato, M., and Lynch, L. (1996). Genetic algorithms and their statistical applications: an introduction. *Computational Statistics and Data Analysis*, 22, 633–651.

Chib, S., and Greenberg, E. (1995). Understanding the Metropolis–Hastings algorithm. *American Statistician*, 49, 327–335.

Collett, D. (1994). *Modelling Survival Data in Medical Research*. Boca Raton, FL: Chapman & Hall/CRC Press.

Commenges, D. (1986). Semi-Markov and non-homogeneous Markov models in medicine. In *Semi-Markov Models: Theory and Applications*, ed. J. Janssen. New York: Plenum Press, pp. 423–436.

Commenges, D., Letenneur, L., Joly, P., Alioum, A., and Dartigues, J. F., (1998). Modelling age-specific risk: application to dementia. *Statistics in Medicine*, 17, 1973–1988.

Cox, D. R. (1955). A use of complex probabilities in the theory of stochastic processes. *Proceedings of the Cambridge Philosophical Society*, 51, 313–319.

Cox, D. R. (1972). Regression models and life tables (with discussion). *Journal of the Royal Statistical Society, Series B*, 34, 187–220.

Cox, D. R., and Miller, H. D. (1977). *The Theory of Stochastic Processes*. London: Chapman & Hall.

Cox, D. R., and Oakes, D. (1984). *Analysis of Survival Data*. New York: Chapman & Hall.

Cox, D. R., and Smith, W. L. (1961). *Queues*. London: Chapman & Hall.

Crowder, M. J. (2001). *Classical Competing Risks*. Boca Raton, FL: Chapman & Hall/CRC.

Crowder, M. J., Kimber, A. C., Smith, R. L., and Sweeting, T. J. (1991). *Statistical Analysis of Reliability Data*. London: Chapman & Hall.

Daniels, H. E. (1954). Saddlepoint approximations in statistics. *Annals of Mathematical Statistics*, 25, 631–650.

Daniels, H. E. (1987). Tail probability approximations. *International Statistical Review*, 55, 37–48.

D'Azzo, J., and Houpis, C. (1981). *Linear Control System Analysis and Design: Conventional and Modern*. New York: McGraw-Hill.

DeFinetti, B. (1972). *Probability, Induction, and Statistics*. New York: Wiley.

DeGroot, M. H. (1970). *Optimal Statistical Decisions*. New York: McGraw-Hill.

DeGroot, M. H., and Schervish, M. J. (2002). *Probability and Statistics*, 3rd edition. Reading, MA: Addison-Wesley.

Diabetic Retinopathy Study Research Group (1981). A modification of the Airlie House classification of diabetic retinopathy. Report 7. *Investigative Ophthalmology and Visual Science*, 21, 210–226.

Doob, J. L. (1953). *Stochastic Processes*. New York: Wiley.

Dorf, R., and Bishop, R. (1995). *Modern Control Systems*. Reading, MA: Addison-Wesley.

Ellis, R., and Rosen, J. (1982). Laplace's method for Gaussian integrals with an application to statistical mechanics. *Annals of Probability*, 10, 47–66.

Erlang, A. K. (1909). The theory of probability and telephone conversations. In *The Life and Works of A. K. Erlang*, Volume 2. ed. E. Brockmeyer, H. L. Halstrom, and A. Jensen (1948). Copenhagen: Danish Academy of Sciences.

Esscher, F. (1932). On the probability function in the collective theory of risk. *Skandinavisk Aktuarist Tidskrift*, 15, 175–195.

Feller, W. (1968). *An Introduction to Probability Theory and Its Applications*. New York: Wiley.

Fieller, E. C. (1941). The biological standardization of insulin. *Supplement to the Journal of the Royal Statistical Society (1940–1941)*, 7, 1–64.

Fix, E., and Neyman, J. (1951). A simple stochastic model of recovery, relapse, and loss of patients. *Human Biology*, 23, 205–241.

Fleming, T. R., and Harrington, D. P. (1991). *Counting Processes and Survival Analysis*. New York: Wiley.

Frydman, H. (1995). Semi-parametric estimation in a three-state duration dependent Markov model from interval censored observations with application to AIDS. *Biometrics*, 51, 502–511.

Gajic, Z., and Lelic, M. (1996). *Modern Control Systems Engineering.* Upper Saddle River, NJ: Prentice Hall.

Garcia, S. (1997). Planning and communication in construction: Impacts on performance. M.S. thesis, University of New Mexico, Albuquerque, NM.

Gelman, A. B., Carlin, J. S., Stern, H. S., and Rubin, D. B. (1995). *Bayesian Data Analysis.* Boca Raton, FL: Chapman & Hall/CRC.

Green, P., and Han, X. (1992). Metropolis methods, Gaussian proposals and antithetic variables. In *Stochastic Models, Statistical Methods, and Algorithms in Image Analysis*, ed. P. Barone, A. Frigessi, and M. Piccioni. Berlin: Springer-Verlag.

Grimmett, G. R., and Stirzaker, D. R. (1992). *Probability and Random Processes.* Oxford: Oxford University Press.

Gross, A. J., Clark, V. A., and Liu, V. (1971). Estimation of survival parameters where one of two organs must function for survival. *Biometrics*, 27, 369–377.

Gross, D., and Harris, C. M. (1998). *Fundamentals of Queueing Theory.* New York: Wiley.

Harrington, H. J., and Tumay, K. (2000). *Simulation Modeling Methods: To Reduce Risks and Increase Performance.* New York: McGraw-Hill.

Hartigan, J. (1964). Invariant prior distributions. *Annals of Mathematical Statistics*, 35, 836–845.

Hougaard, P. (1999). Multi-state models: a review. *Lifetime Data Analysis*, 5, 239–264.

Hougaard, P. (2000). *Analysis of Multivariate Survival Data.* New York: Springer-Verlag.

Hoyland, A., and Rausand, M. (1994). *System Reliability Theory: Models and Statistical Methods.* New York: Wiley.

Huggins, W. (1957). Signal-flow graphs and random signals. *Proceedings of the Institute of Radio Engineers*, January, 74–86.

Huzurbazar, A. V. (1999). Flowgraph models for generalized phase type distributions with non-exponential waiting times. *Scandinavian Journal of Statistics*, 26, 145–157.

Huzurbazar, A. V. (2000a). Modeling and analysis of engineering systems data using flowgraph models. *Technometrics*, 42, 300–306.

Huzurbazar, A. V. (2000b). Modeling time-to-event data using flowgraph models. In *Advances on Methodological and Applied Aspects of Probability and Statistics*, ed. N. Balakrishnan. New York: Taylor & Francis, pp. 561–571.

Huzurbazar, A. V. (2005). A censored data histogram. *Communications in Statistics: Simulation and Computation*, 34(1), to appear.

Huzurbazar, A. V. (2004a). Modeling survival data using flowgraph models. In *Advances in Survival Analysis, Volume 23 of Handbook of Statistics*, ed. N. Balakrishnan and C. R. Rao. Amsterdam: Elsevier, pp. 729–746.

Huzurbazar, A. V. (2004b). Flowgraph models: A Bayesian case study in construction engineering. *Journal of Statistical Planning and Inference*, to appear.

Huzurbazar, S. (1999). Practical saddlepoint approximations. *American Statistician*, 53, 225–232.

Huzurbazar, S., and Huzurbazar, A. V. (1999). Survival and hazard functions for progressive diseases using saddlepoint approximations. *Biometrics*, 55, 198–203.

Huzurbazar, V. S. (1955). Exact forms of some invariants for distributions admitting sufficient statistics. *Biometrika*, 42, 533–537.

Huzurbazar, V. S. (1976). *Sufficient Statistics*. New York: Marcel Dekker.

ISO 8402 International Standard: Quality Vocabulary, International Standards Organization.

Jaynes, E. T. (1980). Marginalization and prior probabilities. In *Bayesian Analysis in Econometrics and Statistics*, ed. A. Zellner. Amsterdam: North-Holland.

Jeffreys, H. (1961). *Theory of Probability*, 3rd edition. New York: Oxford University Press.

Jensen, J. L. (1995). *Saddlepoint Approximations*. New York: Clarendon Press.

Johanssen, F. (1907). Waiting times and number of calls. Reprinted in *Post Office Electrical Engineers' Journal (London)*, October 1910 and January 1911, as cited by Kendall (1951).

Johnson, N., Kotz, S., and Balakrishnan, N. (1994). *Continuous Univariate Distributions*, Volumes 1 and 2, 2nd edition. New York: Wiley.

Joly, P., and Commenges, D. (1999). A penalized likelihood approach for a progressive three-state model with censored and truncated data: Application to AIDS. *Biometrics*, 55, 887–890.

Kalbfleisch, J., and Prentice, R. (1980). *The Statistical Analysis of Failure Time Data*. New York: Wiley.

Kaplan, E., and Meier, P. (1958). Nonparametric estimation from incomplete observations, *Journal of the American Statistical Association*, 53, 457–481.

Karlin, S., and Taylor, H. (1975). *A First Course in Stochastic Processes*. New York: Academic Press.

Karlin, S., and Taylor, H. (1981). *A Second Course in Stochastic Processes*. New York: Academic Press.

Kass, R., and Wasserman, L. (1996). The selection of prior distributions by formal rules. *Journal of the American Statistical Association*, 91, 1343–1370.

Kelley, A., and Pohl, I. (1995). *A Book on C: Programming in C*, 4th edition. Reading, MA: Addison-Wesley.

Kendall, D. G. (1951). Some problems in the theory of queues. *Journal of the Royal Statistical Society, Series B*, 13, 151–185.

Kendall, D. G. (1953). Stochastic processes occurring in the theory of queues and their analyses by the method of the imbedded Markov chain. *Annals of Mathematical Statistics*, 24, 338–354.

Klein, J., and Moeschberger, M. (1997). *Survival Analysis: Techniques for Censored and Truncated Data*. New York: Springer-Verlag.

Klein, J., and Shu, Y. (2002). Multi-state models of bone marrow transplantation studies. *Statistical Methods for Medical Research*, 11, 117–139.

Klein, J., Keiding, N., and Copelan, E. (1993). Plotting summary predictions in multistate survival models: probabilities of relapse and death in remission for bone marrow transplantation patients. *Statistics in Medicine*, 12, 2315–2332.

Klein, R., Klein, B., Moss S., Davis, M., and Demets, D. (1984a). The Wisconsin Epidemiologic Study of Diabetic Retinopathy: II. Prevalence and risk of diabetic retinopathy when age at diagnosis is less than 30 years. *Archives of Ophthalmology*, 102, 520–526.

Klein, R., Klein, B., Moss S., Davis, M., and Demets, D. (1984b). The Wisconsin Epidemiologic Study of Diabetic Retinopathy: III. Prevalence and risk of diabetic retinopathy when age at diagnosis is 30 or more years. *Archives of Ophthalmology*, 102, 527–532.

Kleinrock, L. (1975). *Queuing Systems*, Volume 1. New York: Wiley.

Kofler, M. (1997). *MAPLE: An Introduction and Reference.* Reading, MA: Addison-Wesley.

Kolassa, J. E., (1997). *Series Approximation Methods in Statistics*, 2nd edition. New York: Springer-Verlag.

Kulkarni, V. G. (1999). *Modeling, Analysis, Design, and Control of Stochastic Systems.* New York: Springer-Verlag.

Lagakos, S. (1976). A stochastic model for censored survival time in the presence of an auxiliary variable. *Biometrics*, 32, 551–559.

Lee, W. (1993). *Mobile Communications Design Fundamentals.* New York: Wiley.

Lee, W. (1995). *Mobile Cellular Telecommunications: Analog and Digital Systems.* New York: McGraw-Hill.

Leonard, T. (1982). Comment on: A simple predictive density function. *Journal of the American Statistical Association*, 77, 657–658.

Leung, A., and Tam, C. (1999). Models for assessing hoisting times of tower cranes. *Journal of Construction Engineering and Management*, 125, 385–391.

Li, H., Cao, J., and Love, P. (1999). Using machine learning and GA to solve time–cost trade-off problems. *Journal of Construction Engineering and Management*, 125, 347–353.

Limnios, N. (1992). Throughput availability in Markov systems. *IEEE Transactions on Reliability*, 41, 219–224.

Limnios, N., and Oprisan, G. (2001). *Semi-Markov Processes and Reliability.* Boston: Birkhauser.

Lindley, D. V. (1952). The theory of queues with a single server. *Proceedings of the Cambridge Philosophical Society*, 48, 277–289.

Lindsay, J., and Ryan, L. (1993). A three-state multiplicative model for rodent tumorigenicity experiments. *Applied Statistics*, 42, 283–300.

Little, J. (1961). A proof for the queueing formula: $L = \lambda W$. *Operations Research,* 9, 383–387.

Lloyd, D., and Lipow, M. (1964). *Reliability: Management, Methods, and Mathematics.* Englewood Cliffs, NJ: Prentice-Hall.

Longini, I. M., Clark, W., Byers, R., Ward, J., Darrow, W., Lemp, G., and Hethcote, H. (1989). Statistical analysis of the stages of HIV infection using a Markov model. *Statistics in Medicine*, 8, 831–843.

Longini, I. M., Clark, W., Gardner, L., and Brundage, J. (1991). The dynamics of CD4+ T-lymphocyte decline in HIV-infected individuals: a Markov modeling approach. *Journal of Acquired Immune Deficiency Syndromes*, 4, 1141–1147.

Lorens, C. S. (1964). *Flowgraphs for the Modeling and Analysis of Linear Systems.* New York: McGraw-Hill.

Lugannani, R., and Rice, S. (1980). Saddlepoint approximation for the distribution of the sum of independent random variables. *Advances in Applied Probability*, 12, 475–490.

MacKay, D. (1998). Introduction to Monte Carlo methods. In *Learning in Graphical Models*, ed. M. Jordan. Norwell, MA: Kluwer Academic, pp. 175–204.

Maindonald, J., and Braun, J. (2003). *Data Analysis and Graphics Using R*. Cambridge: Cambridge University Press.

Marshall, G., and Jones, R. (1995). Multi-state models for diabetic retinopathy. *Statistics in Medicine*, 14, 1975–1983.

Martz, H., and Waller, R. (1982). *Bayesian Reliability Analysis*. New York: Wiley.

Mason, S. J. (1953). Feedback theory: some properties of signal flow graphs. *Proceedings of the Institute of Radio Engineers*, 41, 1144–1156.

Mason, S. J. (1956). Feedback theory: further properties of signal flow graphs. *Proceedings of the Institute of Radio Engineers*, 44, 920–926.

Medley, G., Anderson, R. M., Cox, D. R., and Billard, L. (1987). Incubation period of AIDS in patients infected via blood transfusion. *Nature*, 328, 719–721.

Meeker, W., and Escobar, L. (1998). *Statistical Methods for Reliability Data*. New York: Wiley.

Multi-state models (2002). *Statistical Methods in Medical Research*, 11(2).

Neal, R. (1997). Markov chain Monte Carlo methods based on "slicing" the density function. *Technical Report* 9722. Department of Statistics, University of Toronto, Ontario, Canada.

Neal, R. (1998). Suppressing random walks in Markov chain Monte Carlo using ordered overrelaxation. In *Learning in Graphical Models*, ed. M. I. Jordan. Cambridge, MA: MIT Press.

Neal, R. (2003). Slice sampling. *Annals of Statistics*, 31, 705–767.

Nelder, J., and Mead, R. (1965). A simplex method for function minimization. *Computer Journal*, 7, 308–313.

Nelson, W. (1969). Hazard plotting for incomplete failure data. *Journal of Quality Technology*, 1, 27–52.

Nelson, W. (1990). *Accelerated Testing: Statistical Models, Test Plans, and Data Analysis*. New York: Wiley.

Neuts, M. F. (1981). *Matrix-Geometric Solutions in Stochastic Models: An Algorithmic Approach*. Baltimore: Johns Hopkins University Press.

Olds, C. D. (1963). *Continued Fractions*. New York: Random House.

Pepe, M. S., Longton, G., and Thornquist, M. (1991). A qualifier Q for the survival function to describe the prevalence of a transient condition. *Statistics in Medicine*, 10, 413–421.

Perks, W. (1947). Some observations on inverse probability including a new indifference rule. *Journal of the Institute of Actuaries*, 73, 285–313.

Prabhu, N. (1965). *Stochastic Processes: Basic Theory and Applications*. New York: Macmillan.

Pritsker, A., Sigal, C., and Hammesfahr, R. (1989). *SLAM II Network Models for Decision Support*. Englewood Cliffs, NJ: Prentice Hall.

Pyke, R. (1961). Markov renewal processes with finitely many states. *Annals of Mathematical Statistics*, 32, 1243–1259.

Raiffa, H., and Schlaifer, R. (1961). *Applied Decision Theory.* Cambridge, MA: Harvard University Press.

Ramaswami, V., and Lucantoni, D. (1979). On the merits of an approximation to the busy period of the GI/G/1 queue. *Management Science*, 25, 285–289.

Ramaswami, V., and Neuts, M. F. (1980). Some explicit formulas and computational methods for infinite server queues with phase type arrivals. *Journal of Applied Probability*, 17, 498–514.

Reid, N. (1988). Saddlepoint methods and statistical inference (with discussion). *Statistical Science*, 3, 213–238.

Reid, N. (2003). Asymptotics and the theory of inference. *Annals of Statistics*, 1695–2095.

Resnick, S. (1992). *Adventures in Stochastic Processes.* Boston: Birkhauser.

Ross, S. (1988). *A First Course in Probability*, 3rd edition. New York: Macmillan.

Samuelson, P. A. (1947). *Foundation of Economic Analysis.* Cambridge, MA: Harvard University Press.

Satten, G. A., and Longini, I. M. (1994). Estimation of incidence of HIV infection using cross-sectional marker surveys. *Biometrics*, 50, 675–688.

Saunders, P. (1993). Epidemiology of mental disorder in old age: GMS-AGECAT studies. *Advances in Old Age Psychiatry*, 4, 1–2.

Savage, L. J. (1954). *The Foundations of Statistics.* New York: Dover.

Shaked, M., and Shantikumar, J. G. (1983). Phase type distributions. In *Encyclopedia of Statistical Science*, ed. S. Kotz and N. Johnson. New York: Wiley.

Silverman, B. (1986). *Density Estimation for Statistics and Data Analysis.* New York: Chapman & Hall.

Sittler, R. (1956). Systems analysis of discrete Markov processes. *Institute of Radio Engineers Transactions on Circuit Theory*, Vol. 3, 257–266.

Skovgaard, I. M. (1987). Saddlepoint expansions for conditional distributions. *Journal of Applied Probability*, 24, 875–887.

Smith, A. F., and Gelfand, A. E. (1992). Bayesian statistics without tears: a sampling–resampling perspective. *American Statistician*, 46, 84–88.

Taylor, H., and Karlin, S. (1984). *An Introduction to Stochastic Modelling.* Orlando, FL: Academic Press.

Temkin, N. (1978). An analysis for transient states with applications to tumor shrinkage. *Biometrics*, 34, 59–66.

Therneau, T., and Grambsch, P. (2000). *Modeling Survival Data: Extending the Cox Model.* New York: Springer-Verlag.

Thomas, R., and Rosa, A. (2003). *The Analysis and Design of Linear Circuits*, 4th edition. New York: Wiley.

Tierney, L., and Kadane, J. B. (1986). Accurate approximations for posterior moments and marginal densities. *Journal of the American Statistical Association*, 81, 82–86.

Voelkel, J. G., and Crowley, J. (1984). Nonparametric inference for a class of semi-Markov processes with censored observations. *Annals of Statistics*, 12, 142–160.

Weibull, W. (1951). A statistical distribution of wide applicability. *Journal of Applied Mechanics*, 18, 293–297.

Welch, B., and Peers, H. (1963). On formulae for confidence points based on integrals of weighted likelihoods. *Journal of the Royal Statistical Society, Series B*, 25, 318–329.

Whitehouse, G. E. (1970). GERT, a useful technique for analyzing reliability problems. *Technometrics*, 12, 33–48.

Whitehouse, G. E. (1973). *Systems Analysis and Design Using Network Techniques.* Englewood Cliffs, NJ: Prentice-Hall.

Whitehouse, G. E. (1983). Flowgraph analysis. In *Encyclopedia of Statistical Science*, Volume 3, ed. D. L., Banks, C. B., Read, and S. Kotz. New York: Wiley, pp. 156–159.

Wichita Business Journal (2002). Boeing planned for strike on west coast docks. Wichita, KS, October 4.

Williams, B. J., and Huzurbazar, A. V. (2004). Posterior sampling with constructed likelihood functions. *Los Alamos National Laboratory Technical Report* LA-UR-04-2792.

Wilson S., and Solomon P. (1994). Estimates for different stages of HIV/AIDS disease. *Computations in Applied Biosciences*, 10(6), 681–683.

Wolstenholme, L. (1999). *Reliability Modelling: A Statistical Approach.* Boca Raton, FL: Chapman & Hall/CRC.

Wood, A., Booth, J., and Butler, R. W. (1993). Saddlepoint approximations to the CDF for some statistics with non-normal limit distributions. *Journal of the American Statistical Association*, 88, 680–686.

Yau, C. L., and Huzurbazar, A. V. (2002). Analysis of censored and incomplete survival data using flowgraph models. *Statistics in Medicine*, 21, 3727–3743.

Yeung, K., and Nanda, S. (1996), Channel management in microcell/macrocell cellular radio systems. *IEEE Transactions on Vehicular Technology*, 45, 601–612.

Zellner, A. (1971). *An Introduction to Bayesian Inference in Econometrics*. New York: Wiley.

Author Index

Flowgraph Models for Multistate Time-to-Event Data, By Aparna V. Huzurbazar
ISBN 0-471-26514-4

Subject Index

Flowgraph Models for Multistate Time-to-Event Data, By Aparna V. Huzurbazar
ISBN 0-471-26514-4

WILEY SERIES IN PROBABILITY AND STATISTICS

ESTABLISHED BY WALTER A. SHEWHART AND SAMUEL S. WILKS

The ***Wiley Series in Probability and Statistics*** is well established and authoritative. It covers many topics of current research interest in both pure and applied statistics and probability theory. Written by leading statisticians and institutions, the titles span both state-of-the-art developments in the field and classical methods.

Reflecting the wide range of current research in statistics, the series encompasses applied, methodological and theoretical statistics, ranging from applications and new techniques made possible by advances in computerized practice to rigorous treatment of theoretical approaches.

This series provides essential and invaluable reading for all statisticians, whether in academia, industry, government, or research.

ABRAHAM and LEDOLTER · Statistical Methods for Forecasting
AGRESTI · Analysis of Ordinal Categorical Data
AGRESTI · An Introduction to Categorical Data Analysis
AGRESTI · Categorical Data Analysis, *Second Edition*
ALTMAN, GILL, and McDONALD · Numerical Issues in Statistical Computing for the Social Scientist
AMARATUNGA and CABRERA · Exploration and Analysis of DNA Microarray and Protein Array Data
ANDĚL · Mathematics of Chance
ANDERSON · An Introduction to Multivariate Statistical Analysis, *Third Edition*
*ANDERSON · The Statistical Analysis of Time Series
ANDERSON, AUQUIER, HAUCK, OAKES, VANDAELE, and WEISBERG · Statistical Methods for Comparative Studies
ANDERSON and LOYNES · The Teaching of Practical Statistics
ARMITAGE and DAVID (editors) · Advances in Biometry
ARNOLD, BALAKRISHNAN, and NAGARAJA · Records
*ARTHANARI and DODGE · Mathematical Programming in Statistics
*BAILEY · The Elements of Stochastic Processes with Applications to the Natural Sciences
BALAKRISHNAN and KOUTRAS · Runs and Scans with Applications
BARNETT · Comparative Statistical Inference, *Third Edition*
BARNETT and LEWIS · Outliers in Statistical Data, *Third Edition*
BARTOSZYNSKI and NIEWIADOMSKA-BUGAJ · Probability and Statistical Inference
BASILEVSKY · Statistical Factor Analysis and Related Methods: Theory and Applications
BASU and RIGDON · Statistical Methods for the Reliability of Repairable Systems
BATES and WATTS · Nonlinear Regression Analysis and Its Applications
BECHHOFER, SANTNER, and GOLDSMAN · Design and Analysis of Experiments for Statistical Selection, Screening, and Multiple Comparisons
BELSLEY · Conditioning Diagnostics: Collinearity and Weak Data in Regression

*Now available in a lower priced paperback edition in the Wiley Classics Library.

* BELSLEY, KUH, and WELSCH · Regression Diagnostics: Identifying Influential Data and Sources of Collinearity
BENDAT and PIERSOL · Random Data: Analysis and Measurement Procedures, *Third Edition*
BERRY, CHALONER, and GEWEKE · Bayesian Analysis in Statistics and Econometrics: Essays in Honor of Arnold Zellner
BERNARDO and SMITH · Bayesian Theory
BHAT and MILLER · Elements of Applied Stochastic Processes, *Third Edition*
BHATTACHARYA and WAYMIRE · Stochastic Processes with Applications
BILLINGSLEY · Convergence of Probability Measures, *Second Edition*
BILLINGSLEY · Probability and Measure, *Third Edition*
BIRKES and DODGE · Alternative Methods of Regression
BLISCHKE AND MURTHY (editors) · Case Studies in Reliability and Maintenance
BLISCHKE AND MURTHY · Reliability: Modeling, Prediction, and Optimization
BLOOMFIELD · Fourier Analysis of Time Series: An Introduction, *Second Edition*
BOLLEN · Structural Equations with Latent Variables
BOROVKOV · Ergodicity and Stability of Stochastic Processes
BOULEAU · Numerical Methods for Stochastic Processes
BOX · Bayesian Inference in Statistical Analysis
BOX · R. A. Fisher, the Life of a Scientist
BOX and DRAPER · Empirical Model-Building and Response Surfaces
*BOX and DRAPER · Evolutionary Operation: A Statistical Method for Process Improvement
BOX, HUNTER, and HUNTER · Statistics for Experimenters: An Introduction to Design, Data Analysis, and Model Building
BOX and LUCEÑO · Statistical Control by Monitoring and Feedback Adjustment
BRANDIMARTE · Numerical Methods in Finance: A MATLAB-Based Introduction
BROWN and HOLLANDER · Statistics: A Biomedical Introduction
BRUNNER, DOMHOF, and LANGER · Nonparametric Analysis of Longitudinal Data in Factorial Experiments
BUCKLEW · Large Deviation Techniques in Decision, Simulation, and Estimation
CAIROLI and DALANG · Sequential Stochastic Optimization
CHAN · Time Series: Applications to Finance
CHATTERJEE and HADI · Sensitivity Analysis in Linear Regression
CHATTERJEE and PRICE · Regression Analysis by Example, *Third Edition*
CHERNICK · Bootstrap Methods: A Practitioner's Guide
CHERNICK and FRIIS · Introductory Biostatistics for the Health Sciences
CHILÈS and DELFINER · Geostatistics: Modeling Spatial Uncertainty
CHOW and LIU · Design and Analysis of Clinical Trials: Concepts and Methodologies, *Second Edition*
CLARKE and DISNEY · Probability and Random Processes: A First Course with Applications, *Second Edition*
*COCHRAN and COX · Experimental Designs, *Second Edition*
CONGDON · Applied Bayesian Modelling
CONGDON · Bayesian Statistical Modelling
CONOVER · Practical Nonparametric Statistics, *Third Edition*
COOK · Regression Graphics
COOK and WEISBERG · Applied Regression Including Computing and Graphics
COOK and WEISBERG · An Introduction to Regression Graphics
CORNELL · Experiments with Mixtures, Designs, Models, and the Analysis of Mixture Data, *Third Edition*
COVER and THOMAS · Elements of Information Theory
COX · A Handbook of Introductory Statistical Methods

*Now available in a lower priced paperback edition in the Wiley Classics Library.

*COX · Planning of Experiments
CRESSIE · Statistics for Spatial Data, *Revised Edition*
CSÖRGŐ and HORVÁTH · Limit Theorems in Change Point Analysis
DANIEL · Applications of Statistics to Industrial Experimentation
DANIEL · Biostatistics: A Foundation for Analysis in the Health Sciences, *Eighth Edition*
*DANIEL · Fitting Equations to Data: Computer Analysis of Multifactor Data, *Second Edition*
DASU and JOHNSON · Exploratory Data Mining and Data Cleaning
DAVID and NAGARAJA · Order Statistics, *Third Edition*
*DEGROOT, FIENBERG, and KADANE · Statistics and the Law
DEL CASTILLO · Statistical Process Adjustment for Quality Control
DeMARIS · Regression with Social Data: Modeling Continuous and Limited Response Variables
DEMIDENKO · Mixed Models: Theory and Applications
DENISON, HOLMES, MALLICK and SMITH · Bayesian Methods for Nonlinear Classification and Regression
DETTE and STUDDEN · The Theory of Canonical Moments with Applications in Statistics, Probability, and Analysis
DEY and MUKERJEE · Fractional Factorial Plans
DILLON and GOLDSTEIN · Multivariate Analysis: Methods and Applications
DODGE · Alternative Methods of Regression
*DODGE and ROMIG · Sampling Inspection Tables, *Second Edition*
*DOOB · Stochastic Processes
DOWDY, WEARDEN, and CHILKO · Statistics for Research, *Third Edition*
DRAPER and SMITH · Applied Regression Analysis, *Third Edition*
DRYDEN and MARDIA · Statistical Shape Analysis
DUDEWICZ and MISHRA · Modern Mathematical Statistics
DUNN and CLARK · Basic Statistics: A Primer for the Biomedical Sciences, *Third Edition*
DUPUIS and ELLIS · A Weak Convergence Approach to the Theory of Large Deviations
*ELANDT-JOHNSON and JOHNSON · Survival Models and Data Analysis
ENDERS · Applied Econometric Time Series
ETHIER and KURTZ · Markov Processes: Characterization and Convergence
EVANS, HASTINGS, and PEACOCK · Statistical Distributions, *Third Edition*
FELLER · An Introduction to Probability Theory and Its Applications, Volume I, *Third Edition,* Revised; Volume II, *Second Edition*
FISHER and VAN BELLE · Biostatistics: A Methodology for the Health Sciences
FITZMAURICE, LAIRD, and WARE · Applied Longitudinal Analysis
*FLEISS · The Design and Analysis of Clinical Experiments
FLEISS · Statistical Methods for Rates and Proportions, *Third Edition*
FLEMING and HARRINGTON · Counting Processes and Survival Analysis
FULLER · Introduction to Statistical Time Series, *Second Edition*
FULLER · Measurement Error Models
GALLANT · Nonlinear Statistical Models
GHOSH, MUKHOPADHYAY, and SEN · Sequential Estimation
GIESBRECHT and GUMPERTZ · Planning, Construction, and Statistical Analysis of Comparative Experiments
GIFI · Nonlinear Multivariate Analysis
GLASSERMAN and YAO · Monotone Structure in Discrete-Event Systems
GNANADESIKAN · Methods for Statistical Data Analysis of Multivariate Observations, *Second Edition*
GOLDSTEIN and LEWIS · Assessment: Problems, Development, and Statistical Issues
GREENWOOD and NIKULIN · A Guide to Chi-Squared Testing

*Now available in a lower priced paperback edition in the Wiley Classics Library.

GROSS and HARRIS · Fundamentals of Queueing Theory, *Third Edition*
*HAHN and SHAPIRO · Statistical Models in Engineering
HAHN and MEEKER · Statistical Intervals: A Guide for Practitioners
HALD · A History of Probability and Statistics and their Applications Before 1750
HALD · A History of Mathematical Statistics from 1750 to 1930
HAMPEL · Robust Statistics: The Approach Based on Influence Functions
HANNAN and DEISTLER · The Statistical Theory of Linear Systems
HEIBERGER · Computation for the Analysis of Designed Experiments
HEDAYAT and SINHA · Design and Inference in Finite Population Sampling
HELLER · MACSYMA for Statisticians
HINKELMAN and KEMPTHORNE: · Design and Analysis of Experiments, Volume 1: Introduction to Experimental Design
HOAGLIN, MOSTELLER, and TUKEY · Exploratory Approach to Analysis of Variance
HOAGLIN, MOSTELLER, and TUKEY · Exploring Data Tables, Trends and Shapes
*HOAGLIN, MOSTELLER, and TUKEY · Understanding Robust and Exploratory Data Analysis
HOCHBERG and TAMHANE · Multiple Comparison Procedures
HOCKING · Methods and Applications of Linear Models: Regression and the Analysis of Variance, *Second Edition*
HOEL · Introduction to Mathematical Statistics, *Fifth Edition*
HOGG and KLUGMAN · Loss Distributions
HOLLANDER and WOLFE · Nonparametric Statistical Methods, *Second Edition*
HOSMER and LEMESHOW · Applied Logistic Regression, *Second Edition*
HOSMER and LEMESHOW · Applied Survival Analysis: Regression Modeling of Time to Event Data
HUBER · Robust Statistics
HUBERTY · Applied Discriminant Analysis
HUNT and KENNEDY · Financial Derivatives in Theory and Practice
HUSKOVA, BERAN, and DUPAC · Collected Works of Jaroslav Hajek—with Commentary
HUZURBAZAR · Flowgraph Models for Multistate Time-to-Event Data
IMAN and CONOVER · A Modern Approach to Statistics
JACKSON · A User's Guide to Principle Components
JOHN · Statistical Methods in Engineering and Quality Assurance
JOHNSON · Multivariate Statistical Simulation
JOHNSON and BALAKRISHNAN · Advances in the Theory and Practice of Statistics: A Volume in Honor of Samuel Kotz
JOHNSON and BHATTACHARYYA · Statistics: Principles and Methods, *Fifth Edition*
JOHNSON and KOTZ · Distributions in Statistics
JOHNSON and KOTZ (editors) · Leading Personalities in Statistical Sciences: From the Seventeenth Century to the Present
JOHNSON, KOTZ, and BALAKRISHNAN · Continuous Univariate Distributions, Volume 1, *Second Edition*
JOHNSON, KOTZ, and BALAKRISHNAN · Continuous Univariate Distributions, Volume 2, *Second Edition*
JOHNSON, KOTZ, and BALAKRISHNAN · Discrete Multivariate Distributions
JOHNSON, KOTZ, and KEMP · Univariate Discrete Distributions, *Second Edition*
JUDGE, GRIFFITHS, HILL, LÜTKEPOHL, and LEE · The Theory and Practice of Econometrics, *Second Edition*
JUREČKOVÁ and SEN · Robust Statistical Procedures: Aymptotics and Interrelations
JUREK and MASON · Operator-Limit Distributions in Probability Theory
KADANE · Bayesian Methods and Ethics in a Clinical Trial Design

*Now available in a lower priced paperback edition in the Wiley Classics Library.

KADANE AND SCHUM · A Probabilistic Analysis of the Sacco and Vanzetti Evidence
KALBFLEISCH and PRENTICE · The Statistical Analysis of Failure Time Data, *Second Edition*
KASS and VOS · Geometrical Foundations of Asymptotic Inference
KAUFMAN and ROUSSEEUW · Finding Groups in Data: An Introduction to Cluster Analysis
KEDEM and FOKIANOS · Regression Models for Time Series Analysis
KENDALL, BARDEN, CARNE, and LE · Shape and Shape Theory
KHURI · Advanced Calculus with Applications in Statistics, *Second Edition*
KHURI, MATHEW, and SINHA · Statistical Tests for Mixed Linear Models
*KISH · Statistical Design for Research
KLEIBER and KOTZ · Statistical Size Distributions in Economics and Actuarial Sciences
KLUGMAN, PANJER, and WILLMOT · Loss Models: From Data to Decisions
KLUGMAN, PANJER, and WILLMOT · Solutions Manual to Accompany Loss Models: From Data to Decisions
KOTZ, BALAKRISHNAN, and JOHNSON · Continuous Multivariate Distributions, Volume 1, *Second Edition*
KOTZ and JOHNSON (editors) · Encyclopedia of Statistical Sciences: Volumes 1 to 9 with Index
KOTZ and JOHNSON (editors) · Encyclopedia of Statistical Sciences: Supplement Volume
KOTZ, READ, and BANKS (editors) · Encyclopedia of Statistical Sciences: Update Volume 1
KOTZ, READ, and BANKS (editors) · Encyclopedia of Statistical Sciences: Update Volume 2
KOVALENKO, KUZNETZOV, and PEGG · Mathematical Theory of Reliability of Time-Dependent Systems with Practical Applications
LACHIN · Biostatistical Methods: The Assessment of Relative Risks
LAD · Operational Subjective Statistical Methods: A Mathematical, Philosophical, and Historical Introduction
LAMPERTI · Probability: A Survey of the Mathematical Theory, *Second Edition*
LANGE, RYAN, BILLARD, BRILLINGER, CONQUEST, and GREENHOUSE · Case Studies in Biometry
LARSON · Introduction to Probability Theory and Statistical Inference, *Third Edition*
LAWLESS · Statistical Models and Methods for Lifetime Data, *Second Edition*
LAWSON · Statistical Methods in Spatial Epidemiology
LE · Applied Categorical Data Analysis
LE · Applied Survival Analysis
LEE and WANG · Statistical Methods for Survival Data Analysis, *Third Edition*
LEPAGE and BILLARD · Exploring the Limits of Bootstrap
LEYLAND and GOLDSTEIN (editors) · Multilevel Modelling of Health Statistics
LIAO · Statistical Group Comparison
LINDVALL · Lectures on the Coupling Method
LINHART and ZUCCHINI · Model Selection
LITTLE and RUBIN · Statistical Analysis with Missing Data, *Second Edition*
LLOYD · The Statistical Analysis of Categorical Data
MAGNUS and NEUDECKER · Matrix Differential Calculus with Applications in Statistics and Econometrics, *Revised Edition*
MALLER and ZHOU · Survival Analysis with Long Term Survivors
MALLOWS · Design, Data, and Analysis by Some Friends of Cuthbert Daniel
MANN, SCHAFER, and SINGPURWALLA · Methods for Statistical Analysis of Reliability and Life Data
MANTON, WOODBURY, and TOLLEY · Statistical Applications Using Fuzzy Sets
MARCHETTE · Random Graphs for Statistical Pattern Recognition

*Now available in a lower priced paperback edition in the Wiley Classics Library.

MARDIA and JUPP · Directional Statistics
MASON, GUNST, and HESS · Statistical Design and Analysis of Experiments with Applications to Engineering and Science, *Second Edition*
McCULLOCH and SEARLE · Generalized, Linear, and Mixed Models
McFADDEN · Management of Data in Clinical Trials
* McLACHLAN · Discriminant Analysis and Statistical Pattern Recognition
McLACHLAN, DO, and AMBROISE · Analyzing Microarray Gene Expression Data
McLACHLAN and KRISHNAN · The EM Algorithm and Extensions
McLACHLAN and PEEL · Finite Mixture Models
McNEIL · Epidemiological Research Methods
MEEKER and ESCOBAR · Statistical Methods for Reliability Data
MEERSCHAERT and SCHEFFLER · Limit Distributions for Sums of Independent Random Vectors: Heavy Tails in Theory and Practice
MICKEY, DUNN, and CLARK · Applied Statistics: Analysis of Variance and Regression, *Third Edition*
*MILLER · Survival Analysis, *Second Edition*
MONTGOMERY, PECK, and VINING · Introduction to Linear Regression Analysis, *Third Edition*
MORGENTHALER and TUKEY · Configural Polysampling: A Route to Practical Robustness
MUIRHEAD · Aspects of Multivariate Statistical Theory
MULLER and STOYAN · Comparison Methods for Stochastic Models and Risks
MURRAY · X-STAT 2.0 Statistical Experimentation, Design Data Analysis, and Nonlinear Optimization
MURTHY, XIE, and JIANG · Weibull Models
MYERS and MONTGOMERY · Response Surface Methodology: Process and Product Optimization Using Designed Experiments, *Second Edition*
MYERS, MONTGOMERY, and VINING · Generalized Linear Models. With Applications in Engineering and the Sciences
NELSON · Accelerated Testing, Statistical Models, Test Plans, and Data Analyses
NELSON · Applied Life Data Analysis
NEWMAN · Biostatistical Methods in Epidemiology
OCHI · Applied Probability and Stochastic Processes in Engineering and Physical Sciences
OKABE, BOOTS, SUGIHARA, and CHIU · Spatial Tesselations: Concepts and Applications of Voronoi Diagrams, *Second Edition*
OLIVER and SMITH · Influence Diagrams, Belief Nets and Decision Analysis
PALTA · Quantitative Methods in Population Health: Extensions of Ordinary Regressions
PANKRATZ · Forecasting with Dynamic Regression Models
PANKRATZ · Forecasting with Univariate Box-Jenkins Models: Concepts and Cases
*PARZEN · Modern Probability Theory and Its Applications
PEÑA, TIAO, and TSAY · A Course in Time Series Analysis
PIANTADOSI · Clinical Trials: A Methodologic Perspective
PORT · Theoretical Probability for Applications
POURAHMADI · Foundations of Time Series Analysis and Prediction Theory
PRESS · Bayesian Statistics: Principles, Models, and Applications
PRESS · Subjective and Objective Bayesian Statistics, *Second Edition*
PRESS and TANUR · The Subjectivity of Scientists and the Bayesian Approach
PUKELSHEIM · Optimal Experimental Design
PURI, VILAPLANA, and WERTZ · New Perspectives in Theoretical and Applied Statistics
PUTERMAN · Markov Decision Processes: Discrete Stochastic Dynamic Programming
*RAO · Linear Statistical Inference and Its Applications, *Second Edition*
RAUSAND and HØYLAND · System Reliability Theory: Models, Statistical Methods, and Applications, *Second Edition*

*Now available in a lower priced paperback edition in the Wiley Classics Library.

RENCHER · Linear Models in Statistics
RENCHER · Methods of Multivariate Analysis, *Second Edition*
RENCHER · Multivariate Statistical Inference with Applications
* RIPLEY · Spatial Statistics
RIPLEY · Stochastic Simulation
ROBINSON · Practical Strategies for Experimenting
ROHATGI and SALEH · An Introduction to Probability and Statistics, *Second Edition*
ROLSKI, SCHMIDLI, SCHMIDT, and TEUGELS · Stochastic Processes for Insurance and Finance
ROSENBERGER and LACHIN · Randomization in Clinical Trials: Theory and Practice
ROSS · Introduction to Probability and Statistics for Engineers and Scientists
ROUSSEEUW and LEROY · Robust Regression and Outlier Detection
RUBIN · Multiple Imputation for Nonresponse in Surveys
RUBINSTEIN · Simulation and the Monte Carlo Method
RUBINSTEIN and MELAMED · Modern Simulation and Modeling
RYAN · Modern Regression Methods
RYAN · Statistical Methods for Quality Improvement, *Second Edition*
SALTELLI, CHAN, and SCOTT (editors) · Sensitivity Analysis
*SCHEFFE · The Analysis of Variance
SCHIMEK · Smoothing and Regression: Approaches, Computation, and Application
SCHOTT · Matrix Analysis for Statistics
SCHOUTENS · Levy Processes in Finance: Pricing Financial Derivatives
SCHUSS · Theory and Applications of Stochastic Differential Equations
SCOTT · Multivariate Density Estimation: Theory, Practice, and Visualization
*SEARLE · Linear Models
SEARLE · Linear Models for Unbalanced Data
SEARLE · Matrix Algebra Useful for Statistics
SEARLE, CASELLA, and McCULLOCH · Variance Components
SEARLE and WILLETT · Matrix Algebra for Applied Economics
SEBER and LEE · Linear Regression Analysis, *Second Edition*
*SEBER · Multivariate Observations
SEBER and WILD · Nonlinear Regression
SENNOTT · Stochastic Dynamic Programming and the Control of Queueing Systems
*SERFLING · Approximation Theorems of Mathematical Statistics
SHAFER and VOVK · Probability and Finance: It's Only a Game!
SILVAPULLE and SEN · Constrained Statistical Inference: Order, Inequality and Shape Constraints
SMALL and McLEISH · Hilbert Space Methods in Probability and Statistical Inference
SRIVASTAVA · Methods of Multivariate Statistics
STAPLETON · Linear Statistical Models
STAUDTE and SHEATHER · Robust Estimation and Testing
STOYAN, KENDALL, and MECKE · Stochastic Geometry and Its Applications, *Second Edition*
STOYAN and STOYAN · Fractals, Random Shapes and Point Fields: Methods of Geometrical Statistics
STYAN · The Collected Papers of T. W. Anderson: 1943–1985
SUTTON, ABRAMS, JONES, SHELDON, and SONG · Methods for Meta-Analysis in Medical Research
TANAKA · Time Series Analysis: Nonstationary and Noninvertible Distribution Theory
THOMPSON · Empirical Model Building
THOMPSON · Sampling, *Second Edition*
THOMPSON · Simulation: A Modeler's Approach
THOMPSON and SEBER · Adaptive Sampling
THOMPSON, WILLIAMS, and FINDLAY · Models for Investors in Real World Markets

*Now available in a lower priced paperback edition in the Wiley Classics Library.

TIAO, BISGAARD, HILL, PEÑA, and STIGLER (editors) · Box on Quality and Discovery: with Design, Control, and Robustness
TIERNEY · LISP-STAT: An Object-Oriented Environment for Statistical Computing and Dynamic Graphics
TSAY · Analysis of Financial Time Series
UPTON and FINGLETON · Spatial Data Analysis by Example, Volume II: Categorical and Directional Data
VAN BELLE · Statistical Rules of Thumb
VAN BELLE, FISHER, HEAGERTY, and LUMLEY · Biostatistics: A Methodology for the Health Sciences, *Second Edition*
VESTRUP · The Theory of Measures and Integration
VIDAKOVIC · Statistical Modeling by Wavelets
VINOD and REAGLE · Preparing for the Worst: Incorporating Downside Risk in Stock Market Investments
WALLER and GOTWAY · Applied Spatial Statistics for Public Health Data
WEERAHANDI · Generalized Inference in Repeated Measures: Exact MANOVA and Mixed Models
WEISBERG · Applied Linear Regression, *Third Edition*
WELSH · Aspects of Statistical Inference
WESTFALL and YOUNG · Resampling-Based Multiple Testing: Examples and Methods for *p*-Value Adjustment
WHITTAKER · Graphical Models in Applied Multivariate Statistics
WINKER · Optimization Heuristics in Economics: Applications of Threshold Accepting
WONNACOTT and WONNACOTT · Econometrics, *Second Edition*
WOODING · Planning Pharmaceutical Clinical Trials: Basic Statistical Principles
WOOLSON and CLARKE · Statistical Methods for the Analysis of Biomedical Data, *Second Edition*
WU and HAMADA · Experiments: Planning, Analysis, and Parameter Design Optimization
YANG · The Construction Theory of Denumerable Markov Processes
*ZELLNER · An Introduction to Bayesian Inference in Econometrics
ZHOU, OBUCHOWSKI, and McCLISH · Statistical Methods in Diagnostic Medicine

*Now available in a lower priced paperback edition in the Wiley Classics Library.

Printed and bound by CPI Group (UK) Ltd, Croydon, CR0 4YY
16/09/2026
14963441-0002